Cities and Nature

Cities and Nature connects environmental processes with social and political actions. The book reconnects science and social science to demonstrate how the city is part of the environment and how it is subject to environmental constraints and opportunities. This second edition has been extensively revised and updated with in-depth examination of theory and critical themes. Greater discussion is given to urbanization trends and megacities; the post-industrial city and global economic changes; developing cities and slums; urban political ecology; the role of the city in climate change; and sustainability.

The book explores the historical relationship between cities and nature, contemporary challenges to this relationship, and attempts taken to create more sustainable cities. The historical context situates urban development and its impact on the environment, and in turn the environmental impact on people in cities. This provides a foundation from which to understand contemporary issues, such as urban political ecology, hazards and disasters, water quality and supply, air pollution and climate change. The book then considers sustainability and how it has been informed by different theoretical approaches. Issues of environmental justice and the role of gender and race are explored. The final chapter examines the ways in which cities are practicing sustainability, from light "greening" efforts such as planting trees, to more comprehensive sustainability plans that integrate the multiple dimensions of sustainability.

The text contains case studies from around the globe, with many drawn from cities in the developing world, as well as reviews of recent research, updated and expanded further reading to highlight relevant films, websites and journal articles. This book is an asset to students and researchers in geography, environmental studies, urban studies and planning and sustainability.

Lisa Benton-Short is Associate Professor of Geography at George Washington University (GWU), Washington, DC. An urban geographer, she has research interests in environmental issues in cities, parks and public spaces, and monuments and memorials. She is also Director of Academic Programs in Sustainability at GWU.

John Rennie Short is Professor of Geography and Public Policy at the University of Maryland, Baltimore County. He has published 35 books and numerous articles and is recognized as an international authority on the study of cities.

Routledge critical introductions to urbanism and the city

Edited by Malcolm Miles, University of Plymouth, UK
and John Rennie Short, University of Maryland, USA

International Advisory Board:

Franco Bianchini	Jane Rendell
Kim Dovey	Saskia Sassen
Stephen Graham	David Sibley
Tim Hall	Erik Swyngedouw
Phil Hubbard	Elizabeth Wilson
Peter Marcuse	

The series is designed to allow undergraduate readers to make sense of, and find a critical way into, urbanism. It will:

- cover a broad range of themes;
- introduce key ideas and sources;
- allow the author to articulate her/his own position;
- introduce complex arguments clearly and accessibly;
- bridge disciplines, and theory and practice;
- be affordable and well designed.

The series covers social, political, economic, cultural and spatial concerns. It will appeal to students in architecture, cultural studies, geography, popular culture, sociology, urban studies and urban planning. It will be trans-disciplinary. Firmly situated in the present, it also introduces material from the cities of modernity and post-modernity.

Published:

Cities and Consumption – *Mark Jayne*
Cities and Cultures – *Malcolm Miles*
Cities and Economies – *Yeong-Hyun Kim and John Rennie Short*
Cities and Cinema – *Barbara Mennel*
Cities and Gender – *Helen Jarvis with Paula Kantor and Jonathan Cloke*
Cities and Design – *Paul L. Knox*
Cities, Politics and Power – *Simon Parker*
Cities and Sexualities – *Phil Hubbard*
Children, Youth and the City – *Kathrin Hörschelmann and Lorraine van Blerk*
Cities and Photography – *Jane Tormey*
Cities and Climate Change – *Harriet Bulkeley*
Cities and Nature, Second Edition – *Lisa Benton-Short and John Rennie Short*

Forthcoming:

Cities, Risk and Disaster – *Christine Wamsler*

Cities and Nature
Second Edition

Lisa Benton-Short
and John Rennie Short

Routledge
Taylor & Francis Group

LONDON AND NEW YORK

First edition published 2008
by Routledge
2 Park Square, Milton Park, Abingdon, Oxon OX14 4RN

Second edition published 2013
by Routledge
2 Park Square, Milton Park, Abingdon, Oxon OX14 4RN

Simultaneously published in the USA and Canada
by Routledge
711 Third Avenue, New York, NY 10017

Routledge is an imprint of the Taylor & Francis Group, an informa business

British Library Cataloguing in Publication Data
A catalogue record for this book is available from the British Library

Library of Congress Cataloging-in-Publication Data

Benton-Short, Lisa.
Cities and nature / Lisa Benton-Short and John Rennie Short. – 2nd ed.
   p. cm.
Includes bibliographical references and index.
1. Urban ecology (Sociology) 2. Urbanization – Environmental aspects. 3. City planning –
Environmental aspects. 4. Urban pollution. 5. Sustainable development. I. Short, John R.
II. Title.
HT241.B46 2013
307.76—dc23   2012039645

ISBN: 978-0-415-62555-5 (hbk)
ISBN: 978-0-415-62556-2 (pbk)
ISBN: 978-0-203-10350-0 (ebk)

Typeset in Times New Roman and Futura
by Cenveo Publisher Services

Printed and bound in Great Britain by
TJ International Ltd, Padstow, Cornwall

To Bonnie and Harriet, again.

Contents

Figures

Tables

Boxes

Preface

The first edition of this book appeared in 2008, with most of the writing and research done in 2006/7. The subject of urban–nature relations has continued to develop robustly since then, with a substantial increase in the literature covering a range of topics from historical urban–environment relations to the recent development of comprehensive sustainability plans for many cities. The emergence and widening of such areas as urban political ecology, cities and climate change, and urban sustainability, as well as the new research in traditional areas such as urban hazards, air and water pollution, have all added a vast body of work to the theme of cities and nature that we considered in the first edition. A second edition allows taking stock of these developments.

The book is much expanded from the first edition. There are now 16 chapters in five sections, expanded from the 12 chapters in three sections. The second edition is organized around five central themes:

1 the urban environment in historical context;
2 contemporary urban environments;
3 urban physical systems;
4 urban environmental issues;
5 realigning urban–nature relations.

In Part I, the urban environment in historical context, we explore the pre-industrial and industrial city. The concerns of the industrial city provide the context for the urban environment regulatory framework that still exists today in many cities. In Part II, the contemporary urban environment, we consider global urban trends and then focus on the postindustrial city and the developing city. Part III focuses on urban physical systems with specific chapters on urban sites, hazards and disasters and urban political ecology. Important contemporary urban environmental concerns are examined in Part IV. A brief chapter introduces the recent emergence of an urban environment regulatory framework. Individual chapters then consider the case of water pollution, air pollution, climate change and garbage. Part V, in a more normative approach, covers issues of urban sustainability and environmental justice. Urban sustainability explores the theories of sustainability and the ways in which cities are practicing sustainability—from light "greening" efforts such as planting trees, to more comprehensive plans that integrate the multiple dimensions of sustainability.

We have moved some material around from the first edition so that there is a cleaner narrative flow. New chapters include the postindustrial city, the developing city and climate change; we've added a new section in Chapter 16 on theories of sustainability as well. In all chapters the latest research is incorporated into the literature review and guides to further reading. The geographical coverage is extended to cast a wider net that includes many more cities in developing countries. The examples and case studies provide a more global coverage.

This second edition allows us to expand theoretical development and add breadth to existing chapters by considering non-US/EU cities; in particular increasing the representation of examples of developing world cities. We feel that the first edition concentrated too much on US/EU cities and the new edition provides an opportunity to offer a more global representation. The second edition also allows us to update data, legislation, examples and trends.

The reorganization of chapters, addition of new chapters and updates and improvements in all chapters allow us to more thoroughly document critical issues and trends that are at the cutting edge of this field of study. We have added new photographs, data sources, data and more recent legislative developments. In summary, the book is much expanded and, we think, much improved with greater theoretical depth and wider geographical coverage.

This second edition was prompted by a suggestion from Andrew Mould and shaped by three anonymous referees who took the time and no small effort to read through the previous edition, making many suggestions for revision, excision and addition. We thank you.

We were heartened by the very positive response to the first edition and wanted to expand, amplify and enlarge one of the most dynamic and exciting areas of intellectual inquiry, political importance and social significance. As we transition into a predominantly urban world the quality, sustainability, vulnerability and opportunities of the urban environment provides one of the most pressing human concerns. Not only a site of problems, the urban environment is also a platform for progressive social change.

Acknowledgments

We would like to first acknowledge that this second edition is the result of years of teaching the topic of "urban environmental issues" to undergraduates and graduates at the George Washington University and the University of Maryland, Baltimore County. In the process of teaching this course, and using the first edition, we found our students were invaluable in asking provocative questions and providing new and inspirational sources of material in their term papers and research projects. Notably, students in Lisa's graduate seminar on "Urban Sustainability" in the spring of 2012 provided sources for updating tables and data, and suggested topics for us to consider including.

We would also like to thank Mike Mussman and Barbara Bridges, research assistants at UMBC, for their valuable contributions in this second edition. Thanks also to Becky Barton, Michele Judd, David Rain, Joe Dymond and Elizabeth Chacko for granting permission to reproduce images in this work.

Thanks go to three anonymous reviewers who provided thoughtful and important suggestions to improve this second edition. The increased size, scope and depth of the second edition are a response to their careful assessments.

Every effort has been made to contact copyright holders for their permission to reprint material in this book. The publishers would be grateful to hear from any copyright holder who is not here acknowledged, and will undertake to rectify any errors or omissions in future editions of this book.

Lastly, Columbus, Cosmo and Mackinder were frequent desktop editors and writers, and reminded us when it was time for dinner.

1 The city and nature

Around four in the afternoon, on January 12, 2010 an earthquake measuring a devastating 7 on the Richter scale occurred in Haiti. Over 300,000 were killed, a similar number were injured, almost one million people were made homeless and much of the capital city of Port au Prince was reduced to rubble. Despite international relief efforts, even by 2012 half a million people still live in temporary shelter, often in appalling conditions. The environmental disaster revealed not only the unstable nature of the underlying geology, but also the fractures in society: the dysfunctional nature of the Haitian state; the shoddy building in the nation's capital; the crippling poverty of most of its citizens; and the marked inequality that allowed the tiny elite to remained unaffected while the poorest people were most negatively impacted.

Another story from a richer country: in the summer of 1995 a heat wave struck the city of Chicago. For over a week in July the temperature reached over 100°F every day. By July 20, over 700 people had died. It was referred to in the press as a "natural disaster," the unfortunate outcome of a freak meteorological condition. In his 2002 book *Heat Wave* Eric Klinenberg questioned this assumption and undertook a social autopsy of the event. He found that deaths were greatest among more elderly people living on their own. The tragedy was not simply a natural disaster, but the outcome of the social isolation of seniors, retrenchment of public assistance and declining neighborhoods. Most victims were seniors who lived alone in neighborhoods that lacked a sense of community, and where there was perception of danger in the streets. Trapped inside their homes, and with few visits from public health officials, many poor isolated seniors overheated and died. The disaster was not the result of high temperatures, but high temperatures as mediated through a complex set of social and political relationships.[1]

We take two points from these stories. First, the city is situated in broader physical processes and entangled in the ecosystem—in these cases, respectively situated on plate tectonic fault lines and part of the rising temperatures of summer warming. Second, these environmental processes, exposed by the extreme nature of the events, are filtered through social arrangements of political and economic differences. There are no such things as "natural" disasters in cities, just as there are no cities independent of nature.

BOX 1.1

Words and definitions

Words are important, slippery, relational things. We need to make clear what we mean when we use the terms nature, environment and city. Nature has many different meanings, but two of the most important are essential quality (as in "it is in his nature to do these things") and the material world that can include or exclude human beings. Words are relational and nature was often counterpoised against technology and/or culture. In this book we will use the term "nature" to refer to the material world that includes human beings. We use it less in opposition to human society and more as the container of physical resources and cultural meanings. We will use nature and environment interchangeably.

There has long been a distinction between nature and city in linguistic usage. The main point of this book is to show that the city is part of the material world and this materiality is shaped and structured by and in cities. Rather than situating the city on one side of a crude binary of nature–society, we suggest understanding the city as a hybrid space that constitutes and reconstitutes the natural and the social, a complex space of physical and socio-economic flows. The city is both an environmental and a social construct. The city is an integral part of nature and nature is intimately interwoven into the social life of cities. The city is a vital node in entangled networks of flows of technology, socio-physical processes and socio-economic relations. The city is a site where the emergent connections between the political and the ecological are revealed and contested.

We need to make the point that "natural" disasters are, on closer inspection, much more closely connected to social processes than we often acknowledge. When people build on eroding hillsides or locate houses in earthquake zones, then the natural disaster turns out to be in part a social construction. And the term "natural" disaster also hides the social-economic implication of their effect. Citizens of poor countries are more affected by floods and storms because they do not have expensive technology to provide as much early warning or as rapid evacuation as richer countries. The same storm will have vastly different human consequences in different places. And even within the same city the experience of disaster by rich and poor residents can be vastly different. In Port au Prince the elite rich who lived in the green suburb of Petionville, up in the hills above the city, escaped the plight of the urban poor trapped in the devastated city below. Here few homes were destroyed and police were quickly mobilized to protect the residents and their property. Haiti's elite, because they were rich enough to live in Petionville, were spared from much of the devastation of the earthquake.[2] In the US a few years earlier, the searing images of those trapped in New Orleans in the immediate aftermath of Hurricane Katrina also reminded us of the gulf that separates the socio-economic status of those able to leave compared to those stranded in the city.

Disasters provide a visible connection between nature and the city. Upon closer inspection the natural appears more social and the social life of cities is more accurately seen as implicated in environmental processes. The city is the center of a society–environment dialectic. We develop this theme more fully in subsequent chapters.

The intellectual background

Cities provide an inevitable contrast to the "natural." A consistent strand of thought has sought to place the city as a human invention in opposition to the "natural," the "pristine," and the "wilderness." Protecting the environment has usually meant halting the encroachment into pristine areas such as rainforests and tundra. Most often, environmental protection has been defined as meaning something outside of, and mostly unrelated to, the concerns and interests of our cities. Cities have been described and understood as somehow separate from the so-called "natural world." This has been reinforced by the increasing separation of life in the city from the wider environmental context. When food is more available in a supermarket aisle instead of in the fields outside our homes and when we can turn up the heating to keep out the cold and/or turn on the air-conditioning to keep out the oppressive heat, there is a tendency to see the city as somehow removed and independent from the physical world.

Urban theorizing for a long time was developed as if a city was on a flat, featureless plain. Urban studies have long ignored the physical nature of cities; instead, the emphasis was on the social, political and economic rather than the ecological. And yet cities are ecological systems, they are predicated upon the physical world as mediated through the complex prism of social and economic power. In recent years there has been a renewal of interest in the city as an ecological system, with emphasis on the complex relationships between environmental issues and urban concerns, and between social networks and ecosystem flows. Take three examples of what we will term the *new urban political ecology*. First, William Solecki and Cynthia Rozenzweig look at the biodiversity–urban society relationships in the greater New York Metropolitan Region.[3] They use such concepts as the "ecological footprint" and vulnerability to global environmental change to analyze the current interactions between biodiversity and urban society. Second, Paul Robbins considers the flows and networks that link the suburban lawn to the chemical industry, pollution and toxic waste.[4] The green strips of lawn in front of the suburban house are a recreation of the "natural" created by heavy doses of the "chemical." Third, Julian Yates and Jutta Gutberlet analyze the flows of food waste in the Brazilian city of Diadema.[5] The flow of food waste is collected by recyclers, then used by urban gardeners as a vital fertilizer of soils in community gardens. The urban connections between physical and political processes are revealed through tracing such flows and examining their effects.

In the emerging and fast-developing field of urban political ecology, cities are viewed as imbrications of the physical and the social, the ecological and the political. The city is implicated with the "natural" world in connections that embody and reflect social, economic and political power. The city is an integral part of nature and nature is intimately interwoven into the social life of cities.

The city as ecosystem

Physical geographer Ian Douglas suggested that the city can be modeled as an ecosystem with inputs of energy and water, and outputs of noise, climate change, sewerage, garbage and air pollutants.[6] Another way to consider the city–nature dialectic then is to consider the city as an ecological system with a measurable amount of environmental inputs and outputs. Table 1.1 gives

Table 1.1 London's environmental inputs and outputs

Inputs	Tons/year
Oxygen	40,000,000
Water	876,000,000,000 (litres)
Food	6,900,000
Paper	2,200,000
Plastic	2,100,000
Construction material (bricks, sand, concrete)	27,000,000
Energy needs (tons of oil)	13,276,000
Outputs	Tons/year
Carbon dioxide	41,000,000
Sulfur dioxide	400,000
Nitrogen oxide	280,000
Sewage and sludge	7,500,000
Industrial/commercial waste	14,029,000
Household waste	3,900,000

Source: The City Limits project: a resource flow and ecological footprint analysis of Greater London at www.citylimitslondon.com

an estimate of the range and amount of inputs and outputs for London. Among the most obvious inputs are energy and water.

Human activity in the city is dependent on large and consistent inputs of energy. When we leave heated buildings to drive in cars to purchase goods we use energy. The commercial activities we pursue and the microclimates we create (heating in winter, cooling in summer) all use energy. In seeking to overthrow the tyranny of nature, cities use prodigious amounts of energy. Cities are deeply dependent on energy sources. In the US, since the beginning of the twentieth century, petroleum has traditionally been very cheap and cities now sprawl across the landscape. In countries where energy is more expensive cities tend to be higher in density and more reliant on public transport. Large-scale suburban sprawl is a function of cheap energy. It is tempting to theorize the impact of a long-term, sustained increase in energy prices on suburban sprawl and urban structure.

Water is an essential ingredient of life. The people and commerce of cities are utterly dependent upon water. One of the largest urban differences

Figure 1.1 Gold Coast, Queensland, Australia. This image evokes the sense of the city as a physical–social artifact, imbrications of the social and the ecological. This transect reveals the interconnections between ocean, beach, built form and river as water surfaces and green spaces intersperse with roads, houses and commercial buildings in a complex space

Source: Photo by John Rennie Short

in the world is between cities and citizens with clean, easily accessible water and others with expensive, inaccessible and polluted water supplies. Immense engineering projects have been undertaken to provide inexpensive and clean water, and as cities have grown the catchment areas have extended outwards and the engineering sophistication of piping in water has grown and deepened. In poorer cities polluted urban water remains a major source of disease and illness, especially for children.

Even in rich countries the availability of fresh water is a determinant of the limits of urban growth. In the arid west of the US, for example, urban growth has been predicated upon massive federal subsidies and expensive engineering projects that have provided fresh water at low cost to the consumers. The ecological limits are always more flexible than the environmental determinists suggest, but they are not infinitely extendable. We may be reaching the "water" limits of urban growth in the arid US.

Cities have an impact on the environment. The notion of the ecological footprint of the city has been developed.[7] It is defined as the amount of land required to meet the resource needs of a city and absorb its waste. The cities of the richer world, with their heavier energy usage and larger waste consumption, have a larger and heavier footprint than cities of the developing world. Recent studies use the metric of global hectares (gha) per capita. A study of the city of Swindon in England, for example, estimated that the footprint was 5.6–5.9 gha.[8] The city of Calgary in Canada also undertook a survey and found that the city's footprint was 9.5–9.9 gha per person.[9] High energy use was the single biggest contributor. As a result of the study, the city government of Calgary delayed greenfield developments and moved toward greater use of renewable sources. The city's light rail system is now powered entirely by wind-generated energy.

Cities modify the environment. The most obvious example of this is the urban heat island. Cities tend to be warmer because of the amount of extra heat produced in the city and the heat absorption of man-made materials such as tarmac, asphalt and concrete. Heat is absorbed by these surfaces during the day and released at night. The net result is for the air around cities to be warmer than surrounding rural areas. One side-effect is to reduce the need for heating in the winter but to increase the need for air-conditioning in the summer. The heat island means you can turn the heating down in London in December but need to increase the air-conditioning in Washington, DC in August. The extra heat causes a thermally induced upward movement of air, and an increase in cloud and raindrop formation. Cities are often cloudier, more prone to thunder and slightly warmer than surrounding rural areas.

Human activity in the city also produces pollutants. Industrial processes and auto engines emit substances that include carbon oxides, sulfur oxides, hydrocarbons, dust, soot and lead. The air in cities has traditionally been very unhealthy, which is part of the reason for the higher urban death rate throughout most of human history. The pall of smog that hangs over many cities is a visible reminder of the effects of concentrated human activity on

BOX 1.2

An early urban ecologist

In 1864, the naturalist and geographer George Perkins Marsh, in his book *Man and Nature*, commented that the likely cause of the decay of once-flourishing civilizations such as the Roman Empire was in part due to acts of environmental neglect. He suggested, for example, that

Rome imposed on the products of agricultural labor in the rural districts taxes which the sale of the entire harvest would scarcely discharge; she drained them of their population by military conscription; she impoverished the peasantry by force and unpaid labor on public works; she hampered industry and internal commerce by absurd restrictions and unwise regulations. Hence, large tracts of land were left uncultivated, or altogether deserted, and exposed to all the destructive forces which act with such energy on the surface of the earth when it is deprived of those protections by which nature originally guarded it.

Source: Marsh, G. P. (1864) *Man and Nature: Or Physical Geography as Modified by Human Action*. New York: Scribner, pp. 10–12.

the environment. The pollutants of cities are not only injurious to the health of individuals, but they also cause more general damage; cities are in part a major cause of global warming and ozone depletion.

A major output of cities is garbage. High mass consumption in association with elaborate packaging has created a rising tide of garbage in cities. Burning it causes air pollution, while burying it leads to massive landfills. The environmental justice literature shows that many environmentally hazardous facilities are generally located in poor, minority and more weakly organized communities. Issues of environmental management are tied in to wider issues of equity and social justice. Patterns of environmental racism are clear when we note that most noxious facilities are located in lower income, more marginal communities.

Cities also emit noise. Cities are noisy places and households who inhabit busy urban streets for more than 15 years are on average likely to experience a 50 percent reduction in hearing capacity. The effects of noise pollution vary from annoyance to deterioration in hearing. A high background noise level leads to a general increase in stress and the lessening of the quality of urban life.

Cities are an integral part of the hydrological cycle. Cities impact the daily and seasonal flows of water. The large amount of impermeable surfaces, for example, means that when it rains run-off levels spike dramatically. Cities thus need to create modified flows through channels and conduits that can cope with the irregular high flow rates. But the large amount of impermeable surfaces in association with the channelization of water courses can lead to distinct surges in water flow after rain and in many cases patterns of flooding. As urbanization increases so too does the overloading of the hydrological cycle. Cities also modify the flow and direction of rivers in order to increase commercial activity. In Chicago, engineers actually reversed the flow of the Chicago River in the nineteenth century to facilitate industrial growth. Cities also tend to pollute water systems, thus reducing the amount of fresh water and in some cases posing major health hazards.

To theorize the city as an ecological unit is to open new possibilities for understanding the complex relations between the environmental inputs necessary for urban growth and the environmental impacts of urban growth.

Nature and the city

Nature is present in cities in often unforeseen and unplanned ways. Wildlife in a variety of forms continues to find ecological niches in the city. Urban tensions can be examined through the narration of the relationship between cities and wildlife. Urban animal geographies can tell us a great deal about the city–nature dialectic, whether it is in the stories of rats in cities or the story of hawks in the city. Consider the case of Pale Male and Lola, two red-tailed hawks that made their nest in the facade of an exclusive high-rise apartment block in New York City's Fifth Avenue. Hawks have been noticed in the area since 1998, and every year the birds return to nest, breed and feed their young. Birdwatchers followed their progress through binoculars, cameras and websites. There is something heroic about the capacity of hawks to thrive in the city. Some residents of the apartment block thought other-wise. The president of the co-op board, wealthy real-estate developer Richard Cohen, unilaterally ordered the nest removed in December 2004. Red-tailed hawks are rare enough to have been protected by a treaty signed in 1918 between several nations, including the US, Canada and Russia. An earlier attempt to evict the birds was blocked when their defenders invoked this international agreement. However, the nest could be removed if it contained no eggs or chicks. The co-op board used this loophole and their decision initiated a major protest. Protesters dressed as birds mounted a vigil across the street from the building. The media publicized the story. One subtext was resistance to the power of the wealthy. Apartments at 927 Fifth Avenue can sell for as much as $18 million, and residents include the wealthy and the famous. The image of very rich residents evicting hawks from their perch was too delicious to ignore. The extremely negative publicity for the apartment building and its residents eventually led to a reversal of the eviction. Pale Male and Lola continued to nest on the building, but for six barren years there were no new chicks after the disturbance of their original nest. In recent years, many more red-tailed hawks have taken up residence in New York City. A 2007 study commissioned by the Audubon Society reported that pairs of red-tails were spotted breeding in nests at 32 locations throughout the city, and hawk watchers say they have spotted hundreds of unattached red-tails across the five boroughs. At the south end of the park, a hawk couple dubbed Pale Male Junior (or just Junior) and Charlotte nested on the Trump Park Hotel on Central Park South in 2005 and successfully raised two eyasses. In 2007 they moved their nest to a building on Seventh Avenue at Fifty-Seventh Street (two blocks south of the park) and raised one eyass. Junior and Charlotte may often be found in the vicinity of the Time Warner Center at the southwest corner of the park. Pale Male and Lola continue to be the most closely watched celebrities; there are three children's books about the birds, and several websites that update sightings and photographs (see www.palemale.com). The birds were even the subject of a 2009 documentary called *The Legend of Pale Male* by Frederic Lilien. But Lola disappeared in December 2010. She is now presumed dead. Pale Male was spotted with a new mate in early January 2011. The new female hawk, given the name "Ginger" because of her dark feathers on her neck and chin, is much younger. In the summer of 2011, Pale Male enthusiasts were rewarded: Ginger and Pale Male were the proud parents of two eyasses.

Not all wildlife that shares the urban environment is as welcome as Pale Male and Lola. Rats, for example, have managed to find a home in most cities. They inhabit the dark tunnels and the hidden recesses of the city and have become symbols of disease and decay; they bring out fear and loathing rather than love and respect. And yet, like the cuddlier animals or the more photogenic birds, they too are urban survivors.

There is also a more self-conscious referencing of nature in cities. Consider urban parks. It is difficult to imagine London without Hyde Park, New York without Central Park or Washington without the National Mall. Landscape architects such as Frederick Law Olmsted have left a permanent legacy in cities. The modern park movement is more closely tied into active participation than the environmental contemplation so beloved of the early park movement. City parks are now developed as much for their recreational opportunities as their aesthetic appeal. Urban planners realize that the successful referencing of nature can be linked with the promise of economic redevelopment. Whether it is in the beaches of southern California, the lakeside shore of Chicago, or the parks of London and Paris, a commonly accepted attractive feature of urban life is the successful (re)incorporation of nature into the urban lifestyle, the city's image and the metropolitan experience.

Sometimes cities reclaim rather than subdue nature. Seoul, South Korea reconnected with an old stream that used to pass through the city, Cheonggyecheon (Figure 1.2). The five mile (8.2 km) stream was an integral part of the early history of the city, a setting for trade and festivities and a marker of divisions between the rich and poor parts of the city. But in Seoul's rapid postwar economic growth the physical environment became

Figure 1.2 Cheonggyecheon, Seoul, South Korea
Source: Photo by John Rennie Short

predominately a site for economic expansion. The river was blocked and covered, disappearing completely in 1978 under highway construction. But in recent years, with a change of emphasis on sustainability, biodiversity and quality of urban life, the river was re-excavated as part of a massive urban ecological renewal project costing $281 million. The river was reclaimed and renewed as the city embraced the waterway. In 2005 the project, essentially a linear water park, was opened to the public as a series of walkways along the river. The aim was to reintroduce nature back into the city and promote more eco-friendly urban designs.[10] The project, despite the criticism of its costs and lack of ecological and historical authenticity, has resulted in a cooling of the urban heat island, an increase in biodiversity, and the provision of an attractive urban public space. It was a reconnection with both nature and history in a city that for 30 years had ignored both. The river was reinvented as well as rediscovered.

Urban sustainability

A central idea in the emerging interest in the relationship between the city and nature is the notion of urban sustainability; the idea that cities can be environmentally sustainable over the longer term. There is a widely shared belief that many cities impose such heavy environmental costs that the long-term future of the city may be undermined. The heavy reliance on fossil fuels, for example, and the increased use of the private auto as the main urban transport mode tend to degrade the environment. These problems are particularly acute in rapidly growing cities and cities in developing economies where environmental regulation is weaker and environmental improvement may not be considered an important political issue.[11] In China, for example, the rapid and often unregulated economic growth of the past 20 years has been purchased with the severe degradation of air, water and land.

We can picture a three-stage model of the relationship between cities and environmental sensitivity; in the early stages urban growth is small and environmental impacts, while strong, are highly localized. As the city moves into a more industrialized mode of production, environmental degradation is more severe as the environmental impacts are heavier and longer lasting. As the economy matures and people become more affluent, a greater premium is often placed on the quality of the urban environment. Environmental reforms are often instituted. In cities around the world, rich and poor, developed and developing, the struggle to live in a better urban environment, with clean water, fresh air and pleasant conditions is an important source of mobilization and platform for action.[12]

Questions of environmental quality are intimately connected to issues of social justice; the worst environmental conditions are imposed on the lower income, most marginal urban residents. Poverty and environmental degradation tend to go hand in hand in a web of multiple deprivation and social exclusion. Graham Haughton presents five principles of sustainable development that are also based on social justice: generation equity, inter-generational equity, geographic equity, procedural equity and interspecies equity.[13] He suggests approaches to achieving sustained development that include creating more self-reliant cities that reduce the environmental impact on the wider bioregion and redesigning cities so that land is used more effectively and rationally. Michael Hough also lays out a road map for urban sustainability that recognizes the importance of maintaining the integrity of urban ecosystems.[14] One oft-touted example of a more suitable city development is Curitiba, Brazil. This city of 1.8 million people developed a master plan in 1965 that limited central city growth and guided development to two north–south running corridors. The concentrated growth allowed the more effective use of public transport. There are now 1,100 buses that carry 1.4 million passengers each day and a network of pedestrian routes that allow people to travel by foot in the central business area. The net result is less demand for private car usage resulting in less pollution and a more pleasant urban environment. Many cities in the developed world can learn from Curitiba. Sustainable development practices can flow from poor to rich countries as well as from rich to poor countries. In many poor countries the need to husband scarce resources, recycle goods and reimagine the city provides a rich context for new urban practices.

A new approach

In the social sciences there is now an emerging body of literature that considers the environmental context of urban life and, in the physical sciences, a growing awareness that cities are environments worthy of serious ecological analysis. Let us end this introductory chapter with some brief examples of such recent work.

There is the exciting and growing field of urban environmental history. A deep historical perspective is provided by Chris Boone and Ali Modarres in their book *City and Environment*.[15] There are also detailed case studies of individual cities that provide a close-up of the urbanization of nature. William Cronon in *Nature's Metropolis* examines the relationship between Chicago and its hinterland from 1850 to 1890.[16] He shows how the physical world was turned into a commodified human landscape as grain, lumber and meat production transformed prairies and woodlands into the physical basis for the city's growth and development. Merchants, railway owners and primary producers transformed the "wilderness" into a humanized landscape that was the basis for the city's impressive economic growth. Cronon's work demonstrates that urban economic growth draws heavily on a physical world. A more recent work by Matthew Klingle looks at the environmental history of Seattle.[17] He tells the story of continual landscape modification. The city was an important site in the intertwined processes of the transformation of nature and the constitution of society.

On the one hand, social scientists are examining the social context of the city–nature dialectic. There is now a large body of work that we can characterize as urban political ecology. Consider the work of Matthew Gandy. His book *Concrete and Clay: Reworking Nature in New York* looks at the urbanization of nature in New York City and explores a series of relationships between nature, the city and social power in his consideration of the creation of the city's water supply, Central Park, the construction of urban parkways, a radical Puerto Rican environmental group in the 1960s and 1970s, and an anti-waste campaign in the Greenpoint-Williamsburg district of Brooklyn. He examines the environmental justice movement in a city where toxic facilities and land uses are consistently concentrated in minority-dominated areas of the city.[18] In a more recent work he explores the reasons behind the dysfunctional water supply system of Mumbai, India: the result of colonialism, rapid urban growth and the dominance of the middle-class interests over the needs of the majority poor.[19] Water supply is not only an engineering problem but also an issue bound up with politics and power. In 2011 a special issue of the *International Journal of Urban Sustainable Development* was entirely devoted to urban water poverty.[20] This theme is evident in Erik Swyngedouw's 2004 book *Social Power and the Urbanization of Water*, which focuses on the city of Guayaquil in Ecuador, where 600,000 people lack easy access to potable water. He shows that flows of water are deeply bound up with flows of power and influence and water provision is not simply about connecting supply and demand but about the interconnections between the physical and the social, the environmental and the political.[21] These works reveal a similarity in theoretical outlook; the city is implicated with the "natural" world in connections that embody and reflect social, economic and political power. The city is an integral part of nature and nature is intimately interwoven into the social and political life of cities.

On the other hand some ecologists are using their techniques and approaches to consider the city as an ecosystem. Mary Cadenasso and colleagues have developed a model of the urban ecosystem as a complex of biophysical, social and built components. Using both watersheds and patch dynamics they seek to model the fluxes of energy, matter, population and capital with the goal of identifying the feedback between ecological information and environmental quality.[22] In a more detailed use of this urban ecology approach Eric Keys and colleagues, working on Phoenix, Arizona looked at the spatial structure of land use from 1970 to 2000. They show how there was a marked change from agricultural to urban land and that remaining areas of desert were increasingly fragmented with implications for the urban ecology and biodiversity. They suggest the land-use changes had an impact on higher levels of carbon emissions, creating a hotter urban heat island and a decline in native plant species.[23] Michael Paul and Judy Meyer show some of the connections between urbanization and stream ecology: the increased amount of impervious surfaces and the greater amount of contaminant run-off reduce the richness of algal, invertebrate and fish communities.[24] We not only have a large and increasing body of literature on the city as ecosystem, we also have some long-term study sites. In the US, for example, there are two urban ecological sites in Baltimore and Phoenix that have provided an array of interesting material for the modeling of urban ecosystems.[25]

There is now a fascinating and important area of interdisciplinary work in the convergence of social scientists considering the nature of cities and ecologists looking at the city. This book will review the various strands of research that show how the city is part of the environment, subject to environmental constraints and opportunities, a shaper as well as a container of environmental processes. The book is written from the assumption that we can only but improve our understanding of the physical environment by considering its many and subtle links with the city and we can only but enhance our social understanding of the city by exploring its many relationships to the physical world.

Organization of the book

The subsequent chapters are organized into five parts that connect to central themes

1   the city and nature in historical context;
2   the contemporary urban context;
3   urban physical systems;
4   the main urban environmental issues;
5   (re)aligning urban–nature relations.

In Part I we show how the emergence of cities fundamentally altered the relationship between nature and society. A major theme in the history of city–nature relations is the rise of the industrial city, in Europe and North America in the nineteenth and early twentieth centuries, and in the developing world since the latter half of the twentieth century. In Chapter 2 we examine how the early, pre-industrial cities modified the physical environment while the physical environment also impacted cities. As cities emerged and grew, new environmental problems such as pollution and disease prompted new systems of regulation and infrastructural modifications that in turn modified the city–environmental relationship. Chapter 3 notes how the rise of the industrial city generated vastly increased amounts of air, land and water contamination, but also brought about subsequent policy reforms and new forms of urban design. Coping with the problems of the urban environments during industrialization is an important stimulus of new knowledge and the creation of new policies and infrastructure.

In Part II we describe the contemporary urban dynamics. In Chapter 4 we introduce contemporary global urban trends and patterns. We identify the major environmental themes linked to the current wave of urban transformations. Chapter 5 takes up the themes of the postindustrial city while Chapter 6 considers the environmental issues in cities in the developing world.

In Part III we look at the urban–nature dynamic in a variety of ways. In Chapter 7 we discuss how the occupancy of specific sites—such as deserts, beaches and flood plains—creates constraints and opportunities. The Haitian earthquake, the 2004 tsunami, Hurricane Katrina in 2005 and 9/11 make it increasingly clear that cities are vulnerable to disasters. In Chapter 8 we explore this vulnerability and argue that there is no such thing as a "natural disaster." We prefer the term environmental hazard/disaster because it highlights the social, economic and political forces that mediate or exacerbate hazards. At the same time we acknowledge that the city is also resilient. Efforts to recover and rebuild after fires and earthquakes, for example, often provide the impetus for urban growth. A discussion of urban hazards and disasters provides an important facet of the social–nature dialectic. In Chapter 9 we document the emergence of an urban political ecology. Looking at cities as ecological systems holds out enormous promise for combining the insights of traditional ecology with the perspective of critical social science. Developments in this field of study allow us to look at cities more precisely as social-biophysical complexes.

In Part IV we focus on particular urban environmental issues. Chapter 10 describes the recent rise of environmental regulation and sets the context for understanding policies and responses to critical environmental issues. In Chapters 11 and 12 we examine water and air issues, respectively. One of the most dramatic environmental changes impacting cities is global climate change, as discussed in Chapter 13. The issue of urban waste is examined in Chapter 14. In some cases cities have made significant improvements and advances. For many megacities in the developing world it is a mixed picture, with some positive developments but also instances of cities being almost overwhelmed by the sheer size of growth in relation to the available resources.

Part V considers theoretical and practical issues of realigning urban–nature relations. In Chapter 15 we raise the issue of environmental justice and investigate how issues of class, race and gender interconnect with environmental issues and social justice. In Chapter 16 we consider the discourses of urban sustainability that aim to not only reconnect cities to their local and regional ecologies, but also to the global commons. We then turn to the practice of sustainability and attempts to redesign, recreate and rethink cities within a larger framework of livability, sustainability, equity and social justice.

The impetus for this book is our belief that environmental issues are increasingly urban-based and that environmental issues are central to the urban condition. We are optimistic that since cities are the ultimate social creation, they can become important sites of positive transformation and pivotal places of social progress.

Guide to further reading

Boone, C. and Moddares, A. (2006) *City and Environment*. Philadelphia, PA: Temple University Press.

Douglas, I., Goode, D., Houck, M. and Wang, R. (eds.) (2010) *Routledge Handbook of Urban Ecology*. New York: Routledge.

Girardet, H. (2008) *Cities People Planet: Urban Development and Climate Change*. Chichester: Wiley.

Grimm, N. B., Faeth, S. H., Golubiewski, N. E., Redman, C. L., Wu, J., Bai, X. and Briggs, J. M. (2008) "Global change and the ecology of cities." *Science* 319: 756–760.

Heynen, M., Kaika, M. and Swyngedouw, E. (eds.) (2006) *In the Nature of Cities*. New York and London: Routledge.

Kaika, M. and Swyngedouw, E. (2011) "The urbanization of nature: great promises, impasse and new beginnings," in *The New Blackwell Companion to the City*, G. Bridge and S. Watson (eds.). Chichester: Wiley-Blackwell, pp. 96–107.

Niemelä, J., Breuste, J. H., Elmqvist, T., Guntenspergen, G., James, P. and McIntyre, N. E. (eds.) (2011) *Urban Ecology: Patterns, Processes, and Applications*. Oxford: Oxford University Press.

Peet, R., Robbins, P. and Watts, M. J. (eds.) (2010) *Global Political Ecology: A Critical Introduction*. New York: Routledge.

Roberts, P., Ravetz, J. C. and George, C. (2009) *Environment and the City*. New York: Routledge.

Short, J. R. (2005) *Imagined Country: Environment, Culture and Society*. Syracuse, NY: Syracuse University Press.

Stefanovic, I. G. and Scharper, S. B. (2012) *The Natural City: Re-envisioning the Built Environment*. Toronto: University of Toronto Press.

Part I
The urban environment in history

2 The pre-industrial city

The first cities

The first, ancient cities began to emerge some 5,000 or 6,000 years ago in
various regions around the world. They began first as a shift from tribal
communities and villages to larger, more complex social, economic and
political systems. The earliest cities were found in Mesopotamia (cities such
as Ur, Erech, Lagash, and Larsa that flourished in the southern portion of the
Tigris–Euphrates river valley areas), Egypt (along the Nile such as Heliopolis,
Memphis and Nekheb) and the Indus Valley (Harappa and Mohenjo-daro).
At its height Ur might have had a population of 25,000. In China, the Huang
Ho Valley appears to have been the region where the first cities of Shang and
Chengchow in eastern Asia emerge. In the Valley of Mexico the Mayan
cities of Tikal and Uaxactun are among the oldest discovered. Grecian cities
such as Thebes and Troy began to emerge around 1200 BCE, while the city of
Rome began as a cluster of villages along the Tiber a few hundred years
later. The early cities share common elements that reveal important nature–
society relationships. The most dramatic is the realization that urban growth
is predicated upon environmental sustainability.

The world's first cities appear to have arisen in regions where climate
and soil allowed the land to provide an abundance of plant and animal life
necessary to support larger populations. However, it was not just an agricultural surplus that created cities, but the implementation of social power
that directed labor and the production of a surplus that in turn allowed the
development of cities. Agricultural surplus did not create cities, cities
created agricultural surplus.

Water was one of the most critical elements. Almost all cities were located
along major rivers and based their power (and that of their rulers) on
the control of irrigation systems that served the surrounding countryside.
The urban historian Lewis Mumford notes that it is no accident that the

first cities began in river valleys.[1] Water management was an important ingredient in the development of centralized power. Large-scale engineering projects were only possible with centralized planning and hierarchical authority.

Cities are, at essence, a transformation of the physical environment to a built environment. Thus the environmental impact of cities is multi-directional: the city transforms the surrounding area, affecting the natural environment. In turn, the natural environment provides critical natural resources that impact the city.

The very construction of the early cities involved an environmental transformation. In Middle America, around 2,500 years ago, the Zapotec Indians began building a great city, possibly the first in the New World. The task involved the reshaping of Monte Albán, a 1,500-foot hill overlooking the Valley of Oaxaca in central Mexico. Cutting into the hillsides, workers constructed hundreds of terraces, stepped platforms with retaining walls designed mainly for plain and fancy residences.[2] In order to create a massive center square, workers toiled to flatten out the entire top of the hill by some 200–400 meters; it was a major feat of engineering and eight times bigger than St. Peter's Square at the Vatican.[3] Monte Albán endured for more than 1,000 years, housing some 20,000–30,000 people at its height.

The early cities referenced nature within the city walls. In Uruk it was said half the city was dedicated to open spaces with a greenbelt of market gardens.[4] The reign of Nebuchadnezzar II (604–562 BCE) is credited with

---

BOX 2.1

The collapse of Mayan cities

The Maya lived in the area in Central America which now consists of Yucatan, Guatemala, Belize and southern Mexico. Mayan cities include Tikal and Uaxactun, Chichen Itza, Mayapan, Copan and Palenque. Tikal's population is estimated to have been around 60,000, which would give it a population density several times greater than the average European city at the same period in history.

The Maya were highly accomplished in astronomy, with an intimate knowledge of the calendar. The cosmology of the Maya permeated their lives and structured their cities. Cities were designed to coincide with astronomical rhythms. At Chichen Itza, on the day of spring and autumn equinox, during sunset a sun serpent rises up the side of the stairway of the pyramid called El Castillo. In Mayan sites in the central and southern lowlands many temples have doorways and other features that align to celestial events.

At the heart of the Maya city existed large plazas surrounded by the most important government and religious buildings. The towering palaces and the emphasis on height and verticality gives an imposing air and reflects the desire to reach the heavens, while also reinforcing a rigidly vertical social hierarchy. The most important religious temples sat atop the Maya pyramids, which were impressively decorated. One theory suggests that these temples might have served as propaganda since they were the only structures that could be seen from vast distances. Outside of the center were structures of lesser nobles or smaller temples. The act of building itself required tremendous manpower and the ability to control labor.

Beginning in the eighth century and continuing for some 150 years, the great Mayan cities were abandoned, as wars raged and people fled. By CE 930, the Mayan heartland had lost 95 percent of its population. This prolonged event, known as the "Mayan collapse" is one of the enduring mysteries of pre-Columbian America. Some speculated an invasion and ensuing war might have led to the decline, but some have cast their eye to environmental factors. In 2003 earth scientists suggested that a 200-year dry spell, starting around CE 750 caused widespread droughts and a significant decline in regional rainfall.

Sources: Coe, M. (2005) *The Maya.* 7th edition. New York: Thames and Hudson. Faust, B, Anderson, E. N. and Frazier, J. (eds.) (2004) *Rights, Resources, Culture and Conservation* in the Land of the Maya. Westport, CT: Praeger. Haug, G., Gunther, D., Peterson, L., Sigman, D., Hughen, K. and Aeschlimann, B. (2003) "Climate change and the collapse of the Maya civilization." *Science* 299 (5613): 1731–5. McKillop, H. (2006) *The Ancient Maya: New Perspectives.* New York: W.W. Norton.

building the legendary Hanging Gardens of Babylon in Persia. The Greek historian Diodorus Siculus provides a descriptive account of the gardens:

> The approach to the Garden sloped like a hillside and the several parts of the structure rose from one another tier on tier.... On all this, the earth had been piled ... and was thickly planted with trees of every kind that, by their great size and other charm, gave pleasure to the beholder.

Another noted:

> the Hanging Garden has plants cultivated above ground level, and the roots of the trees are embedded in an upper terrace.... Streams of water emerging from elevated sources flow down sloping channels.... These water irrigate the whole garden saturating the roots of plants and keeping the whole area moist. Hence the grass is permanently green and the leaves of trees grow firmly attached to supple branches.... This is a work of art of royal luxury.[5]

While our million-plus cities seem dauntingly large today, early cities also boasted large populations. Athens, at its most successful in 431 BCE contained 300,000 people, while five centuries later Rome boasted more than one million. Chang'an in China reached one million and Teotihuacán in Mexico reached 200,000. Some were small in physical size but densely populated. The great Buddhist city of Taxila, situated at the foot of the Himalayas, covered just a few hectares, while the harbor at Carthage was not much larger than a soccer field.[6] By the thirteenth century, Paris, Milan and Venice contained populations of at least 100,000; by the end of the sixteenth century London had reached 250,000. All of these cities were challenged by a variety of urban environmental issues such as food and water supply, disease and poverty, traffic congestion, limited housing and energy supplies.

Humans transformed the environment by creating the earliest cities. In the rest of this chapter we will select just three of the many environmental impacts: urban design, disease, and pollution controls.

BOX 2.2

The rise and fall of Angkor Thom or the 'Great City' in Cambodia

Angkor Thom was the great capital of the Khmer Empire, which controlled a large area of what is today Cambodia, Thailand, Laos, and Vietnam from the ninth to the fourteenth century. The city area encompassed more than 385 square miles; it was, in terms of area, the largest urban complex in the pre-industrial world. At its peak, it might have housed up to one million people. Today, the ruins of more than 1,000 temples in the area of Angkor remain a testament to its size and extent.

The city's extensive and advanced hydrological systems supporting large-scale agriculture and fresh water delivery have often been cited as the reason for the city's large population. Banks, channels and reservoirs were part of a large-scale water management network built in the ninth century. Water collected from the hills was stored and distributed for a wide variety of purposes, including flood control, agriculture and ritual. A system of overflows and bypasses carried surplus water away to the lake to the south. The network of reservoirs was extensive, and supplied tremendous amounts of water to the city.

The reason for the fall of the Khmer Empire in the fourteenth and fifteenth centuries has been debated. Recent studies suggest that it was the combination of a dramatic decrease in rainfall in the region in the fourteenth and fifteenth centuries, and the impact of the built environments on the surrounding environment. Kathy Marks has noted:

> The water management system, in particular, had the potential to create some very serious environmental problems and radically remodel the landscape. You can see the city pushing into forested areas, stripping vegetation and re-engineering the landscape into something that was completely artificial.

Combined, the stress of climate change and conflict from neighboring empires created an untenable situation for the Khmer Empire. As researcher Damian Evans concludes:

> The city was certainly big enough, and the agricultural exploitation was intensive enough, to have impacted on the environment. Angkor would have suffered from the same problems as contemporary low-density cities, in terms of pressure on the infrastructure, and poor management of natural resources like water. But they had limited technology to deal with these problems and failed to, ultimately, perhaps.

Sources: Evans, D., Pottier, C., Fletcher, R., Hensley, S., Tapley, I., Milne, A. and Barbetti, M. (2007) "A comprehensive archeological map of the world's largest preindustrial settlement complex at Angkor, Cambodia." *Proceedings of the National Academy of Sciences of the United States of America* 104 (36): 14277–82. Fletcher, R., Penny, D., Evans, D., Pottier, C., Barbetti, M., Kummu, M. and Lusting, T. (2008) "The water management network of Angkor, Cambodia." *Antiquity* 82: 658–70. Marks, K. (2007, August 15) "Metropolis: Angkor, the world's first mega-city." *The Independent*, accessed May 2012 at: http://www.independent.co.uk/news/world/asia/metropolis-angkor-the-worlds-first-megacity-461623.html

Urban design

We can deconstruct ways in which the physical design and construction of the city reflect larger environmental discourses. The layout and design of city spaces convey messages about how people view the natural world. Urban design and urban planning also connect to wider social issues of power: most urban design resulted from the desires and decisions of a few powerful individuals to reflect and embody their power.

Urban land-use patterns are ways in which urban society utilizes and defines its relationship to the physical environment. And while the physical patterns—the arrangement of streets and parks and piazzas—vary widely from Berlin to Kolkata to Constantinople, many cities share common land-use differentiations that include the preeminence of the central areas, the marginality of the periphery and the location of particular crafts and merchant activities. There are two land-use patterns that have endured since the earliest cities: walls and grids. Beyond their function, both symbolize deeper perceptions about the urban environment.

BOX 2.3

Water distribution and aqueducts in ancient Rome

Beginning in the fourth century BCE the city of Rome began construction on a series of aqueducts and established plans for the public distribution and management of water. Although the Tiber River flows through the heart of Rome, as the city grew, it needed more water. The historian E. J. Owens notes "a good supply of water was rightly regarded as one of the essential commodities of the maintenance of urban life in the ancient world." The aqueducts were a system of bridges, arcades, and ducts that carried water from a remote source through an enclosed conduit (sometimes running underground, or above ground on an arcade bridge). Along the lines were settling tanks to remove foreign matter or filters that would keep the water free from debris, a factor that safeguarded public health. Once in or near Rome, water from the aqueducts passed into large, covered catch basins. The catch basins then distributed water through free-flowing canals, lead pipes and terracotta pipes to storage reservoirs and public fountains. One of the more interesting aspects of the water supply systems was that most Romans retrieved their domestic water from public fountains; few had private supplies. The tradition of communal civic usage of public works in part shaped the delivery system of water in Rome. With increased urbanization and population growth and rising issues of public health, Roman engineers separated different aqueducts and arranged for some to be used for drinking, and other lines would be assigned other functions, according to water quality. Water was provided for a variety of uses, including fountains and latrines, public baths and street cleaning.

Roman engineers became renowned for their hydraulic technology in general and the construction of aqueducts in particular, bringing water from great distances. Through the development of water management techniques and the engineering of structural forms to transport, divert and store water, populations were able to settle and develop areas of Rome such as the site of the Roman Forum, which was previously waterlogged swamp. In some regards, the feats of Roman engineering and the construction techniques employed in the building of aqueducts and delivery systems marks the urban foundation of Rome just as much as the Capitol or the Forum.

Sources: Evans, H. (1994) *Water Distribution in Ancient Rome: The Evidence of Frontinus.* Ann Arbor, MI: University of Michigan Press. Hodge, A. T. (1992) *Roman Aqueducts and Water Supply.* London: Duckworth. Owens, E. J. (1991) "The Kremna aqueduct and water supply in Roman cities." *Greece and Rome* 38 (1): 41–58.

Fortifications are found in almost any old city of significant size and importance. Such fortifications ranged from wooden palisades and stone walls to ditches and moats. Citadels, or forts, were often incorporated as part of the walled fortification as was the case in Copenhagen and St. Petersburg. Many cities were walled for defense against the "outside" and people could enter or leave only by the gates.[7] The wall served as a military device, a way to protect a city's market privileges, and a way to control the urban population. Encircling moats and canals added to the defense of the city. But these defensive structures also symbolically provided a clear demarcation line between urban and wilderness, between the civilized and the savage, between the insider and the outsider, between people and wild animals. Contact with the world outside was focused at specific access points, often gates. Those cities without walls often ended ambiguously with a few straggling buildings, and then fields. City walls are important for they show an exercise of control over the physical environment, one that attempts to segregate space into distinctive spheres. Walls might have also served as status markers, reflecting the display of power by political, religious or economic elites. In many cities, walls dominated the visual landscape and were the first structures seen when outsiders approached the city from the countryside. Walls inherently convey the message that urban space is separate and apart from the countryside and the wilderness.

Another common feature shared by cities around the world is the imposition of the rectangular street grid on urban space. The grid plan dates from antiquity; some of the earliest planned cities were built using grids and it is by far the most common pattern found in a variety of political societies from absolutionist powers to monarchies to democratic societies. The grid is a simple, rational order of packing the land, setting streets at right angles to one another. As early as 2600 BCE, Harappa (in North India) and Mohenjo-daro (in Pakistan) were built with blocks divided by a grid of straight streets, laid out at perfect right angles, running north–south and east–west. In ancient Egypt, cities like Giza also used a common orientation: a north–south axis from the royal palace and an east–west axis from the temple, meeting at a central plaza. In Babylon the streets were wide and straight and intersected at right angles. Teotihuacán, near modern-day Mexico City, is one of the largest grid-plan sites in the Americas, covering some eight square miles. Its grid system is also aligned astronomically (Figure 2.1).

The rise of the Roman Empire standardized the grid plan, which became a common tool of Roman city planning. The Roman grid is characterized by a nearly perfect orthogonal layout of streets, all crossing each other at right angles. Typically, gates were set at the midpoint on each of the four sides of the rectangle. The grid system allows for easy navigation and better

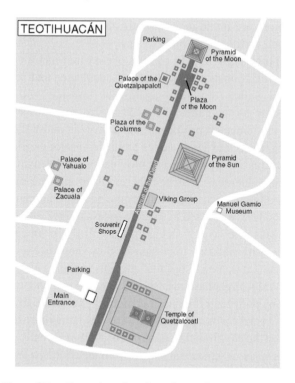

Figure 2.1 Map of Teotihuacán. The city of Teotihuacán is meticulously laid out on a grid, seemingly aligned astronomically. Teotihuacán's main thoroughfare is Calzada de los Muertos (Avenue of the Dead). The Avenue of the Dead connects Piramide de la Luna (Pyramid of the Moon) at the northern end with La Ciudela (the Citadel) at the southern end. The Pyramid of the Sun is directly oriented to a point 15.5° north of west, which is exactly perpendicular to the point on the horizon where the sun sets on the equinoxes

Source: http://www.advantagemexico.com/mexico_city/teotihuacan.html

flowing traffic; however, we can also see that this type of design imposes order on what would otherwise be an organic, chaotic physical environment. The grid system is an imposition of power that determines the shape of living and working spaces. And, in some measure, the grid system inherently denies the importance or the existence of topography. The grid is the triumph of geometry over topography, and its use has become widespread as Figure 2.2, a nineteenth-century map of Cape Town, South Africa shows.

Perhaps one of the most striking examples of the use of the grid system is found in the Forbidden City, located at the exact center of ancient Beijing. The Forbidden City is surrounded by a large area called the Imperial City. As the imperial palace for both the Ming and Qing dynasties (CE 1368–1911), the

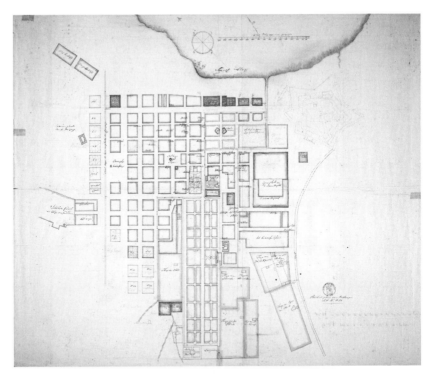

Figure 2.2 Map of Cape Town. The grid system is clear in the 1785 map of the city

Source: Nationaal Archief in The Hague

Forbidden City covered an area of 720,00 square meters and contained more than 9,000 rooms. Figure 2.3 shows the map of the Forbidden City. Emperor Yongle of the Ming Dynasty began building the Forbidden City in 1406; it took one million laborers and 100,000 craftsmen 15 years to complete.

A few features are worth noting. The rigid, rectangular layout was purposefully aligned to the cardinal directions, north, south, east and west. The entire palace, rectangular in shape, is surrounded by walls ten meters high and a moat 52 meters wide. The walls were constructed to withstand attacks by cannons and are thick and squat. The wall has four gates with towers above them: East Magnificent Gate, Meridian Gate to the south, West Magnificent Gate and Gate of Obedience and Purity to the North. On each of the four corners stand four turret towers. The layout within the walls is similarly geometric: the city is divided into northern and southern parts. The southern parts served as the emperor's work area; the northern parts contained his living quarters. These structures were arranged along a central axis at the exact midpoint and are symmetrical on either side.

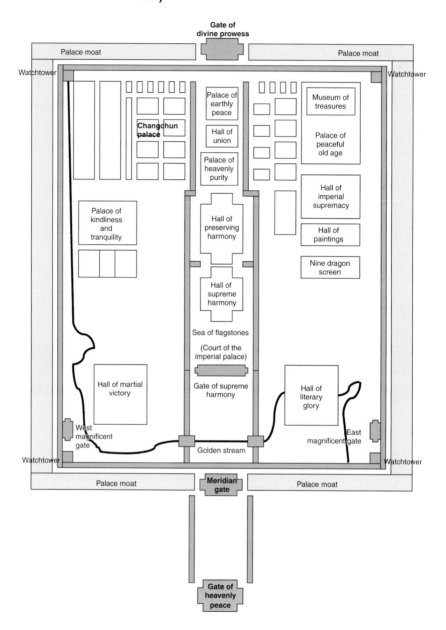

Figure 2.3 Map of Forbidden City

Three structures in the southern area include the Hall of Supreme Harmony, the Hall of Central Harmony and the Hall of Protective Harmony. The Hall of Supreme Harmony was the site of ceremonies, including the emperor's coronation to the throne, his marriage and other official state events. The Hall of Central Harmony was where the emperor rested and received officials; the Hall of Protective Harmony was the site of banquets.

In addition to the use of the grid and the wall, the Forbidden City is replete with symbolic meaning that reflects Chinese culture that connects to wider conceptions about the cosmos. The city itself was patterned after the Heavenly Palace. In ancient Chinese astrology, the Heavenly Palace centered around the North Star and was considered the center of heaven. The number nine received special inclusion in the city design. The number of houses in the Forbidden City is 9,999, and on every door are patterns of nine nails in vertical and horizontal lines. Ancient Chinese regarded nine as the biggest number, to which only emperors were entitled.

The gridiron has prevailed in cities around the world—from Europe to Asia to the Americas. In US cities, the grid plan was nearly universal in the construction of new towns and cities. In some cases the grid softened over time. Boston, set on a grid in the mid-1600s, becomes more organic and curvilinear as it approaches the harbor. In San Francisco a grid was imposed on what must be considered among the most dramatic topography of US cities. Figure 2.4 shows Savannah Georgia in 1734. Note that the entire area has been cleared of trees and vegetation: the natural world has been obliterated and the city (re)constructed from a *tabula rasa*. As US cities grew outward, particularly after the twentieth century, the grid becomes less prevalent.

Despite a multitude of geographies and topographies, of altitudes and latitudes, many cities on the grid share common design features: a lack of sensitivity to the physical environment, the imposition of the grid regardless of topography, a focus on the geometric (geometry over geography) and an underlying sense of the control of people-made space. The earliest palimpsest for many cities is the not-so-subtle attempt to conquer the physical environment, if not by axe or stone, then by design.

Most cities have gardens, small parks or large parks and other "green" amenities. These can also tell us something about how society viewed the natural world. A garden is an arrangement of nature, whose plant materials and ordering principles are determined by prevailing ideas about the relationship between nature and society.

In Florence, Italy, the Boboli Gardens were constructed on a grand scale during the Italian Renaissance. Part of Renaissance culture saw the universe

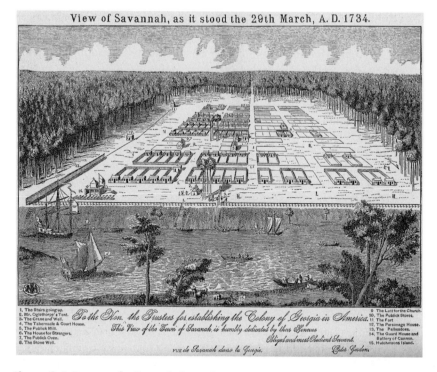

Figure 2.4 Savannah, Georgia in 1734

Source: "View of Savannah, as it stood the 29th March, A.D. 1734." From Report on the Social Statistics of Cities, compiled by George E. Waring, Jr., United States. Census Office, Part II, 1886. Courtesy of the University of Texas Libraries, the University of Texas at Austin. http://www.lib.utexas.edu/maps/historical/savannah_1734.jpg

as a series of hierarchies—with God at the top, humans at the center and nature at the bottom. This larger metanarrative, or environmental discourse, thus informed urban design. Wealthy Italians referred to the gardens of their private residences as "villa gardens." "Villa" meant that all the formal parts of the ground were arranged in direct relation to the house; they were seen as extensions to the palaces and were treated as a living space. The historian Claudia Lazzaro, in her book *The Italian Renaissance Garden*, notes that "Gardens in Renaissance Italy are witness to the attitude of contemporaries toward nature, much of which was inspired by classical culture."[8] The goal of many Renaissance garden designers was to implicitly recreate the gardens of classical antiquity, which were characterized by planting trees in ordered rows, clipping boxwoods into ornamental animals and shapes, decorating grottoes and placing statutes of the gods of pagan mythology around the gardens (see Figure 2.5). In this way, many Renaissance gardens were as much "art" as they were nature; as much contrived and manipulated as they were organic. Many gardens required significant earth-moving and water-powered devices. The gardens were dominated by this formal approach to design, although some aspects of the view that nature was "uncontrollable" also informed design. For example, many small bed areas were allowed to grow at will; groves of trees were not always planted in orderly rows.

The Boboli Gardens are a sixteenth-century Medici garden, located on the grounds behind the Palazzo Pitti, which was purchased by Eleanor de Toledo, wife of Cosimo I Medici in 1549. The Medicis' wealth originated from the wool trade; later they became influential international bankers. The Medicis were among the most powerful families in Florence during the Renaissance. Eleanor was instrumental in the creation of the gardens, hiring Niccolo

Figure 2.5 Boboli Gardens highlights the themes of formal garden design of Renaissance Florence. The imposition of order can be seen in these rigid geometric hedges

Source: Photo by Lisa Benton-Short

Pericoli, known as Tribolo, a famous architect, to design the gardens between 1550 and 1558. Tribolo's plan was to center the gardens on a large fountain, framed by vegetation. But the plan had to incorporate a U-shaped hillside behind the palace, which Tribolo initially planned to plant with a "boschetti" of evergreen trees planted in rows. Dwarf fruit trees were planted in large beds and near the orchards was a fishpond. There was also an extensive botanical garden that did not survive to the present.

Tribolo's plan was the basis for many of the royal gardens in Europe, including Versailles. The palace and grounds of Versailles remains one of the most famous gardens in the world, although the term "garden" is an understatement. Similar to the Boboli Garden, Versailles was as much a political statement about power and the ability to display it on a monumental scale as it was about engaging with nature. The park and garden were designed by Andre Le Norte between 1661 and 1700. His plan included magnificent features, parterres, great basins, an orangery and even a canal. Avenues project from the palace toward different horizons, bringing together forest, garden, palace and city. The most visually obvious design element is the use of geometry—circles, diagonals, squares and rectangles both formalize and define different planting beds and spaces which were immaculately clipped and maintained (see Figure 2.6). A vast collection of outdoor sculpture and fountains add to the Baroque sense of grandeur. Importantly, the gardens of Versailles spoke of the power of an aristocracy; this was not a public garden in the twenty-first-century definition.

The design of Versailles is intricately linked to wider intellectual developments of the time. The historian Carolyn Merchant has argued that the rise of the Scientific Revolution of the sixteenth and seventeen centuries was a time when the cosmos ceased to be viewed as an organism and became instead a machine.[9] She argues that this was a dramatic shift in a broader worldview of nature–society relationships. Advancements in physics, mathematics and technology impacted politics, literature, art, philosophy, religion and, we argue, even urban design. This crucial intellectual shift, which Merchant calls "the death of nature," is characterized by accelerated exploitation of human and natural resources in the name of culture and progress.[10] This new worldview that saw the earth as a machine not only sanctioned exploitation, but also subjugation. We see this reflected in the plans for Versailles in which the defining design elements are rationality and symmetry, revealing an orderly, subdued nature contained within pots and beds, shaped and clipped to repress the organic growth patterns. Nature had become an object to be observed, manipulated and ordered. Even statuary reminded the viewer of human handiwork, not nature's. While this era produced many of the famous

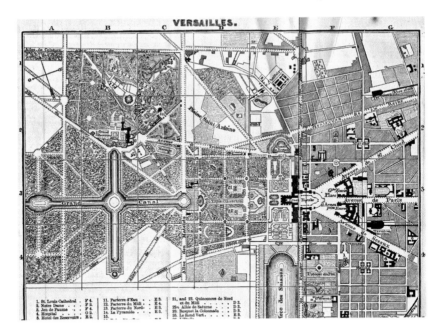

Figure 2.6 Versailles

Source: This image comes from the 4th edition of Meyers' Konversationslexikon, published in 1885

gardens of Europe, we must also see these as products of aristocratic government and broader statements about changing views toward nature, particularly as society appears to be reducing the mysteries of the cosmos and gaining a sense of control over the natural world.

The development of another political capital, Washington, DC, underscores the attempt to embody political ideals with a developing sensitivity to the natural world. The 1791 design of the new federal capital reflected the struggle to create a space representative of American ideals and aspirations. The city was from the beginning rich in the physical and symbolic expression of democratic ideals and a reflection of the formulation of national identity.

Pierre Charles L'Enfant was born in France and trained as an artist and painter under his father at the Royal Academy of Painting and Sculpture in Paris over 1771–6. In 1777, at the age of 23, L'Enfant came to America to volunteer in the Continental Army. L'Enfant rose to the rank of major by the end of the Revolutionary War. George Washington knew of L'Enfant's artistic skills from the sketches and portraits of officers he had done while at Valley Forge.[11] After the war, L'Enfant established himself as an architect and worked in New York and Philadelphia. President Washington asked L'Enfant to design the new federal city.

L'Enfant arrived at Georgetown in March of 1791 intent on designing a capital that would befit a great new country. The concept of a planned city was not new in America. L'Enfant himself knew of the plans for American cities of Annapolis, Savannah, Williamsburg, Philadelphia and New York. He was also well versed in European city planning.

The neoclassical spatial order tradition from which the capital city was created was partly inspired by monumental planning trends in France. Common in France was the use of radial patterns imposed upon orthogonal streets, which offered an urban typology for expedient circulation, enchanting entry vistas and possibilities (space) for defining neighborhoods.[12] Within these monumental plans were symbolic spaces (from which emanated the radial streets). Designing celebratory public spaces in squares and semicircles was supposed to give the city an air of grandeur, delight and edification. All of these planning elements were well known to L'Enfant and eventually incorporated into his design.

In 1791, L'Enfant drafted a comprehensive plan for the new city of Washington, DC. The plan consisted of both a large map and a series of descriptions (Figure 2.7). L'Enfant's broad vision of a capital of buildings, public squares and promenades reflected the new country's optimistic outlook.[13] It was a design intent on celebrating a ceremonial city, the center of national government and culture. The plan was on an immense scale, far beyond the size or even expectation of the government at the time.[14] Two of the most unique characteristics of L'Enfant's plan were its grand scale and the fact that it was laid out so as to utilize to the fullest extent the natural topography.

L'Enfant's plan is based on a template of a grid system of streets on a square block pattern, within which broad diagonal avenues would link the main hills. The resulting circles and squares provided public "reservations" or public space throughout the city. The diagonal avenues were named after the states. Central to the plan were two hills—one called Jenkins Hill, which would be the site of Congress, and the other to the west, which would be the President's Palace (later called the White House). L'Enfant intended the two critical government buildings to be separated by space, a metaphor for the need to separate the branches of government. The broad diagonal street connecting the two was designated as Pennsylvania Avenue. The two elevations were to be linked also by two large parks. One would stretch south from the President's Palace, and from the Congress House westward would be another he termed the Grand Avenue (eventually called the National Mall).[15]

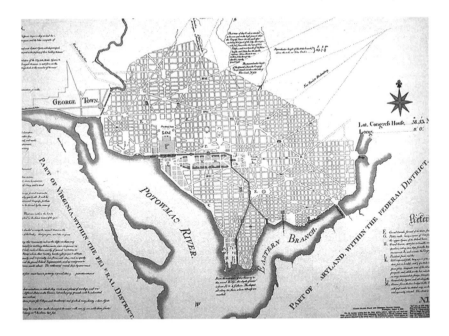

Figure 2.7 Map of Washington, DC, 1791

Source: McMillan Commission Report, 1901. Image courtesy of the National Coalition to save Our Mall. http://www.savethemall.org/mall/resource-hist02.html

Thus a central design element was the axial alignment of the city. L'Enfant envisioned a primary axis east-to-west stretching from the Capitol to the banks of the Potomac. The secondary axis going north-to-south would be perpendicular, crossing from the President's House to the river to the south. Where the two axes crossed, the convergence point, L'Enfant intended to place the statue of President Washington on horseback.

A unique feature of the plan was the inclusion of an intentionally large green space. The National Mall was the centerpiece of the plan. L'Enfant referred to it in his notes as a

> Grand Avenue, 400 feet in breadth and about a mile in length, bordered with gardens, and ending in a slope from the houses [of diplomats] on each side. This Avenue leads to Monument A [an equestrian figure of George Washington], and connects the Congress garden with the President's park.

It would be composed of a tree-lined walkway perhaps planted in the natural, picturesque style of landscape gardening which was gaining popularity in

France and Britain, or the more formal landscaping that was common in European cities.[16] He noted that the stretch of land would remain "a vast esplanade" and the "sort of places as may be attractive to the learned and afford diversion to the idle." His proposal that the major axis of the new city should be two great parks meeting at a central point (the Washington statue) was unusual in that most cities were built around commercial streets.

L'Enfant's plan for the Mall reflected his vision that a democracy should have public open space. In addition, the placement of the Mall as the center-piece of the design represents that open space conferred visible power and strength and symbolized an open, not closed, society. These were important values in the fledgling republic, which sought to distinguish itself from the British monarchy. It was intended that the presidency be viewed physically and symbolically as "open," and accountable to the public. By including green space—a vast open space—at the center of the city, L'Enfant provided an important legacy. His Washington plan predates by more than 70 years the urban park movement.

Today, the Washington Monument is the defining feature of the Washington skyline and the centerpiece of the nation's most symbolic public open space. But at ground level the vast expanse surrounding the obelisk remains, remarkably, unfinished. Seeing this as an opportunity to elicit thoughtful and visionary ideas, in 2010 a coalition of Washington-area historians, architects, professors and others groups organized a competition to begin a dialogue about the underused expanse at the heart of the Mall. The purpose of the Competition was to provide a vision for the monument grounds that responds to twenty-first-century needs while also giving expression, at long last, to the great legacy of the historic visions, including L'Enfant's. The National Ideas Competition for the Washington Monument Grounds repre-sented an important effort to reclaim the Mall for the public, one worthy of further discussion. You can view the different design submissions, the design winners, as well as numerous historic plans for the Washington Monument at: www.wamocompetition.org.

Disease and cities

From their beginnings cities have been places of disease. The gardens of Babylon and the temples of Egypt were emblems of urban glory, but the alleys in their shadows were choked with garbage, vast amounts of human and animal wastes and decaying food.[17] The city was an environment that promoted and disseminated a range of diseases. Each of the four major types of disease transmission—airborne, waterborne, direct contact and by insects or other vectors—was enhanced by urban life.

The transition from nomadic existence to urban life was not without its consequences. The historian Arno Karlan notes that the emergence of agriculture and the domestication of animals generated an epidemiological crisis.[18] For the first time, people lived in continual intimate contact with other species; inevitably, they exchanged pathogens. Pathogens made their way to humans who breathed the same air and dust, who touched animal wastes or butchered animal carcasses, used their wool and hides and consumed their milk, eggs or flesh. Each new domesticated species—dogs, birds, pigs, goats, cows, chickens, cats and even rats—exposed human populations to viruses and bacteria that, over time, transmitted to humans. It is speculated that measles is related to the viruses causing canine distemper, influenza from pigs and horses, and smallpox related to those causing vaccinia in cows. Typhoid, often fatal to humans, may have originated in rodents and birds. Other vector-borne diseases include those spread by rodents, birds, snails, but perhaps the most varied and numerous vectors are insects. Diseases such as malaria, yellow fever and dengue came with greater exposure to mosquitoes due to human-induced changes in the landscape brought about by agriculture and settlement, such as ditches and puddles of irrigated fields, which became ideal breeding grounds for mosquitoes. Portrayed by some as the pinnacle of human ingenuity and civilization, cities also became a breeding ground, transmission point and laboratory for a toxic cocktail of infectious diseases. Cities were very dangerous places and until urban public health reforms in the late nineteenth century, living in cities was always a hazardous activity. So, at second glance the so-called triumph of the rise of agriculture and the development of cities turns out to have unleashed a public health hazard.

The first epidemic of a waterborne disease was probably caused by an infected caveman relieving himself in waters upstream of his neighbors. Perhaps the entire clan was infected, or perhaps the clan fled from the "evil spirits" ravaging their camp. But as long as people lived in small groups, isolated from each other, such incidents were sporadic. Once people began clustering in cities, they shared communal water, handled unwashed food, stepped in excrement from horses, or came into contact with urine used for dyes, bleaches and antiseptics. As farmers and villagers moved into cities, germs lurking in animals, wastes, filth and scavengers were offered feasts, and countless people were sickened and killed by previously unknown epidemics: smallpox, measles, mumps, influenza, scarlet fever, typhus, bubonic plague, and the common cold. As cities grew, they become breeding grounds for waterborne, insect-born and skin-to-skin infectious diseases. Typhus was most common, but typhoid, plague, smallpox, cholera and dysentery often broke out unchecked and seemingly at will.

Fragmented stories and accounts of urban epidemics are found in ancient writings of the Sumerians, Babylonians, Egyptians, Greeks, Romans, Indians and Chinese. With the emergence and growth of cities the element of population density became a crucial factor in the spread of disease, particularly those infections spread via human contact such as coughing, sneezing or casual contact with chamber pots. Thus "crowd diseases" such as smallpox, typhus, leprosy and tuberculosis were a consequence of urbanization.

Poor sanitation along with population density were among other factors that amplified the spread of disease. Street life in medieval England, for example, was dangerous and uncomfortable:

> butcher and poulteres were by no means alone in their careless disposal of animal refuse; fishmongers and cooks and the ordinary households were all guilty. In Chester women carrying entrails of animals from the butchers carried them uncovered and threw them out near the gates, to the public nuisance…. The private citizen was only too ready to dispose of dead dogs and cats by dropping them in the river or just over the town wall.[19]

These conditions facilitated the spread of plague from rats to people in crowded urban areas; there are recorded instances in early urban history. An outbreak of plague (bubonic or perhaps measles) in Athens in 430 BCE was said to have destroyed the city's "Periclean golden age," unraveling its military strengths and its civic and moral fabric. About 100,000 people or 25 percent of the Athenian population was killed. The epidemic, highly contagious and indiscriminate in its victims, was likely exacerbated by the fact that the city was, at the time, under siege during the Peloponnesian War. It was also a major factor in the Athenians' defeat during that war. The plague is a well-known part of history thanks to an eyewitness account by Thucydides, which inspired numerous literary works.[20] Scientists have debated the cause of the epidemic; two of the most common possible causes are typhus (which is unrelated to typhoid) and smallpox.

For the next thousand years in Europe, the plague was a periodic, recurring disease. The cycle of outbreaks in the fourteenth century would be known as the Black Death. The first of several waves of plague descended on Europe starting in 1347 and lasted through 1450. The Black Death consisted of both bubonic and pneumonic forms of the plague. Bubonic plague, caused by flea

BOX 2.4

Disease in ancient Rome

The ancient Roman Empire experienced a number of health epidemics. One of the more devastating examples was the Antonine Plague of CE 165. Troops returning to Rome brought back a disease, most likely smallpox, that killed an estimated five million people. Although the Antonine Plague led to millions of deaths, it was not the only outbreak to cause such casualties. A number of diseases flourished due to deforestation and crowded urban centers. Rome was built around the Tiber River basin. The land proved favorable for intensive farming, which then allowed the city to grow. As the city continued to grow, more and more land was stripped of its timber. The deforestation resulted in frequent flooding and stagnant water which led to a proliferation of insect-borne diseases. Ostia, Rome's main port, had to be abandoned as the land turned to stagnant marshland, an ideal breeding ground for malaria-carrying mosquitoes. A similar fate befell the Pontine territory. Although it began as a very fertile region it too turned into marshland that was unsuitable for farmland. Malaria also plagued this region and the population collapsed.

As in many instances, malaria disproportionately impacted the poor. Although malaria was not then understood to be caused by mosquitoes, people recognized the danger of the marshy valleys. The elite Romans would leave during the hot summers and move to homes on the hills. The poor, however, were stuck in the densely populated valleys. The dense population and the flooded streets served to worsen the impacts of malaria.

These examples underscore that deforestation, poor sanitation and lack of appropriate planning for the growth of the city can lead to significant problems. Rome, however, was able to implement some measures that decreased the prevalence of malaria. The Cloaca Maxima, a series of channels originally built by Tarquinius Priscus in the sixth century BCE to drain rainwater away from the heart of Rome, eventually came to serve as the primary sewerage canal. The Cloaca Maxima helped to remove the enormous volume of graywater produced within the city. Although the canal system helped to mitigate a small amount of the environmental impact of the city on the portion of Tiber used for drinking purposes, all of the graywater transported in the canal returned to the Tiber just outside the city, polluting much of the water supply for the region. The Cloaca Maxima and an expanding overall cloacae system reduced some of Rome's drainage issues. In the fifth century BCE, a sewer was added to the drain. The combination of a drain and a sewer system reduced standing water, therefore reducing malaria breeding grounds; it separated sewage, thus limiting a number of other communicable diseases.

Sources: Murphy, V. (2005, July 11) "Past pandemics that ravaged Europe." *BBC News.* http://news.bbc.co.uk/2/hi/health/4381924.stm (accessed January 19, 2012). O'Sullivan, L., Jardine, A., Cook, A. and Weinstein, P. (2008) "Deforestation, mosquitoes, and ancient Rome: lessons for today." *BioScience* 58 (8): 756–7. Sura, A. (2010) "The Cloaca Maxima: draining disease from Rome." *Vertices Duke University Journal of Science and Technology,* spring: 23–4. http://issuu.com/ritzness/docs/vertices_spring2010 (accessed May 2012).

bites spread by lice on host rats, caused swelling of the lymph nodes (buboes) and fever; the pneumonic form (spread person to person) invaded the lungs and was more deadly. It has been estimated that nearly 25 percent of Europe's population, some 25 million people, died in the 1347–1450 epidemic. As the Black Death spread across the cities of Europe it created panic, death and despair. The impacts were profound. The Italian writer Boccaccio wrote of his experiences living in Florence during the Black Death (1347–9) in the introduction to the *Decameron*:

Whenever, fairest ladies, I pause to consider how compassionate you all are by nature, I invariable become aware that the present work will seem to you to possess an irksome and ponderous opening. For it carries at its head the painful memory of the deadly havoc wrought by the recent plague, which brought so much heartache and misery to those who witnessed, or had experience of it.

What more remains to be said, except that the cruelty of heaven (and possibly, in some measure, also that of man) was so immense and so devastating that between March and July of the year in question, what with the fury of pestilence and the fact that so many of the sick were inadequately cared for or abandoned in their hour of need because the healthy were too terrified to approach them, it is reliably thought that over a hundred thousand human lives were extinguished within the walls of the city Florence.[21]

The Plague would inspire other writers such as Albert Camus, who wrote "The Plague." One of the more well-documented outbreaks of plague occurred in London between 1665 and 1666, which at that time had a population of some 500,000. During the height of the outbreak, some 7,000 people died each week. More than 100,000 people died by its end. The city was nearly deserted. The historian Walter George Bell chronicled the great London Plague and noted that it retreated in 1666 primarily due to the Great London Fire, which burned more than four-fifths of the city.[22]

The Plague was not the only infectious disease to ravage urban populations. In the nineteenth century, cholera became one of the world's first truly global diseases. Cholera did not kill as many people as endemic diseases like typhus and tuberculosis, but its sudden onset, obscure causation and dramatic effects struck fear into the entire population. Today we know that cholera is caused by ingesting water, food or any other material contaminated by the feces of a cholera victim. Contact with a contaminated chamber pot, soiled clothing or bedding can spread the disease. The onset of disease transpires within 12–48 hours of infection and is often characterized by extreme diarrhea, sharp muscular cramps, vomiting and fever and— sometimes—death.

The disease spread along trade routes and into cities, where it spread from port to port (sometimes the germ lived in contaminated kegs of water or in the excrement of infected victims, and was transmitted by travelers). In the early 1830s, cholera entered New York through infected ships. Quarantine regulations which sought to "contain" cities were ineffective. The disease hit worst where poor drainage and human contact came together: the crowded slums.

BOX 2.5

Mapping cholera: the ghost map

In the middle of the nineteenth century, London, a city of two million people crammed into just 30 square miles, was just emerging as one of the first modern cities in the world. But it lacked the public health infrastructure to support its exploding population. The most serious problem confronting the city was what to do with the enormous quantities of raw sewage generated by so many people and animals packed together. In the last week of August 1854, many residents of London's Golden Square, home to some of London's poorest, suddenly took sick and began dying. Their symptoms included upset stomach, vomiting, abdominal cramps, and diarrhea. The local water pump, which was public, free and previously considered a safe source of drinking water, drew from a well beneath Golden Square.

Steven Johnson's book *The Ghost Map*, provides a riveting account of the epidemic, part detective story, part biology lesson. Johnson writes

> Hundreds of residents had been seized by the disease within a few hours of one another, in many cases entire families, left to tend for themselves in dark, suffocating rooms.... You could see the dead being wheeled down the street by the cartload.

Seventy fatalities occurred in a 24-hour period, most within five square blocks. It was cholera. But naming the disease was far more easy than stopping it. No one understood what cholera was or how it spread. At the time, it was considered a miasmal disease. Miasma referred to the spread of disease through odor, and its sources included pools of sewage, animal carcasses, decaying vegetation and poor ventilation. The production of these vapors and orders also reflected prejudices society had about the irresponsible poor and their living habits. Health authorities rejected the notion of waterborne contagions.

Dr. John Snow, a practicing physician, took a different approach to the disease. He suspected a link between the disease and drinking water. His opinion was rejected by the scientific establishment. Yet ultimately Snow made the biggest contribution to solving the mystery of cholera. By plotting the residential locations of those infected with cholera, Snow demonstrated how the cases of cholera in central London were clustered around a single source of contaminated drinking water, a neighborhood water pump on Broad Street (Figure 2.8). The Broad Street hand pumps received untreated water from the Thames. Although some health officials resisted, he managed to convince authorities to break the arm of the pump; the outbreak then subsided.

Figure 2.8 John Snow's map of cholera

Source: Drawn by Dr. John Snow about 1854; shown in L. D. Stamp (1964) A Geography of Life and Death

Although Snow is most famous for his analysis of the Broad Street outbreak, his work confirmed a theory he had been developing: that contaminated water contained disease-causing organisms. Cholera had never impacted Britain until crowded urban conditions exposed people to drinking one another's sewage. Snow's map of cholera detailed the spatial pattern to the geography of the spread of disease and ultimately led to the cause of disease, and a fundamental change in our understanding of disease and its spread. By solving the problem of cholera, Snow revolutionized the way we think of the spread of disease and the rise of the modern city. His legacy was to inform public health reforms of the late nineteenth century that would make cities livable through improvements in sewage and sanitation, thus eliminating conditions that generate epidemics of disease.

Source: Johnson, S. (2006) *The Ghost Map: The Story of London's Most Terrifying Epidemic and How it Changed Science, Cities, and the Modern World.* New York: Penguin/Riverhead Books.

The connection between disease and cities is not simply a physical phenomena limited to epidemiological dimensions. Diseases are very much linked to social and political elements. For example, many early policies and approaches to sanitation characterized slums as places of crime, where moral and social decay threatened disorder and disease.[23] The equation of immigrants, minorities and other slum dwellers with disease became a powerful image used in attempts to control the urban environment. In Victorian London, endemic poverty and disease and the potential violence of the factory laborer created anxiety among the Victorian elite, who viewed these as a threat to progress and social order. The historian Maynard Swanson has argued that in the South African cities of Cape Town and Port Elizabeth, authorities used public fears of disease to justify residential racial segregation. Swanson noted that the fear of cholera, smallpox and plague rationalized efforts to remove or even segregate various elements of society. He concluded that

medical officers and other public authorities in South Africa were imbued with the imagery of infectious disease as a social metaphor, and that this metaphor powerfully interacted with British and South African racial attitudes to influence the policies and shape the institutions of segregation.[24]

Outbreaks of diseases, and particularly an outbreak of bubonic plague in 1901, became an opportunity for those who were promoting segregationist solutions to social problems; thus concern for sanitation is one major factor in the creation of urban apartheid.

Similarly, an 1864 sanitation report on Bombay, India detailed "noxious matters, poisonous gases and accumulated filth," arguing that "filthiness was the worst of Bombay's many Evils." Night-soil collectors struggled to collect ever-increasing amounts of human waste from the streets. Geographer Colin McFarlane has noted that colonial law and regulation was often at variance with the practices of local society; for example, Hindu cremations or Parsi death rituals (where bodies are left to vultures) were considered dangerous and offensive.[25] His work has shown that the idea of the contaminated city reflected different cultural understandings of public and private, appropriate and inappropriate practices and behaviors. While there were some attempts to undertake widespread reforms, most officials tried to create *cordons sanitaires* to protect elite groups (white, British colonials) from the threat carried by the poor and poor places. These two examples show that issues of disease and the fear of the Other went hand in hand.

In Wellington, New Zealand disease also had an impact on land-use change. When Wellington became the nation's capital in 1865, many businesses relocated their head offices or established branches in the city to gain the advantages of being close to government. However, the city barely expanded beyond its early boundaries, and as a result, very high densities of housing formed slums with little or no sanitation, leading to outbreaks of cholera and typhus. The threat of disease and the development of transport infrastructure in the city hastened the shift of more affluent residents outwards to new suburbs.[26]

That cities survived the continued onslaught of diseases is remarkable. Because cities experienced both epidemics and endemics, they became "population sinks"; where death rates exceeded birth rates. Many catastrophic epidemics were followed by a resurgence in population growth as the inflow of migrants from the countryside replenished the urban populations. As Arno Karlen observed, "it is a testimony to human vigor and adaptability that city life flourished despite plagues, famines, wars and migrations."[27]

Cities and pollution controls

As long as there have been cities, there has been pollution and attempts to deal with the issue. Aristotle (384 BCE–322 BCE) mentions in *Athenaion Politeia* (the Constitution of Athenians), a rule that manure should be placed outside the town, at least 2 km away from the town walls. Smoke caused marble to turn gray, and Athens introduced laws about when and where open fires could burn. Similarly, the Roman Senate introduced a law about 2,000 years ago, according to which: "*Aerem corrumpere non licet*," meaning "Polluting air is not allowed."[28] One law stipulated that cheese-making manufactures should be located so that their smoke not pollute other houses.

In 1231 Emperor Frederick II Hofenstaufen, concerned about air quality in Sicily, decreed a new law to clean up the air and went as far as to impose monetary penalties:

> We are disposed to preserve by our zealous solicitude, insofar as we are able, the salubrious air which divine judgment has provided. We therefore command that henceforth no one be permitted to place linen or hemp for retting [soaking] in any waters within the distance of one mile from any city or castle, lest from this, as we have learned certainly happens, the quality of the air is corrupted. We order that the bodies of the dead, not placed in coffins, should be buried to a depth of one-half a rod. If anyone does the contrary, he shall pay our court one augustalis. We further order that those who take the skins of animals should put the carcasses and wastes which create an order outside the territory [of a city] by a fourth part of a mile, or throw them into the sea or river. If anyone does the contrary, he shall pay to our court one augustalis for dogs and animals which are larger than dogs, and one-half an augustalis for smaller animals.[29]

Similarly, the bailiff of the medieval city of York posted an order to deal with both air and water pollution in a city where the stench was pronounced:

> Whereas it is sufficiently evident that the air is so corrupted and infected by the pigsties situated in the king's highways and in the lanes of that town and by the swine feeding and frequently wandering about … and by dung and dunghills and many other foul things placed in the streets and lanes, that great repugnance overtakes the king's ministers staying in that town and also others dwelling and passing through,

BOX 2.6

Early pollution control in Paris

Paris has a long history of attempting to solve problems associated with pollution.

For hundreds of years, the basic rule for dealing with Parisian garbage was "tout-a-la-rue"—all in the street—including household waste, urine and feces. Larger items were frequently thrown into the "no-man's-land" over the city wall or into the Seine. Feces, however, was often collected to be used as fertilizer. Parisian dirt streets were even worse when frequent rain and heavy pedestrian and cart traffic spread the refuse around. The edible muck was often consumed by pigs and wild dogs, and the rest was consumed by microorganisms. The smell of the rotting, decomposing matter was terrible.

Rapidly accumulating garbage in the streets of Paris finally prompted in 1348 Phillipe VI de Valois to pass an ordinance requiring the citizens to sweep in front of their doors and to transport their garbage to dumps or risk fines and imprisonment. He established the first corp of sanitation workers to clean the streets. Even with ordinances issued every few years, these brought little relief and were difficult to enforce. Garbage piled up in the streets, making some completely inaccessible.

In the fifteenth century, Charles IV created official dumps outside the city walls, but the situation inside the city walls still did not improve much. In the late sixteenth century, these dumps became so tall and large that they were fortified, fearing that enemies would use them to point their canons down on the city. Despite these efforts at controlling house waste, however, little was accomplished. Yet another ordinance of 1780 again forbade people from throwing water, urine, feces or household garbage out the window.

Garbage wasn't the only pollution problem. Contaminated water has long been an issue for Paris. In 1539 Francois I ordered property owners to build cesspools, for the collection of human sewage, into each new dwelling. Those who would not comply would have their houses confiscated, and rents were collected to pay for the cesspools. Unfortunately, most of these cesspools leaked, as water tightness was not required. In 1674, another decree stated that feces must be separated from other wastes at the dump, in order to begin manufacturing poudrette or human guano—a greasy, powdery, flammable substance made by open-air fermentation of human sewage. Valued as a fertilizer, it was toxic to breathe and incredibly foul smelling.

Source: Krupa, F. (n.d.) "Paris: urban sanitation before the 20th century—a history of invisible infrastructure." http://www.translucency.com/frede/parisproject/garbage1200_1789.html (accessed June 2012).

the advantage of more wholesome air is impeded, the state of men is grievously injured, and other unbearable inconveniences ... the king, being unwilling longer to tolerate such great and unbearable defects there, orders the bailiffs to cause the pigsties, aforesaid streets and lanes to be cleansed from all dung and to cause them to be kept thus cleansed hereafter.[30]

In 1900, a pandemic of bubonic plague, which had originated a few years earlier in China, spread to Australia and New Zealand. In Sydney, Australia, a citizens' committee formed to raise funds for the Sydney Hospital for Sick Children in order to deal with the increasing number of patients. In New Zealand, public fear of the plague stimulated the passage of a Public Health Act which set up a Department of Health with effective powers. The country was divided into six health districts, each with a medical officer in charge. In addition, each district had a senior medical officer for the Maori population specifically. These measures led to the initiation of effective action against plague, typhoid fever and tuberculosis, to improved quality of food and drugs, to the establishment of boards for the control of hospitals for infectious and general diseases.

Throughout history there are many examples of ordinances such as local pollution control or public health initiatives. However, we can identify two major shifts that prompted changes in national policies that impacted cities. The first, which we will explore in more detail in the next chapter, is the advent of the industrial city, which prompted a variety of public health/pollution abatement regulations. The second is the environmental revolution of the mid-to-late twentieth century, which created more stringent regulations to limit pollution and introduced national standards.

Take the case of air pollution, which had an increasing impact with industrialization and population growth. Air pollution was a problem in British cities for more than 800 years. Until the twelfth century, most Londoners burned wood for fuel. But as the city grew and the forests shrank, wood became scarce. Large deposits of coal provided a cheap alternative. By the thirteenth century, Londoners were burning soft, bituminous coal to heat their homes and fuel their factories. In 1257 Queen Eleanor of Provence visited Nottingham Castle and was forced to move because of fouled air, full of heavy coal smoke. Numerous attempts to control coal burning and to punish offenders were made during the thirteenth and fourteenth centuries but were largely ineffective. In 1661 John Evelyn wrote an anti-coal treatise *Fumifungium: or the Inconvenience of the Aer and Smoake of London Dissipated*, in which he pleaded with the king and Parliament to do something about the dark yellow mist with a pungent smell that hung over the city. By the middle of the nineteenth century London's air was so highly polluted that it figured in popular novels. Here is an extract from a Dickens novel, *Bleak House*, first published in 1853:

> [A]s he handed me into a fly after superintending the removal of my boxes, I asked him whether there was a great fire anywhere? For the streets were so full of dense brown smoke that scarcely anything was to be seen.
> "Oh, dear no, miss," he said. "This is a London particular."
> I had never heard of such a thing.
> "A fog, miss," said the young gentleman.

Episodes such as Charles Dickens describes were common in mid and late nineteenth-century London, a result of the burning of coal laden with high levels of sulfur released into the air when combusted. Smoke particles from industrial plumes would mix with fog, giving it a yellow-black color. Air pollution was long accepted as simply an unfortunate fact of London life. Increasing pollution does not lead automatically to environmental regulations; only when the problems reach high on the political agenda is action taken. In December 1952 an anticyclone settled over London. The wind dropped and a thick fog began to form. Londoners burned more coal to combat the winter cold. The Great London Smog, as it was named, darkened the streets of London for five days and levels of sulfur dioxide increased seven-fold and levels of smoke increased three-fold. The smog killed approximately 4,000 people immediately and caused another 8,000 premature deaths. Most of the 4,000 deaths occurred due to pneumonia, bronchitis, tuberculosis or heart failure, with the peak in the number of deaths coinciding with the peak in both smoke and sulfur dioxide pollution levels. The British government, initially reluctant to admit that coal smoke was the cause, blamed the deaths on a flu epidemic. But pressure mounted and in 1956 the government passed its first Clean Air Act, which created smokeless zones in the city where only clean-burning fuels were allowed. The Clean Air Act differed from previous legislation because it controlled not only domestic sources of air pollution, but industrial. Within a few years after the legislation, smoke emissions declined by more than one-third; and the city experienced an overall decrease of smoke concentrations by some 70 percent. Since the regulation, the reduction of sulfur dioxide has made London's infamous "peasoupers" a thing of past.

Conclusions

This chapter has considered the environmental impacts of cities as well as the important interdependence of cities on their physical environment. By necessity, the emergence of cities altered the relationship between nature and society in profound ways. We have seen how the earliest cities modified the physical environment in the creation of specific urban designs and growing environmental impacts. The physical environment also impacted the cities as new incubators for disease. The case of pollution controls shows the example of a feedback system when growing urbanization created perceived environmental problems that in turn engendered new systems of regulation and infrastructural modifications that in turn affected changes to the city–environment dynamic.

Guide to further reading

Bell, W. (1994) *The Great Plague of London*. London: Bracken Books.

Berg, S. (2007) *Grand Avenues: The Story of the French Visionary who Designed Washington, D.C.* New York: Pantheon Books.

Craddock, S. (2004) *City of Plagues: Disease, Poverty and Deviance in San Francisco*. Minneapolis, MN: University of Minnesota Press.

Daunton, M (ed.) (2001) *The Cambridge Urban History of Britain*. Cambridge: Cambridge University Press.

Falck, Z. J. S. (2010) *Weeds: An Environmental History of Metropolitan America*. Pittsburgh, PA: University of Pittsburgh Press.

Hempel, S. (2007) *The Strange Case of the Broad Street Pump: John Snow and the Mystery of Cholera*. Berkeley, CA: University of California Press.

Hope, V. (2000) *Death and Disease in the Ancient city*. London and New York: Routledge.

Johnson, S. (2006) *The Ghost Map: The Story of London's Most Terrifying Epidemic and How it Changed Science, Cities, and the Modern World*. New York: Penguin/ Riverhead Books.

Massard-Guilbaud, G. and Rodger, R. (2011) *Environmental and Social Justice in the City: Historical Perspectives*. Cambridge: White Horse Press.

Mays, L. (ed.) (2010) *Ancient Water Technologies*. London and New York: Springer.

Mukerji, C. (1997) *Territorial Ambitions and the Gardens of Versailles*. Cambridge: Cambridge University Press.

Mumford, L. (1989) *The City in History*. San Diego, CA and New York: Harvest Books.

Penna, A. N. and Wright, C. E. (2009) *Remaking Boston: An Environmental History of the City and its Surroundings*, Pittsburgh, PA: University of Pittsburgh Press.

Porter, Y. (2004) *Palaces and Gardens of Persia*. Paris: Flammarrion.

Reps, J. (1965) *The Making of Urban America: A History of City Planning in the United States*. Princeton, NJ: Princeton University Press.

Schott, D., Luckin, B. and Massard-Guilbaud, G. (eds.) (2005) *Resources of the City: Contributions to an Environmental History of Modern Europe*. Aldershot and Burlington, VT: Aldershot.

Spary, E. (2000) *Utopia's Garden: French Natural History From Old Regime to Revolution*. Chicago, IL: University of Chicago Press.

Spirn, A. W. (1984) *The Granite Garden: Urban Nature and Human Design*. New York: Basic Books.

An excellent website on John Snow is The John Snow Archive and Research Companion at matrix.msu.edu/~johnsnow/maps.php.

3 The industrial city

The City is of Night, perchance of Death,
But certainly of Night; for never there
Can come the lucid morning's fragrant breath
After the dewy dawning's cold grey air
                James Thomson, *The City of Dreadful Night*, 1880

When James Thomson wrote *The City of Dreadful Night* in 1880, he was referring to the dirty, gritty city of London in the midst of rapid industrialization. Thomson's London is beset by disease and doom. His image was accurate for many of Europe's and America's industrializing cities in the eighteenth and nineteenth centuries. The onset of the industrial revolution profoundly and irrevocably shifted human relationships with their physical environment. And while it is probably true that life in the cities was not necessarily worse than that in rural areas, the problems of pollution and poverty and distress were more evident, more massed, and less easy to ignore. For the most part, when cities were smaller and density low, pollution was perceived more as a nuisance than a threat to human health. The industrial era generated new agents in the city: the factory and the railroad. Rapid urbanization and increasing population density created a strained, hazardous and degraded physical environment that had visible and often significant health impacts. But the industrial city was also the cauldron in which new environmental–social relations were forged. New public health measures were introduced and the urban parks movement as well as the garden cities movement are just some of the responses that reshaped, both in imagination and practice, the urban–nature dynamic.

In this chapter we focus on the nineteenth-century industrial city that emerged primarily in Europe and North America. This history is important because the increasing and often visibly noticeable environmental contamination of water, land and air in the nineteenth-century industrial city became the catalyst for sanitation and public health reforms. These policy reforms continue to influence twenty-first-century approaches to environmental reform. The chaos of the nineteenth-century industrial city also brought about new forms of urban design and the rise of modern urban planning. We will draw our examples largely on the experience of the US and to a lesser extent the UK, but a similar tale can be told for most industrial cities around the world.

Pollution in Coketown

In 1844 Friedrich Engels published *The Condition of the Working Class in England*, based in part on his stay in Manchester in England. His description of the "irregular cramming together of dwellings," the streets of "filth and disgusting grime," the "coal-black, foul-smelling stream," and the general "stench of animal putrefaction" vividly conveyed his impression of the new urban form of the industrial city. He noted:

> If any one wishes to see in how little space a human being can move, how little air—and such air!—he can breath, how little of civilization he may share and yet live, it is only necessary to travel hither. Everything which here arouses horror and indignation is of recent origin, belongs to the industrial epoch.[1]

Later, Charles Dickens, in his novel *Hard Times*, would coin the phrase "Coketown" to refer to cities like Manchester, a new type of city where industrialism produced the most degraded urban environment the world had yet seen.

The urban historian Lewis Mumford said Coketown was where

> the immense productivity of the machine, the slag heaps and rubbish heaps reached mountainous proportions, while the human beings whose labor made these achievements possible were crippled and killed almost as fast as they would have been on a battlefield.[2]

The economic foundations of the industrial city were the exploitation of the coal mine, the vastly increased production of iron, and the use of steady, reliable mechanical power—the steam engine. These functions had a devastating effect on the physical environment.

The industrial city was characterized by mechanization and the intensification of the use of resources. Coal replaced wood as the leading source of energy. Mines were built to extract both coal and iron. New industries emerged such as smelting and large-scale manufacturing. Coal, iron ore, lumber and petroleum were abundant raw materials. It was the age of steam

Table 3.1 Urbanization in Manchester, England

Date	Estimated urban population
1685	6,000
1760	30,000
1801	72,275
1850	303,382

Source: Lewis Mumford (1989) *The City in History*. San Diego, CA: Harvest Books, p. 455

engines, factories and smokestacks. A new working class of factory workers was created. The factory system encouraged the centralization of production in cities. Factories had to be near or have access to sources of raw material, a sufficient labor force and sizable markets.[3] But, as more and more factories concentrated in urban areas, the result was both increased urban population growth and the discharge of pollution in massive amounts. In Britain, people moved to the new burgeoning industrial cities like Leeds, Halifax, Rochdale, Swansea as well as Bristol, Liverpool and London where jobs flourished. Consider the rapid transformation of Manchester, England from small mill city to a large industrial one, as shown in Table 3.1.

Although the cities of Britain were at the forefront of the industrial revolution, the rest of Europe and North America were not far behind. Cities like Essen, Cologne, Toronto and Melbourne began to grow rapidly by the late nineteenth century. Toronto's population increased from 30,000 in 1851 to 56,000 in 1871, 86,400 in 1881 and 181,000 in 1891. Swift economic change and population growth also occurred in cities such as Boston, Pittsburgh, Cleveland, Milwaukee, Cincinnati and Philadelphia. Industrialization generated unprecedented levels of urbanization. For example, in 1830 Cincinnati's population was 24,800; by 1850 it had quadrupled to 115, 400 and by 1870 it was 216,000. The city's early factories and industries grew out of Ohio's agricultural roots and became known as "Porkopolis" during the 1800s, as the city became the pork processing capital of the United States. Pittsburgh's rapid urbanization and industrialization was based on manufacturing of steel, creating a very smoky city. Figure 3.1a shows the smokestacks of Pittsburgh.

The advent of the industrial economy and the trend toward specialized labor resulted in many trades and crafts that carried distinctive health risks. Laborers who skinned leather and worked with hides were exposed to anthrax all day, every day. Potters and miners were poisoned with toxic metals such as mercury, lead and arsenic.[4]

Figure 3.1a Industrial Pittsburgh view of the Point, circa 1900
Source: The Carnegie Library of Pittsburgh

Figure 3.1b View of the Strip District at 11:00 a.m. in June of 1906
Source: Library of Congress

BOX 3.1

Recycling slag heaps

Red-hot and spitting fire, the crackling of molten slag was a common sight and sound in industrial Pittsburgh. Like most industrial cities, Pittsburgh's riverfront was home to mills and warehouses, and lots of slag dumps. When the furnaces "bake" ore to obtain steel, slag is what is left over. It is made up mostly of aluminum and silica. When slag cools, it becomes a hard, chunky compound of silicon, phosphorus, manganese and limestone. For every ton of steel produced, at least a quarter ton of slag is left over. It was a common sight in many industrial cities to see railroad cars pour molten slag down hills, creating slag heaps. But Pittsburgh was home to some of the biggest.

One of the largest slag dumps was Brown Dump, eight miles south of Pittsburgh. Carnegie Steel bought the land and opened the dump in 1913. The slag was transported by rail cars from the mills of Pittsburgh to this once-remote valley. The pile grew until it became an artificial mountain, as hard as concrete. In 1947 a local reporter called it one of the best, continuous shows of free fireworks. By the time the dump closed in the late 1960s, it stood 200 feet high and covered the approximate area of 130 square city blocks. Carnegie Steel was considering ways to diversify their holdings and realized they had a valuable piece of real estate. They decided to lease out the land. For several decades, Lafarge, a Virginia-based company, removed up to 300,000 to 500,000 tons of slag to sell for reuse in roads, parking lots and other commercial projects. A portion of the mountain was leveled and a strip mall was built. Opened in 1979, and remodeled in 1997, the three-level mall contains 1,290,000 square feet (120,000 m²) of retail space and approximately 100 stores. By the late 1990s the Century III Mall was the fourth largest shopping mall in the Greater Pittsburgh Area. But by 2011 the shopping mall was in decline; certain sections were closed.

Another slag dump was along the Monongahela River, just five miles from downtown Pittsburgh. In early 2001, real-estate developers began clearing the site for "Summerset," a residential community intended to rejuvenate the community. The $243 million project intended to produce 700 homes and apartments on 240 acres. Summerset is considered one of the best examples of smart growth in a central city.

Finally, another slag pile of some 238 acres is located in a creek area known as Nine Mile Run. In 1922 Duquesne Slag Company bought an initial 94 acres. Although there were local efforts to protect this area as urban open space, the economic imperatives of the industrial city led to the sacrifice and destruction of Nine Mile Run. Over the years, the Duquesne Slag Company expanded the slag heap. For over 90 years the creek was a desolate moonscape, buried under as much as 120 feet of industrial byproducts and wastes. The last slag was dumped on the site around 1972; by then approximately 17 million cubic yards of slag filled the valley nearly from end to end. In 1995 the Urban Redevelopment Authority of Pittsburgh purchased the land for $3.8 million. The city's master plan for the site envisions an extension of Frick Park, over the slag, to the Monongahela River and construction of several hundred units of mixed-income housing. The Nine Mile Run aquatic ecosystem restoration was completed by the Army Corps of Engineers in July 2006. The restoration work included stream channel reconfiguration, wetland reconstruction, native wildlife habitat enhancement, and native tree, shrub and wildflower plantings. It was the largest urban stream restoration in the US at the time of its completion.

Sources: (1997, November 18) "Giant slag dump experiences rebirth in suburban Pittsburgh." *The Observer-Reporter.* McElwaine, A. (n.d.) "Slag in the park." http://www.ninemilerun.org/slag-in-the-park (accessed June 2012). Swaney, C. (2001, January 28) "Houses are to replace a Pittsburgh slag heap." *New York Times.* http://www.nytimes.com/2001/01/28/realestate/houses-are-to-replace-a-pittsburgh-slag-heap.html?pagewanted=all&src=pm (accessed May 12, 2012).

In Manchester, high-density areas were associated with high mortality rates; in Liverpool a 1871 survey showed that a large proportion of deaths were of young children; similarly in New York City, the mortality rate for infants in 1810 was 120 per 1,000 live births, but rose to 180 per 1,000 live births by 1850 and to 240 in 1870.[5] The causes were poor housing, lack of sanitation, lack of clean water, poor diet and endemic disease. Industrial cities experienced an environmental crisis on a scale not encountered previously. In short, the industrial city witnessed widespread environmental deterioration of water, land and air.

Water pollution

> I wander through each dirty street
> Near where the dirty Thames does flow
> And mark in every fact I meet
> Marks of weakness, marks of woe
> William Blake

Water was a critical environmental element in production and manufacturing during the industrial revolution. It was an ingredient in many industrial processes and a convenient dumping ground for industrial byproducts. In the industrial era, cities faced two pressing water issues. The first was water quality. The second was locating sufficient water supplies for a rapidly urbanizing population.

Water pollution in the industrial city originates from two main sources. The first is residential: human and animal waste, which is composed of organic compounds. The second is commercial: factories and businesses. Commercial sources included both organic byproducts and, increasingly, inorganic byproducts as advances in technology helped to fabricate new inorganic materials such as plastics, dioxins and other heavy metals. Factories, especially those involved in textile, chemical and iron and steel industries, were often sited by rivers or bays because they needed large quantities of water for production processes. Factories used water for steam boilers and to cool engines, and in the making and disposing of chemical solutions and dyes. Water, once used, was dumped back into rivers, creeks and tributaries. These resources became open sewers, poisoning aquatic life. By the 1870s, for example, the Passaic River in New Jersey was so polluted the city had to abandon it as a water supply, and commercial fishing in the area was effectively ended.[6] In Chicago, the center of slaughtering and meat processing, nearly one million cattle and pigs were slaughtered each year. Meat plants routinely flushed carcasses and unwanted parts into the river, while glues, gums, dyes, fertilizer, sausage casings, brushes from processing and packaging plants nearby were also dumped into water sources. The Chicago River was so polluted that ice cut from the river released a disgusting stench as it melted.[7] Similar problems affected Canadian cities such as Toronto. Despite possessing a beautiful lake and miles of beaches, the city's beaches quickly became far too polluted to use, largely a side-effect of dumping garbage directly into the lake. In addition, many of the city's smaller rivers and creeks such as Garrison Creek and Taddle Creek were used as open sewers. Officials decided to cover these creeks up to prevent disease outbreaks.

The poor condition of London's Thames made a great impression on poets and others. In one poem written in 1859, *State of the Thames*, the Thames is decried by a London satirist:

> River, river, reeking river!
> Doomed to drudgery foule and vile;
> Noisome, noxious fumes distilling,
> Fumes which streets and housing filling
> Harpy life, defile[8]

In Australia, the city once called "Marvelous Melbourne" was given the infamous title "Marvelous Smellbourne" for the stench caused by uncollected garbage, and its huge, open sewers and lack of water treatment. The River Yarra was as polluted as any urban European water course by the late nineteenth century.[9]

Human wastes were disposed of in cesspools and privies (although sometimes they were tossed out on the street or in vacant lots); hence most of the residential waste was put onto the land.[10] Before 1850 no US city had sewer systems for the collection of human waste or wastewater. In Europe some cities had sewer systems built of masonry dating back as early as the thirteenth or fourteenth century, but many had fallen into disrepair. In Paris, for example, the sewers were decayed and poorly maintained and were considered dangerous places to go. In Victor Hugo's 1862 *Les Miserables*, the sewer was politically symbolic: sewers harbored enemies of the state, and outcasts such as thieves and prostitutes.

Prior to the nineteenth century, water supplies were obtained from mostly local sources such as wells, nearby ponds, streams and rivers. But for many industrial cities, by the 1860s population had outgrown water supplies. New York, Philadelphia and Boston were unable to provide an adequate supply of clean water to a growing population. This supply issue meant there was a shortage of clean water for drinking, bathing or cooking, and it also hampered efforts to fight fires.

BOX 3.2

The sewers of Paris

"Paris above ground is an ever-changing panorama, which anyone can view by paying for it; sometimes the coin is simply money, or cheaper and better yet, a little enterprise or exercise; but too often it is a sight draft upon either health or morals.... Few, however, think of glancing at subterranean Paris; that mighty labyrinth of streets beneath ground, seen but rarely by human eyes, but without which Paris above ground would be an inhabitable mass, or a generator of pestilence. There is nothing here for show, but all for use. Built to endure for ages, and to subserve the necessities of millions of human beings, performing in the material economy of social life functions as important and as indispensable as the veins and arteries in physical life, they are worthy of a glance, at all events, that we may learn the labor and expense in lighting, watering, and cleaning a modern capital. These indispensable offices are all moving quietly on in their prescribed paths, unseen and almost unknown by the millions of noisy feet above them. Yet, should any derangement ensue, the health and comfort of the city is at once in jeopardy. Were the Tuileries consumed by fire, and the Arc of Triumph engulfed in an earthquake, the Parisians would simply have two fine monuments the less. But were the drains, water, and gas of Paris to be suddenly arrested, the city would become uninhabitable, and the ancient marshes of Lutèce would regain their lost empire. It was not, however, until the commencement of the last century that a regular system of drainage was established ... the system has been continually improved upon, until it has rendered Paris the cleanest and best lighted capital in the world."

(1854) "Life in Paris: sketches above and below ground." *Harper's Bazaar*, February: 306–7. http://www.sewerhistory. org/articles/whregion/1854_ah01/index.htm (accessed June 28, 2012).

Land pollution

Industrial cities generated new pollutants that were discarded on land. Factories sought the easiest methods of disposal for non-liquid wastes. Rubbish, garbage, ashes, scrap metals and slag (formed during iron smelting and other metallurgical processes) were often disposed of on open or vacant lots around the city.

Many industrial cities were ill-prepared to provide adequate sanitation services. As we noted in the previous chapter, in Paris, the basic rule for dealing with garbage was "*tout-a-la-rue*," translating as "all in the street," including household waste, urine, feces and animal carcasses. This was common practice in many cities. Collection and disposal of solid waste was uncoordinated, sometimes privatized so that only the wealthy residents could afford to have their waste collected. In New York, street teams collected garbage, and loaded it on barges to be dumped at sea; in St. Louis, Boston, Baltimore and Chicago, refuse was hauled to open dumps near the city's edge.[11] As late as the 1870s, many cities had no public provision for the collection and disposal of household garbage. Ever-increasing mounds of waste became a major issue for the industrial city, but the debate was who was ultimately responsible for providing such services—private interests or the public authorities of the city?

Not all land pollution came from the factories. The horse was a principal cause of dirty streets. In 1830 the first horse-drawn buses appeared in New York; at their peak, horses would number 170,000. For the next century, the horse would remain the primary form of transportation for both people and goods, pulling wagons, trolleys and horse cars. Most of the horses were commercial or work animals. They hauled freight and made deliveries. The environmental impacts of the horse were significant. A single horse discharged more than several gallons of urine and nearly 20 pounds of manure each day. The evidence of the horse was everywhere—piles of manure littered the streets attracting swarms of flies and generating stench, and the occasional discarded carcass of a horse that dropped dead while performing its job (see Figure 3.2).[12] Streets could turn into cesspools when it rained; in Paris ladies and gentlemen were assisted in their navigation through streets littered with horse droppings by "crossing-sweepers."[13] The historian Joel Tarr has called the horse the predecessor to the auto, noting that the horse generated many of the same problems attributed to the car: air contaminants, noxious odors and noise. Even more disturbing was the fact that while the horse created environmental problems for the city, urban conditions made life incredibly difficult for the horse. The average streetcar horse had a life expectancy of two

Figure 3.2 Dead horse in the streets of New York City, circa 1910

Source: Library of Congress

years; many horses were whipped and abused by their drivers spurring them on to haul heavy loads. Many horses died in the open street.

Air pollution

> Here was the very heart of industrial America, the center of its most lucrative and characteristic activity, the boast and pride of the richest and grandest nation ever seen on earth—and here was a scene so dreadfully hideous, so intolerably bleak and forlorn that it reduced the whole aspiration of man to a macabre and depressing joke.
>
> H. L. Mencken writing on Pittsburgh in a
> 1927 essay, "The Libido of the Ugly"

The opening paragraphs of Charles Dickens' 1853 novel *Bleak House* speaks of "smoke lowering down from the chimney-pots, making a soft black drizzle, with flakes of soot in it as big as full-grown snowflakes—gone into mourning, one might imagine, for the death of the sun." By the mid-1800s more than one million Londoners were burning soft coal

BOX 3.3

Garbage in the city

For much of New York City's history, trash—or refuse—was tossed out onto the street, allowed to rot in ditches along the thoroughfares, and left in huge heaps that became play-grounds for vermin and poverty-stricken children. Even the stately mansions along Central Park threw out their house-hold waste like everybody else. For much of the nineteenth century, trash removal was a private, not municipal service, which made garbage an issue of social class. In more densely populated areas of town such as the filthy, over-crowded slums of the Five Points on Manhattan's Lower East Side, garbage run-off and sewage contaminated public drinking water, giving rise to disease and sky-high mortality rates for young and old alike.

By the 1800s, the filth in lower Manhattan had accumulated to a depth of two to three feet in the wintertime, when house-hold waste and horse manure combined with snow. Even when Manhattan's population was less than one million, in the mid-nineteenth century, city horses dumped 500,000 pounds of manure a day on its streets, in addition to 45,000 gallons of urine. For most of the nineteenth century, waste collected from the streets of New York City was dumped into the ocean. The East River was a stinking mess, choked with all the refuse of downtown New York, which citizens tossed off the end of a pier built expressly for garbage dumping.

New York City was so filthy rumor has it that sailors could smell the city six miles out to sea. And all of this filth exacer-bated a public health crisis. A cholera epidemic in the 1830s killed 3,515 people, which was roughly 12 percent of the population at the time. That same percentage would mean about 100,000 people today. The mortality rate in 1860 New York was equal to that of medieval London.

Before 1872, responsibility for street cleaning and waste collection was assumed by a succession of public and pri-vate ventures. Political ties figured strongly in the awarding of contracts to carting operations, and the city often took over for contractors who performed inadequately. Robin Nagle's work on the history of garbage in New York shows that despite public outrage about garbage-strewn streets, corruption meant that money set aside for street cleaning was going into the pockets of the Tweed and Tammany politicians. Eventually, it was so dirty for so long, no one thought that it could be any different. In an online interview with OnEarth, Nagle summarized,

> Imagine, on your own block, that you can't cross the street, even at the corner, without paying a street kid with a broom to clear a path for you, because the streets were layered in this sludge of manure, rotting vegetables, ash, broken up furniture, debris of all kind. It was called "corporation pudding" after the city government. And it was deep—in some cases knee-deep.

Political patronage and corruption, however, remained an obstacle to effective service until 1895, when George Waring Jr. was appointed commissioner.

By 1898, the city's Metropolitan Board of Health issued a proclamation forbidding "the throwing of dead animals, garbage or ashes into the streets," and New Yorkers were ordered to stop dumping their garbage off the East River pier.

Sources: Nagle, R. (2013) *Picking Up: On the Streets and Behind the Trucks with the Sanitation Workers of New York City*. New York: Farrar, Straus and Giroux. OnEarth (2010) "Interview with Robin Nagle the New York City Department of Sanitation's anthropologist-in-residence, and professor of anthropology at New York University." http://www.onearth. org/article/digging-into-new-york-citys-trashy-history (accessed June 2012). Tenement Museum (2010, October 8) "Manure, rubish, slops and waste." http://tenement-museum. blogspot.com/2010/10/questions-for-curatorial-manure-rubish.html (accessed June 2012).

and winter "fogs" became more than a nuisance. In 1873 a coal smoke-saturated fog hovered over the city for several days, causing 268 deaths from bronchitis. By 1905 the term "smog" had been coined to describe London's combination of natural fog and coal smoke. By then, the phenomenon was part of London history and the dirty, smoke-filled "peasoupers" were legendary.

After the 1830s, many industrial cities relied primarily on coal as a major source of energy. Two types of coal are readily found. The first is bituminous or soft coal, which is high in sulfur; the second type is anthracite coal which is harder and burns more cleanly. In early decades, bituminous was more plentiful and used more frequently, but when it is consumed, much of its residue goes directly into the air. The severity of smoke and particulate matter that is released when coal is burned left its mark on buildings, on laundry and in the lungs of city dwellers. Cities such as Pittsburgh and St. Louis relied primarily on bituminous coal. Although Pittsburgh faced water and land pollution issues, it surpassed most other cities in terms of air pollution. Smoke pollution was the most visible byproduct of coal consumption—the industrial engine of Pittsburgh—and atmospheric inversions in the city and in the region exacerbated conditions. However, smoke's perceived link with industrial prosperity made control of the problem difficult.

Air pollutants in the industrial city included soot from smokestacks and locomotives; factories belched chemicals including chlorine, ammonia, carbon monoxide, carbon dioxide, hydrosulfuric acid, and methane. Although there were concerns about air pollution, it tended to be regarded as a nuisance and was lower on the environmental agenda for two important reasons. First, the impacts of smoke were the dirtying of buildings and high cleaning prices for clothing, but scientists at the time were unable to link smoke with health problems. In contrast, water pollution had a more immediate and direct link to public health through the spread of infectious diseases; the rise of bacterial science had shown cause and effect scientifically.[14] Second, smoke was equated with progress, growth and jobs and the smoky skies were a constant reminder of economic growth and prosperity.

There were some attempts to control smoke. Some US cities banned the use of bituminous coal-burning locomotives from their streets. In 1869 Pittsburgh attempted to ban the construction of beehive coke ovens within the city limits. While some smoke-control efforts had limited success in reducing smoke, they basically failed to make substantive inroads into the problem.[15]

Reforming the industrial city

> The time has arrived when manure heaps, slaughter houses, fat and bone boiling establishment, glue manufactures, outdoor or unsewered privies and all kindred occupations and nuisances cannot be much longer tolerated within the built-up portions of New York or Brooklyn.
>
> New York City Metropolitan Board of Health, March 1866

In the industrial age, few people attached importance to preserving the quality of the environment. But there were consequences to unlimited and unregulated growth. Short-term consequences of environmental pollution and the industrial city were numerous: malformation of the bones due to lack of sunshine and a poor diet; skin disease from dirty water; small pox, typhoid and scarlet fever spread through dirt and human excrement in land and water; tuberculosis from bad diet and overcrowded conditions in tenements. Long-term consequences were little understood at the time, but we now know that these include health problems associated with long-term exposure to pollutants and chemicals. City authorities and citizens appeared to tolerate a great deal of filth in their cities and to ignore questions of environmental quality. Many had accepted environmental degradation as the price for economic growth.

For many people the fetid, squalid and segregated industrial city was a difficult place to live and work. Wages were low, hours were long and many people lived in poor to atrocious housing, where overcrowding, high rents and poor conditions were the norm. Environmental conditions included coal-filled air, polluted waterways, and garbage-strewn streets. Disease, frequently deadly, was rampant as housing squalor, poor nutrition, expensive health care, and the absence of decent public infrastructures combined to make death a frequent visitor to most households.

By the last quarter of the nineteenth century, however, an urban reform movement emerged seeking to find solutions to these problems. The movement had four main strands. Structural reformers believed that modifying local government and installing "good government" would make the problems disappear. Moral reformers pushed legislative change aimed at controlling individual actions (such as Prohibition, 1919–33 in the US) and limiting immigration (Johnson-Reed Act of 1924). Social reformers promoted legislative change to a range of social and economic issues, including child labor, housing bylaws, bank regulation, and public health. Finally, a growing environmental consciousness and increased concern for sanitation galvanized public opinion. Pollution became no longer simply a nuisance—it was an unwanted and sometimes dangerous byproduct of industrialization.

Table 3.2 Industrial reforms and the creation of public utilities

Pure water supply developed
Street cleaning and garbage pick up
Improved water closet
Stationary bathtubs with water pipes
Water system with running water for every house and apartment
Collective sewage system
Establishment of Board of Health
Establishment of Sanitation Department

Source: by Lisa Benton-Short.

Major reforms during this time would dramatically change environmental quality (Table 3.2).

Public health reforms developed out of the dangers of the industrial city. The cholera epidemics of the 1830s and 1840s had social and political ramifications beyond public health, raising fundamental questions about medical treatment, social policy and the livability of the urban environment. In Europe, the 1832 cholera epidemic killed 20,000 Parisians, convincing an enlightened few of the need for a systematic public health system.[16] In 1842 Edwin Chadwick, a commissioner on the British Royal Commission, published his *Report on the Sanitary Condition of the Laboring Population of Great Britain*. It was preceded by various alarming medical reports on Manchester and London and followed a major outbreak of cholera in 1931–2.[17] Chadwick advocated bringing health and sanitation under central control, but this threatened the power of many vested, private interests. Like many of his contemporaries, Chadwick conceived the city as an organic system, analogous to the human body. A healthy city, like a healthy body, depended on the free circulation and exchange of vital fluids. Sanitary reformers sought to imitate the economy of the human body by integrating the supply of water, the disposal of sewage and the production of food in a single self-regulating system.[18] Ultimately Chadwick's work inspired new sanitary laws in England and Australia, and even inspired American health officials to see refuse as a health issue.

What guided thinking before the 1880s was what the environmental historian Martin Melosi has called the "Law of Purification." "Clean" meant whatever the observer could touch, taste, smell or see with the naked eye. If water was clear, odorless and tasteless, it was pure. And moving water was thought to quickly dilute any pollutants, resulting in "harmless" discharges. This "what you see is what you get" approach to pollution analysis lasted until the late 1800s. If the water looked clean it was. If it was cloudy it was impure. That invisible organisms could live and even thrive in a watery environment was beyond imagination. Although the microscope was invented in 1674, it took another 200 years for scientists to isolate and identify microbes and their link to particular diseases. Finally, in the 1890s, chemists and sanitary engineers identified sewage-polluted water as the carrier for infectious disease.

The Law of Purification worked when populations were low and the discharge of pollutants were predominately organic rather than inorganic. A large body of water can naturally purify small quantities of sewage and waste, but a large urban population produces so much sewage that bacteria can no longer cope with the heavy load, thus depleting the oxygen levels that fish and plants need to survive. Eventually a river, lake or stream may become eutrophic.

In the US, cities where disease was common began to lose business to those which were healthier. New Orleans, which suffered ongoing outbreaks of cholera, lost out to Chicago; in Memphis, outbreaks of typhoid and dysentery almost destroyed its economy.[19] With more than just civic pride at stake, cities began to consider how to reform sanitation and deal with pollution. The historian M. Christine Boyer commented

> the environmental chaos of the American city became linked in the minds of the improvers to the social pathologies of urban life. Long before poverty, poor housing and slums were thought of as economic and political symptoms, improvers saw a link between environmental conditions and the social order, between physical and moral contagion. [20]

As cities outgrew their local water supplies, they were compelled to develop more extensive public water supplies. Many municipalities attempted to solve the problem through private companies, but there were uneven results. Private companies were often reluctant to provide water for civic purposes like flushing streets clean, and the working class and poor could not afford their high prices and so had no access to water. New York City established an ambitious water works program in the 1830s, as they constructed a dam on the Croton River, some 40 miles north of the city, and built aqueducts to convey the water on its long journey.[21] By the 1860s the capacity of the system was 72 million gallons per day, but by the 1880s even this was not sufficient, because population growth was so rapid. In 1885 New York City began construction on the New Croton Aqueduct, which would provide 300 million gallons per day. Further urban growth would put strains on even this capacity and so New York decided to access water in the Catskill Mountains, 100 miles to the north. Other cities began to discuss the development of urban water supplies. In addition to the building of new water supply systems, new technologies such as bathtubs, shower bath and water closets were becoming standard items in middle-class homes. This placed new demands on municipal water supplies. By the late 1880s many US cities developed municipally owned supplies, bringing in water from nearby rivers, lakes and groundwater sources. Cities, once reluctant to spend money on public services, now realized that providing a reliable source of water was both a social and economic imperative. For example, the Citizen's Association of Chicago supported water works projects on the basis that it would improve Chicago's business climate and thus ensure the prosperity of their members.[22]

In cities around the Western world, it became a matter of prestige and pride to construct large-scale water and sanitation infrastructure projects. In London, for example, eliminating the "great stink" of the Thames became a national priority. In Paris, sanitizing the city was one of Baron Haussmann's main concerns when he was Prefect of the Seine in 1853–70. In Athens, bringing adequate fresh water to the city became an urgent political and social issue. However, such large-scale engineering projects required major capital investment, an adequate supply of labor, social consensus and political commitment.[23] In Australia between 1880 and 1910 both Sydney and Melbourne built underground sewers as the main means of disposing of human wastes. As cholera had hastened the arrival of London's sewers, so the outbreak of typhoid epidemics in both cities during the 1880s had hastened the change. In addition, the new infrastructure had a social connotation: Sydneysiders welcomed the coming of the water closet as at last removing a shameful blot on Australian civilization (this was in spite of the fact that very few British cities other than London were any further advanced).[24] Indeed, water supply systems, sewerage infrastructure, and indoor toilets became indispensable markers of civilization in Europe, North America and Australia.

In addition to new developments in water supply, many advances took place with regard to water quality (Table 3.3). In 1882 Robert Koch was able to demonstrate with a microscope that tiny microorganisms (*tubercle bacillus*) were responsible for the transmission of tuberculosis. In 1884 he identified the cholera bacillus, *Vibrio cholerae*. He found the bacteria causing bubonic plague, leprosy and malaria and helped pioneer techniques that would allow others to find the germs causing cholera, typhoid fever and anthrax. Finally, the link between organic pollutants—primarily fecal contamination—was understood, and new technologies in filtration emerged to clean the water of such pathogens. Bacterial researchers such as Koch and Louis Pasteur not only established germ theory, but helped to clarify the etiology of various waterborne diseases such as typhoid fever. By 1900 more than 21 germs that caused disease had been identified.

Table 3.3 Outline of water treatment milestones

Date	Event
312 BC	First Roman aqueduct
Eighth century	Distillation
1582	First pump (London)
1652	First American supply work (Boston)
1685	Slow sand filtration
1761	First steam pump (London)
1829	First large municipal sand filter supply (London)
1849	New York City begins building sewers (125 miles of lines built 1849–73)
1855	London sewers modernized after cholera outbreak
1858	Year of the "Great Stink" of the Thames River (London)
1885	Water softening
1885	Water bacteriology
1891	Aeration
1895	Iron and manganese removal
1907	Chlorine disinfection

Source: adapted from Weinstein, M. (1980) *Health in the City*. New York: Pergamon Press, p. 28.

BOX 3.4

Reforming Victorian London

By the late nineteenth century, the Thames was a river so full of human, animal and industrial waste of unprecedented volume, it was called the Great Stink. How to clean up the river, remove the refuse in the streets, and confront the squalor of slums became a primary concern for engineers, scientists and public officials. As a result of water sanitation efforts and improvements in water supply and water removal, the death rate in England declined from 20.5 per 1,000 in 1861 to 16.9 in 1901. London was a much cleaner place to live at the end of the nineteenth century than at the beginning.

Sanitary reforms in Victorian London certainly revolved around improving public health, but as historian Michelle Allen has pointed out, such reforms also connected to broader social concerns. At their most ambitious, sanitary reforms sought to lift up the poor and working classes, to moralize the population, and to create a more harmonious social order. Clean water, applied inwardly and outwardly, was both an instrument and symbol of Victorian morality. Cleanliness was an outward sign of inward purity. Yet she also notes that sanitary reforms met some surprising resistance. Despite the general assumption that sanitary reforms were widely popular and largely uncontested, many hygienic improvements were actually disparaged, ridiculed and resisted with increasing vehemence throughout the nineteenth century.

While reforms introduced physical changes—such as slum clearance, road building and the building of sewer infrastructure—they not only altered the physical space of the city, they also altered the social and symbolic meanings of the city as well. Challenges to the "Law of Purification" caused social anxiety, in part because reforms ushered in a new paradigm of "cleanliness." That this new paradigm of "cleanliness" generated ambivalent responses reflected an unease that accompanied modernization. For example, large-scale slum clearances became associated with "purification" and the removal of the urban poor, displacing thousands of working residents. Some began to lament the destruction of old areas.

Allen's approach to the reforms of Victorian London provides a more detailed analysis of urban space and the human experience of that space, and what she calls a "critical geography" of sanitary reform. This critical geography uncovers the social, spatial and textual discourses that inform the social order and the built environment and allows us to see the connection between ideas about filth and dirt, environmental issues, and the process of modernization in the city. Reforms—whether in the nineteenth, twentieth or twenty-first century—are often complex transformations that create contradictions and conflicts.

Source: Allen, M. (2008) *Cleansing the City: Sanitary Geographies in Victorian London*. Athens, OH: Ohio University Press.

In Europe and North America scientists, chemists and biologists pioneered sewage treatment processes. Slow sand or mechanical filters removed many of the pathogens. By the early 1900s water treatment plants were using chlorination to disinfect the water supply. Both the new water works and sewage facilities resulted in striking improvement in the city's public health. By the late 1890s, cities in Europe and North America saw dwindling outbreaks of cholera and dysentery. Commitment to full sewage treatment, however, lagged, in part because engineers and politicians believed that the risks involved in using streams for sewage disposal were not sufficient to justify the costs of constructing sewage-treatment plants, and continued to argue for the natural dilution power of waterways.[25]

An important point is that the focus of most of the reforms from the 1890s to 1930s to drinking water and sewage systems focused primarily on the connection between organic pollution, human wastes and human health, resulting in a concentration on sanitary wastes. The historian Joel Tarr notes that the shift from filth theory to germ theory caused public health authorities to reduce their interest in industrial wastes because they did not normally contain disease germs.[26] Although there were organic industrial wastes (such as wastes from dairies, canneries and meat packing), industrial wastes also included metals (such as lead, zinc and arsenic), and inorganic pollutants such as dyes, phenol wastes and cyanides. However, at that time metals and inorganic wastes appeared to have no impact on human health. By placing their highest priority on the potential dangers to water supplies from sewage, public health professionals drew attention away from a concern with industrial wastes. This meant that little, if anything, would be altered until the middle of the twentieth century.

Ironically the reforms dealing with human waste shifted the sink for the wastes from the land to the water. Some cities, particularly cities downstream from other cities, saw an increase in infectious diseases such as diarrhea and typhoid.[27] Thus, for a while new reforms in water and sewage sanitation actually increased organic pollutants, which in turn increased the risk of pathogens. By the early twentieth century, researchers and sanitary engineers solved the problem of organic sewage-polluted water with the development of new technologies such as water filtration and chlorination.

In addition to undertaking water reform projects, cities also turned their attention to the refuse problem. Many cities had rudimentary regulations for the disposal of waste. Pigs were used to scavenge; teams of "rakers" were often employed to remove garbage from the city. One of the first problems to be addressed was that of street cleaning. The environmental historian Martin Melosi argues that street cleaning was an easier problem to address than household garbage, because streets were clearly in the "public" domain, whereas households were a question of individual responsibility.[28] In New York City, the Edict of 1866 evicted 299 piggeries, mandated the cleaning up of some 4,000 refuse-heaped backyards, introduced the first watertight garbage cans, established regular rounds for garbage clean up, and noted that "the well-known capacity of the city to drown in sewerage is more than matched by its talent for smothering under a blanket of garbage and refuse."

In the US, in 1887 the American Public Health Association appointed a committee on Garbage Disposal to study the refuse problems in US cities. In addition, civic awareness also played a critical role. In many cities, local civic organizations created pressure for reforms, such as the establishment of departments of public health and other sanitary services and new laws. The Ladies' Health Protective Association of New York City became a leading force in the fight to bring about sanitation reform in US cities, undertaking projects that included refuse reform, street cleaning improvements and school sanitation.[29]

Perhaps one of the more influential reformers was Colonel George E. Waring, Jr., an engineer and former Civil War officer. He was appointed the first street cleaning Commissioner of New York City in 1895. Waring created an efficient street cleaning operation. He proposed that household refuse be separated into different receptacles. He also helped to build the first municipal rubbish-sorting plant in the US, where salvageable materials were picked out and resold. Waring increased the pay of street cleaners, improved their working conditions and issued them white uniforms. The sweepers were called "White Wings" (see Figure 3.3). His program for New York City indicates an important shift from viewing the refuse problem only as a question of health to a wider more multifaceted urban problem. Sanitary engineering had now become a formal branch of engineering, and the growth and complexities of urban problems such as sewage, water supply, street cleaning and garbage collection required special training and education.

Reforms to air pollution lagged far behind those of water and land reforms. In part this was because many key policymakers considered economic growth essential, and the toleration of polluted air was so entrenched. For many a smoky sky was a sign of progress and prosperity. The urban historian Lewis Mumford once said of the mindset of the industrial city: "Smoke makes prosperity, even if you choke on it." Opposition to a new policy for clean air centered around the argument that smoke meant wealth and jobs.

Figure 3.3 The White Wings, circa 1890s
Source: Library of Congress

By the turn of the century, however, US cities were reporting "Londoners"—
the combination of smoke and fog that had long plagued London. Smoke
appeared to be responsible for physical problems such as the increases in
pulmonary disease.

In the US, the Progressive Era brought many local attempts to regulate smoke
and air pollution. In particular, local civic leagues and women's organizations voiced their concerns and pressured for policy reform. Many women's
clubs were composed of upper-middle class socially prominent women who
had the leisure time to devote to the smoke abatement reform movement.
The Ladies Health Association of Pittsburgh galvanized support for new
smoke ordinances in 1892; in other cities like St. Louis, Cincinnati, Chicago,
and Baltimore similar clubs were established. These organizations were
often supported by engineers who faced mechanical problems caused by the
residual effect of smoke and advocated for technical advances to abate
smoke. Civic groups such as Chambers of Commerce also took up the cru-
sade as part of "civic pride." Ironically, in 1898 steel magnate Andrew
Carnegie delivered a speech to the Pittsburgh Chamber of Commerce urging
it to take up the issue of smoke control.[30]

By 1912, many cities had smoke-control ordinances, but many of the ordinances were lax and stiff fines were uncommon as reformers resisted penalizing industry with costly burdens. The equation of smoke with progress caused many political leaders to move cautiously. One of the more far-reaching reforms occurred in Pittsburgh. In 1941 the Pittsburgh Smoke Control Ordinance set a policy goal to eliminate dense smoke as well as other components of air pollution, such as fly ash. This applied to both industrial and domestic sources. Consumers would have to burn either smokeless fuel or use smokeless technology if they continued to use bituminous coal.[31] Implementation proved difficult but Pittsburgh's air improved considerably by the late 1940s and the city benefited from improved air quality, more sunshine and improved health, as well as savings on cleaning costs and laundry bills. After years of living with so much smoke that the sun was rarely seen, the joke went that improved air quality caused Pittsburghers to point upwards to the sun in the sky in confusion and ask "what is that?"

This cursory discussion of pollution and reforms in the industrial city allows general conclusions to be drawn. First, solutions to one form of pollution often generate new pollution problems in different localities or in different media. Second, both cultural values and scientific knowledge influence pollution policy. Third, nineteenth- and early twentieth-century pollution reforms relied upon a technological solution or fix. There was little effort to question or prohibit the production of pollution—rather, efforts and technology focused on how to clean it up once it was in the environment.

The urban public parks movement

The deteriorating industrial city prompted numerous reformers, including landscape architects and planners, to revision the city and nature–society relationships. The urban public parks movement, which emerged in both Europe and the US during the mid-to-late nineteenth century was a reaction to the problems of the industrial city as planning "visionaries" offered designs of a reconstructed relationship to nature through urban space. Although most cities had open land within city limits (vegetable gardens, squares, commons) these were not often formalized as public space or recreational areas.[32] In both Europe and the US, the ideology of the public park began with a recognition of the importance of open spaces to the health and vitality of the urban population.[33] A constellation of factors provided the genesis of the urban public parks movement in the US: worsening urban conditions and the link with public health, a concern for intellectual and moral improvement, and a challenge to the nation's self-esteem. Urban parks would be aligned with democratic ideals and would be accessible for all socio-economic classes: parks were to be for the public.[34]

One of the most important of these visionaries is Fredrick Law Olmsted, designer of Central Park in New York City and many other parks and park systems, residential developments and college campuses in numerous North American cities. His legacy continues to impact many cities and his planning ideology informed a generation of park planners and landscape architects.

Olmsted was born on April 26, 1822 in Hartford, Connecticut. He was the son of a dry goods merchant. As a young man, he studied civil engineering but struggled until his mid-forties to define his professional identity, shifting from farming, to journalism and then to landscaping.[35] In 1850, when he was 28, he went to England and wrote a book about farming methods, *Walks and Talks of an American Farmer in England*. In 1852, having returned to the US, he went to work as a reporter for the *Times*, and was assigned to write about the South. His experience there sparked his interest in learning more about the democratic condition of society. His transition from writer to landscape architect was improvised; he was hired as the New York City park superintendent before anybody had any idea what the park might look like.[36] New York City had set aside some 750 acres of land on the outskirts of town as the site of a new city park. The site itself was a mixture of swampy areas and rocky outcroppings, and contained numerous heaps of cinders, bricks and other rubbish. The park would be made, not found. As one report noted, "never was a more desolate piece of land chosen for a pleasure ground."[37] In 1858 Olmsted and fellow landscape architect Calvert Vaux, a trained architect, worked together for several months and won the prize for the design for Central Park.

The design for Central Park reveals a very different perspective on nature–society relationships than Versailles or the Boboli Gardens (Figures 3.4 and 3.5). While Olmsted was influenced by European gardens, he also sought an American-derived model that sprang mainly from an anti-urban ideal. Several design elements are worth mentioning. First, Olmsted and Vaux moved away from the formalized, rationalized geometric design of European gardens and beds and instead stressed curvilinear paths with more native plantings. There would be no manicured, clipped animal topiaries in an Olmsted design. Parks were not to provide pictures or allegories; plantings should be permanent and local. Olmsted resisted regimentation and a central *allée*, as characterized Versailles or Luxembourg Gardens. Central Park belonged not to a palace, but to the people. The plan is perhaps best characterized as seemingly disordered and unstructured. Lakes and ponds are nestled in their own places and are not part of a grand canal or waterway.

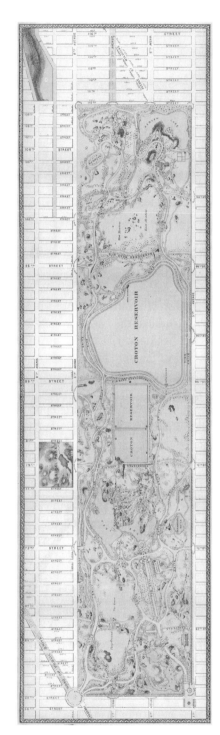

Figure 3.4 Olmsted's Map of Central Park. The winding paths offer a contrast against the rigid grid of the city

Source: Wikimedia: http://commons.wikimedia.org/wiki/File:1868_Vaux_%5E_Olmstead_Map_of_Central_Park,_New_York_City_-_Geographicus_-_CentralPark-CentralPark-1869.jpg

Main spaces are not set apart, but merge with other areas. The design reflected a more organic view of the world and sought to offer opportunities for urban dwellers to "reconnect" with nature, to find spirituality in the green trees, grass and vegetation that contrasted with the bleak industrial city. Curvilinear pathways contrast with the "wearisome rectangularity of cities."[38] It was intended to be an adaptable place. The park was to be the place of numerous activities—roller-skating, ice-skating, rowing, puppet shows, walking and carriage driving, which had both separate and intersecting points. Indeed, almost as soon as the park opened in 1858 the skating pond produced a skating craze.

Olmsted believed parks were the "lungs of the city" and criticized the industrial city for its lack of sunlight, the absence of trees and open spaces. Urban residents, he argued, needed an opportunity to "supply the lungs with air screened and purified by trees, and recently acted upon by sunlight, together with the opportunity and inducement to escape from conditions requiring vigilance, wariness and activity toward other men."[39] Parks were therefore not only aesthetically important, they were linked to public health. Olmsted continues

> We want a ground to which people may easily go after their day's work is done, and where they may stroll for an hour, seeing, hearing, and feeling

Figure 3.5 The informal landscape of Central Park
Source: Photo by Lisa Benton-Short

nothing of the bustle and jar of the streets, where they shall, in effect, find the city put far away from them. We want the greatest possible contrast with the restraining and confining conditions of the town which will be consistent with convenience and the preservation of good order and neatness.[40]

Olmsted's perspective on nature–society relationships is also connected to wider social ideals and the intellectual movement of Romanticism and Transcendentalism associated with writers Henry David Thoreau and poets Ralph Waldo Emerson and William Cullen Bryant. Many had traveled to Europe, visiting parks and gardens. Although most of these parks were originally restricted for use by royalty, by 1850 many began to open for the enjoyment of all the people. This was in contrast to many US cities, where only a few cities—Philadelphia, Washington, DC, Savannah, Boston and New York— had set aside public spaces. A number of American writers became proponents for public parks. Their Romantic idealism looked to the natural world to inspire and to fill the spiritual void present in the dirty, gritty, industrial, mechanical city. Romantic idealism stressed a return to "harmony" with nature and a new-found appreciation for it. Transcendentalists attributed virtues to trees, meadows and lakes and believed that these virtues could be duplicated by human ingenuity and design, thus influencing park design theory. In a sense, Central Park "transcended the city." Ironically, this intellectual perspective emerged in an industrial urban context.

Olmsted's design embraced Romantic ideals—rejecting the grid system in favor of wandering paths. Olmsted and Vaux created informal naturalness. Recognizing that recreating the wilderness would be difficult if not impossible within the urban boundaries of the park, Olmsted opted instead for the pastoral and picturesque. This style called for a composition of smoothness, harmony, serenity and the occasional reminder of the awesome grandeur of a mountain, crevasse, waterfall or lake. The design was to *suggest* nature in places where nature was not actually provided. He favored irregular planting, which suggested the feeling and idea of distance, and tried to avoid the use of flowers, which he felt revealed the hand of man.[41] And he rejected straight lines, hard edges and right angles—all of which were associated with machines and regimentation. Instead, he opted for softer, rounder, more fluid geometries associated with the contours of landscapes and the organic structure of vegetation: it was a romantic rejection of the rational.

As much as Olmsted was influenced by Transcendentalism, he also embraced Progressivism. He was concerned with creating a more democratic social order and his earlier writings were influenced, to some degree, by communitarian thinking. His park design was connected to his political views. In failing to provide opportunities for public recreation, American society had set up barriers of class and caste. Thus his urban public park would not only foster social freedom, it would also act as an agent of moral improvement by providing organized activities and sports.[42] In park design, this translated into building a park as a way of helping to shelter commonplace civilization. Parks were more than scenery: they were social spaces. He believed that the immigrant factory worker or unattended child, with no access to parks or recreation, or much leisure time on the weekends, could become so disillusioned as to threaten the political system or the social order. Hence the parks provided an "outlet" for all social classes. His designs attempted to provide space for all classes. He viewed recreational opportunities in the parks as providing outlets that would "counteract the evils of town life." Thus, in the Olmstedian ideology, parks served as democratic playgrounds. And they were. The parks built during the nineteenth century were, by and large, used mostly by children and the urban working classes.

Olmsted's legacy is widespread throughout the US and Canada. He designed Prospect Park in Brooklyn and Boston's Emerald Necklace—linking various small parks together along a connecting trail, created a model suburban village at Riverside Illinois, planned Mount Royal Park in Montreal and Bell Isle Park in Detroit, created the campus plan for Stanford University in Palo Alto, California, designed the space around the US Capitol, and laid out George W. Vanderbilt's mountain estate, the Biltmore, near Asheville, North Carolina. In the mid-1870s Olmsted began work on Mount Royal Park in Montreal. He considered it a grander natural canvas than Central Park and sought to more fully combine the sublime and the beautiful. His firm designed about 100 projects stretching from Victoria, British Columbia, on the west coast to Truro, Nova Scotia, on the east. He also contributed to preservation plans for both Yosemite and Niagara Falls. His legacy is found not only in the actual parks and other places he built, but in an ideology that helped to generate the broader urban park movement during the Progressive Era in America. Often spearheaded by local elites and civic organizations, the movement would bequeath to US cities such parks as San Francisco's Golden Gate Park, Chicago's South Park system, and parks in Philadelphia, Baltimore, Boston, Buffalo and New Orleans. While these parks have evolved over time to accommodate new demands and needs, the urban parks movement set aside vast areas for parks and recreational spaces that would likely have succumbed to the relentless march of the gridiron. The park movement, born in the city, eventually expanded to preserve threatened scenery and helped to establish a constituency for preservation, eventually providing both intellectual and popular support for the creation of state park systems and the national park system.

The emergence of an urban parks movement was not limited to North American cities. During the nineteenth century, large, naturalistic urban parks began to appear in cities around the world. According to Heath Schenker, author of *Melodramatic Landscapes: Urban Parks in the Nineteenth Century,* urban parks became a stage for the fundamental reworking of civic society, particularly the idea of equity in public affairs and public space. Her book examines the creation of three major city parks: Buttes Chaumont in Paris; Chapultepec Park in Mexico City; and Central Park in NYC. All three were products of complex social, cultural and political forces: Plans for these new parks were informed by notions of melodrama, with the landscape designed to communicate clear moral messages about social harmony and the benefits of contact with nature for the edification of the lower classes. She notes that all three parks represent broader social and political transformations underway.[43]

BOX 3.5

Chapultepec Park

Chapultepec Park, more commonly called the "Bosque de Chapultepec" (Chapultepec Forest), in Mexico City is the largest city park in Latin America, at just over 1,600 acres (686 hectares). The park sits west of the city. Centered on a rock formation called Chapultepec Hill—which translates as grasshopper hill—it is considered the first and most important of Mexico City's "lungs."

Chapultepec is rich in historical significance. Having once been a sacred site for the Aztecs, in 1530 it was designated as public property and a key source for Mexico City's water supply. By the nineteenth century, its springs were running dry, and the viceroy's former pleasure villa had been converted into a military academy. The Castle of Chapultepec saw a major battle in the Mexican–American War, where in 1847 the young cadets sacrificed themselves in a suicidal action against the invading Americans. A monument to the young fighters was erected and the site became the locus of nationalistic celebrations.

Thus the site of Chapultepec was already infused with political meaning when Emperor Maximilian I, in a pact with

Napoleon III, came to power in 1864 and dreamed of recreating Mexico City in the image of Paris. He chose the castle as his residence and the seat of the court (instead of the national palace in the center of the city) and hired architects to redesign the space and grounds into a modern park. At that time, the castle was on the outskirts of the city. Planners designed a wide new boulevard modeled after the Champs-Élysees in Paris to connect the imperial residence with the city center, and named it *Paseo de la Emperatriz* ("Promenade of the Empress"). Following the reestablishment of the Republic in 1867 by President Benito Juarez and the end of the Reform War, the boulevard was renamed Paseo de la Reforma. The boulevard became the spine of modern Mexico City.

In 1876 Porfiriato Diaz won the presidential election and set the city on the path to modernity. Urban improvements included new potable water systems, sewage systems, hospitals and the paving of streets and boulevards. In the last decade of the nineteenth century, the Diaz regime also accelerated improvements to the park. New facilities were built, thousands of trees were planted, new gardens were added; statues, marbles and other artistic objects were placed around the park. As Schenker noted, Chapultepec became the backdrop to a scene where the state articulated its aspirations to modernize the city and where Mexican identity reflected a progressive bourgeoisie that wanted to show that Mexico was no longer a wilderness but a developed, modern nation. But in this version of modernity, the park was a showplace for the rich and middle classes; the poor, who were viewed as a threat to health and morality, were excluded. The history of the park's evolution is also a history of changing politics and national narratives.

Today, Chapultepec Park offers all of the amenities associated with nineteenth-century parks: curving paths and drives, shade trees, serpentine lakes, and vistas that offer respite from the gritty city. Activities in Chapultepec Park are similar to the activities in Central Park in New York or the Bois de Boulogne in Paris on a Sunday afternoon. People of all ages and various walks of life are strolling, chatting, eating, playing games, boating, and generally enjoying themselves. The park boasts several museums, a zoo, numerous restaurants and an amusement park. More than 15 million people visit each year. The park accounts for more than 57 percent of all green space in Mexico City.

Source: Schenker, H. (2009) *Melodramatic Landscapes: Urban Parks in the Nineteenth Century.* Charlottesville, VA: University of Virginia Press.

In the eighteenth century, parks were the exclusive property of aristocrats and monarchs. They were not open to the public. Many of the first parks in Europe were actually hunting parks and private gardens. During the nineteenth century, however, these private properties were transformed into public parks, reflecting social and political transformations such as the establishment of democratic governments. The transformation of London's Hyde Park, Saint James' Park and Kensington Gardens are such examples. Those cities that lacked large estates to convert created parks by appropriating various properties from different owners. By the end of the nineteenth century, nearly every modernizing city had added a large, naturalistic park, both to signify modernity and to attract members of the new business and professional class. The rise of urban public parks in the nineteenth century can be seen not only as a new discourse on urban–environmental relationships, but also a symbol of the wider concepts of equity as a governing principle.

The emergence and development of an urban parks movement represents an important shift in how society values the natural world within the urban context. It was a powerful force in the evolving redefinition of urban form and culture and it laid the conceptual foundation for twentieth-century park development and evolution in cities around the world. And yet it is also important to realize that all urban parks (of any size and function) are cultural landscapes: they are "managed" nature that has been engineered, planted and continually recreated. Under Olmsted's supervision, for example, an army of laborers blasted tons of rock, moved thousands of cartloads of earth, refashioned swamps into lakes and ponds, planted turf, trees and shrubs. As the environmental historian David Schulyer concludes, "it is thus a subtle irony … that, like the city that surrounds it, Central Park is totally a man-made environment."[44]

Garden cities

In 1898 Ebenezer Howard published his book *Garden Cities of To-morrow*. According to Howard, cities in Europe and America had become too densely populated and suffered from "foul air, murky skies, slums and gin palaces." He was fearful of the consequences for society if old cities—and the social conflicts and miseries they embodied—continued. His vision did not seek the amelioration of the old industrial city, but a wholly transformed urban environment.[45] His vision of garden cities sought to create new, self-contained cities of some 500,000 on some 12,000 acres that combined the best features of urban life (industrial employment, social opportunity, places of amusement) with that of the "country" (land, fresh air and an abundance of water and sunshine). This was an ideal model, a blueprint for a new type of city. He would reorient the spatial balance between the built form and green space/open space, thus reforming the physical environment and revolutionizing society. This would "restore the people to the land" as he believed that "human society and the beauty of nature are meant to be enjoyed together."[46] His vision contained two important design elements that would have a lasting legacy on urban planning: zoning and greenbelts.

Howard's map of a garden city is highlighted by the division of land into different uses (Figure 3.6). This is one of the earliest articulations of "zoning."

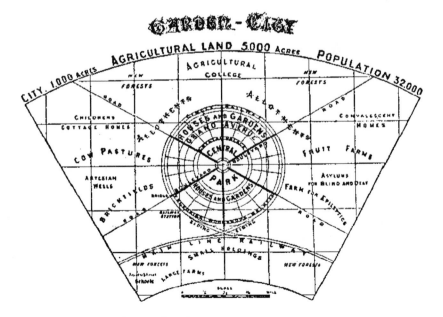

Figure 3.6 Howard's garden city

His plan locates all residential areas away from the belching smokestacks of the factory. In addition, he has incorporated space for urban farms to grow food to supply the city. From the center node out to the radiating smaller nodes, he has planned for railway transportation to link all nodes together. His plan also called for "buffer zones" between each of the distinct land uses. This, too, is one of the earliest articulations of a "green belt" and has become widely adopted in many metropolitan areas in the US and Europe.

In England, Howard and his supporters founded two cities, Letchworth (1903) and Welwyn (1920). In the US more than a dozen "greenbelt cities" were undertaken, including Greenbelt, Maryland, and several were established in Australia. However, Howard's actual garden cities failed to inspire the urban revolution he hoped.

Howard looked at cities in a regional rather than an isolated framework and his vision of a garden city model represents a major step forward in terms of organizing urban space. His vision spoke of a city where people were reunited the nature. Geographer and planner Robert Freestone has observed that it is possible that Howard's physical innovations of garden

Figure 3.7 The garden city of Greenbelt, Maryland, featured walking paths between apartment buildings and to the town center

Source: Photo by John Rennie Short

cities had little impact on the way in which people actually lived; neverthe-less, his ideas and theories influenced a generation of planners throughout the twentieth century.[47] As with the parks of Frederick Law Olmsted, the ideas behind the new urban forms would resonate beyond the US and Britain as planners in cities around the world adopted and adapted the ideals of garden cities.

One such example is Tapiola in Finland. Tapiola was one of the first postwar "new town" projects in Continental Europe. Tapiola was *not* a program of Finland's national government but was constructed by a company acting as a private, nonprofit business, Asuntosäätiö. This firm was established in 1951 by six social and trade organizations, which bought an area of 670 acres in the then-rural county of Espoo about six miles outside Helsinki. Their plan was to create a new community—a working town in a garden setting that would accommodate a cross-section of the population. The Housing Foundation set out to create an ideal garden city. The planners of Tapiola created design teams that included experts in architecture, sociology, civil engineering, landscape gardening, domestic science and youth welfare. The aim of the Housing Foundation was to create a garden city which would be a microcosm of Finnish society: all social classes would live there and there would be different types of buildings, ranging from detached houses to terraced and multi-story blocks. Unlike most traditional European city centers, Tapiola has five separate city centers. Clearly influenced by Howard, the planning team divided the land into four neighborhood units, separated by greenbelts; the center area would house a shopping and cultural center. The planners strived to develop a self-contained community, and aimed to have as many jobs as possible provided nearby. Tapiola's success generated interest in Finnish urban planning, and the city itself gained both a national and international reputation for its high-class architecture and landscaping, as well as an ideological experiment. The Tapiola experience highlights efforts to develop policies to guide urban growth in a more sustainable way.[48]

In the century or so since Howard articulated his garden city idea, many have been planned and built around the world, including in Australia, New Zealand and Latin America.

The urban parks movement and the intellectual legacy of Ebenezer Howard's garden cities provided a counterbalance to the pervasive environmental degradation of the industrial city. These new forms of urban design offered a positive revisioning of the city–nature dialectic.

BOX 3.6

Garden cities in Latin America

Vedado in Havana, Cuba has been characterized as embodying elements from the garden city, as well as a "garden suburb." Designed by José Yboleón Bosque, Vedado developed as a suburb of the Cuban capital in the late 1850s. The original design included the use of square blocks surrounding open space, tree-lined avenues, integration of public parks, green space in front of residences, and the inclusion of grounds for sports and recreation. Vedado became known as "the first Garden City of Latin America."

Vedado was not necessarily an "anticipation" of Howard's model, but rather it synthesized different residential development designs of the late nineteenth century, including the garden city, the organic design of Frederick Law Olmsted, and the combination of blocks and activities of Ildefonso Cerdá. Vedado represents a derivative concept of the "garden city" with diverse suburbs that looked to better conditions and further integration with nature in different Latin American capitals.

Another example is found outside of Buenos Aires, Argentina. Designs for the planned community of Ciudad Jardin Lomas del Palomar began in the late 1930s. It was completed in 1944 and laid out on a pedestrian scale with tree-lined streets. It has the features of a classically planned garden city. Three leafy avenues form a trivium that terminates at a central park. Like other garden cities, the layout was developed to place community facilities, schools, churches, clubs and parks within walking distance of housing areas. Colonnaded shops and community buildings such as a church and school adjoin the central park and create the hub of the neighborhood. Additionally, there are small family homes and a few apartment buildings along the roads. Lomas del Palomar is a condensed community that today has a population of approximately 17,000 residents.

The town radiates out from the central plaza along the tree-lined streets, where all amenities are within walking distance of the home. This has the benefits of reducing congestion as well as leaving less environmental impact as there is less reliance on cars to get around town. It is also situated between two train lines, making public transport easily accessible to other parts of Buenos Aires.

Figure 3.8 Map of Ciudad Jardin Lomas del Palomar
Source: http://en.wikipedia.org/wiki/Ciudad_Jard%C3%ADn_Lomas_del_Palomar

Sources: Almandoz, A. (2004) "The garden city in early twentieth-century Latin America." *Urban History* 31 (3): 438–52. Segre, R. (2000) "Cerdá en el Mar Caribe." *Ciudad y Territorio. Estudios Territoriales* 32 (125): 573. Segre, R. and Baroni, S. (1998) "Cuba y La Habana: Historia, población y territorio." *Ciudad y Territorio. Estudios Territoriales* 30 (116): 370. Hutchings, A. (2011) "Garden suburbs in Latin America: a new field of international research?" *Planning Perspectives*, 26 (2): 313–17.

Guide to further reading

Allen, M. (2008) *Cleansing the City: Sanitary Geographies in Victorian London*. Athens, OH: Ohio University Press.

Beinart, W. (2005) *Environment and History*. London: Routledge.

Brimblecome, P. (1987) *The Big Smoke: A History of Air Pollution in London since Medieval Times*. London: Methuen.

Conan, M. and Wangheng, C. (2008) *Gardens, City Life, and Culture: A World Tour*. Washington, DC: Dumbarton Oaks.

Cronon, W. (1992) *Nature's Metropolis*. New York: W.W. Norton.

Hall, P. (2002) *Cities of Tomorrow: An Intellectual History of Urban Planning and Design in the Twentieth Century*. 3rd edition. Oxford: Blackwell.

McShane, C. and Tarr, J. (2011) *The Horse in the City: Living Machines in the Nineteenth Century (Animals, History, Culture)*. Baltimore, MD: Johns Hopkins University Press.

Meller, H. (2001) *European Cities, 1890–1930s: History, Culture and the Built Environment*. Chichester and New York: Wiley.

Melosi, M. V. (2005) *Garbage in the Cities: Refuse, Reform, and the Environment*. Revised edition. Pittsburgh, PA: University of Pittsburgh Press.

Melosi, M. V. (2008) *The Sanitary City: Environmental Services in Urban America from Colonial Times to the Present*. Abridged edition. Pittsburgh, PA: University of Pittsburgh Press.

Miller, B. (2000) *The Fat of the Land: Garbage in New York, the Last Two Hundred Years*. New York: Four Walls Eight Windows.

Mosley, S. (2001) *The Chimney of the World: A History of Smoke Pollution in Victorian and Edwardian Manchester*. Isle of Harris: White Horse Press.

Schenker, H. (2009) *Melodramatic Landscapes: Urban Parks in the Nineteenth Century*. Charlottesville, VA: University of Virginia Press.

Tarr, J. (1996) *The Search for the Ultimate Sink: Urban Pollution in Historical Perspective*. Akron, OH: University of Akron Press.

Tarr, J. (2003) *Devastation and Renewal: An Environmental History of Pittsburgh and Its Region*. Pittsburgh, PA: University of Pittsburg Press.

Part II

The contemporary urban context

4 Global urban trends

We are in the midst of the third urban revolution. The first began over 6,000 years ago with the first cities in Mesopotamia. These new cities were less the result of an agricultural surplus and more the reflections of concentrated social power that organized sophisticated irrigation schemes and vast building projects. The first urban revolution, independently experienced in Africa, Asia and the Americas was a new way of living in the world. We discussed some of the environmental impacts in Chapter 2. A second urban revolution began in the late eighteenth century with the linkage between urbanization and industrialization that inaugurated the creation of the industrial city and unleashed unparalleled rates of urban growth and environmental transformations. Aspects of this transformation and resultant changes in ideas and practices were discussed in Chapter 3. We are currently in a third urban revolution, a complex phenomenon that began in the twentieth century and is marked by a massive increase in urban populations, the development of megacities and the growth of giant metropolitan regions (see Figure 4.1). We are in the throes of a revolution that we are only just beginning to see, name and theorize. The new lexicon which has emerged to describe cities—"post-modern," "global," "networked," "hybrid," "splintered"—offers some purchase on the rich complexity and deep contradictions of this third urban revolution, but much remains to be said and done before we can make any sense of the new forms of urbanism that characterize the twenty-first century.[1]

In this chapter we will consider some of the major environmental resonances of this dramatic shift associated with three main aspects of global urban change: rapid urbanization; the emergence of megacities; and the development of dispersed metropolitan regions.

Rapid urban growth

The world is increasingly urban. In 1900, only 10 percent of the world's population lived in cities. By 2010 it was more than half, and by 2050 more

Figure 4.1 Rapid large-scale urbanization in the city of Incheon, South Korea
Source: Photo by John Rennie Short

than two out of every three people on the planet will live in cities. Table 4.1 shows the percentage of urban population in 1950 and 2010 and estimates for 2050. The faster rates of urban growth will occur in the developing regions of the world. The respective absolute figures for the urban population are shown in Table 4.2. The world's urban population is increasingly living in cities in the developing world.

This global trend unfolds in different ways. Rates of urban growth have declined in developed economies because they started off with higher levels

Table 4.1 Percentage urban population, 1950–2050

	1950	*2010*	*2050*
World	29.4	51.6	67.2
More developed regions	54.5	77.5	85.9
Less developed regions	17.6	46.0	64.1

Source: Population Division of the Department of Economic and Social Affairs of the United Nations Secretariat, *World Population Prospects: The 2010 Revision* and *World Urbanization Prospects: The 2011 Revision.* http://esa.un.org/unpd/wup/index.html (accessed May 12, 2012).

Table 4.2 Urban population in billions, 1950–2050

	1950	*2010*	*2050*
World	0.74	3.55	6.25
More developed regions	0.44	0.95	1.12
Less developed regions	0.30	2.60	5.12

Source: Population Division of the Department of Economic and Social Affairs of the United Nations Secretariat, *World Population Prospects: The 2010 Revision* and *World Urbanization Prospects: The 2011 Revision.* http://esa.un.org/unpd/wup/index.html (accessed May 12, 2012).

and are slowly approaching their urban maximums. In Australia, more than 90 percent of the population already lives in the six urban areas: Adelaide, Brisbane, Canberra, Melbourne, Perth and Sydney. Urbanization is occurring more rapidly in the developing world. In 1900 only 10 percent of Mexicans lived in cities; by 2010, 77 percent did. In Sudan just 6.8 percent of the population was urban in 1950; by 2010 more than one-third of the population was urban. Even within the category of developing countries, there are substantial differences. Table 4.3 highlights the example of four developing countries. Afghanistan and Kenya are relatively poor countries with limitations on rapid economic expansion; so, while the urban growth rates are impressive, on current trends the urban population will not even reach 50 percent by 2050. Rapidly growing Brazil and China, meanwhile, have ratcheted up phenomenal rates of growth to much higher levels of urbanization. Large-scale urbanization and rapid industrialization continue to go hand in hand.

Urban growth continues apace because of powerful economic and political forces. Cities, especially large cities in the developed world, provide a "thick market" (a large pool of labor and specialized firms), market access and

Table 4.3 Percentage urban population, 1950–2050

	1950	*2010*	*2050*
Afghanistan	5.8	23.2	43.4
Brazil	36.2	84.3	90.7
China	11.8	49.2	77.3
Kenya	5.6	23.6	45.7

Source: Population Division of the Department of Economic and Social Affairs of the United Nations Secretariat, *World Population Prospects: The 2010 Revision* and *World Urbanization Prospects: The 2011 Revision.* http://esa.un.org/unpd/wup/index.html (accessed May 12, 2012).

savings in public goods provision. Cities, because they concentrate things in the same place, make markets more efficient. There are efficiency gains from the scale effects of urban centers. And in turn this spatial ensemble of benefits, through cumulative causation, creates path-dependent development.[2]

Cities bias development not only through the efficient operation of the market but through political processes such as the desire by economic actors to be close to political leaders and administrators, and in much of the developing world, the state's need to buy off the urban masses with public spending. The two arguments are often referred to as "productivity effects" and "rent seeking," respectively. They are not mutually exclusive as they refer respectively to the creation of income and its distribution.

Cities also generate costs, including congestion, pollution and the generation of streams of rural–urban migration that can create large pools of unemployed and underemployed. The costs and benefits vary by city size. Chun-Chung Au and Vernon Henderson suggest an inverted-U relationship between real income and city size. Income increases and then tails off as costs outweigh benefits. In their detailed empirical work on Chinese cities they suggest that real incomes quickly rise with increasing city size and only begin to fall gradually when there are more than half a million to one million workers in a city. They come to the intriguing conclusion that Chinese cities are too small to maximize income increases; they are stunted by government restrictions on rural to urban migration.[3]

There are other factors that lead to high levels of urbanization in the developing world. In the postwar years (1950–80), large-scale development projects, funded in part through the World Bank, were often channeled to create urban infrastructure for an industrial economy. Development projects were centered in cities, including the building of water and sanitation systems, electrical systems, roads, factories and warehouses, port facilities, and the construction of government infrastructure such as courthouses and parliaments. Although there were many rural development projects such as dams and large highways, significant amounts of development loans and funds ended up concentrated in urban areas. The high levels of urban primacy in many developing countries also reinforced growth in selected urban centers. The net result was the encouragement of massive rural to urban migration. To those living in rural areas, cities were places of economic opportunity, and since many rural areas remained without water, electricity or sanitation, cities also seemed to be better places to live. Mass urbanization has important consequences for environmental change and environmental management. Some of the more obvious effects on environment change include pronounced land-use changes and increased environmental impact, especially along the frontiers of urban expansion. Paul Yankson and Katherine Gough looked at the environmental impacts along the peri-urban growth areas of Accra in Ghana. They documented loss of forest cover, shortage of land for farming and increased erosion. As the population increased there was increased need for piped water and adequate sanitation, but service provision varied.[4] More specific changes also include an increase in the number of urban heat islands, an increase in the amount of impermeable surfaces and more polluted run-off. Rapid and large-scale urbanization puts extra pressure on the physical systems such as air, land and water. In many cities in the developing world rapid urban growth was and is still often associated with the creation of a more toxic urban environment.

We now have a large set of case studies that document some of the many changes associated with rapid urbanization, including:

- land-use cover changes;[5]
- ecosystem change;[6]
- ecosystem fragmentation;[7]
- increased resource use;[8]
- water quality changes;[9]
- increased global climate change;[10]
- air quality;[11]
- public health implications.[12]

Many analysts view urban growth as detrimental to environmental quality. There are numerous studies that confirm this finding. However, there is often an implied assumption that the shift from rural to urban in itself involves a downward slide in environmental quality. But this argument ignores the often disastrous effects of agriculture, even traditional agricultural techniques such as slash and burn. The connection between urbanization and environmental deterioration is not a given; it is especially marked in a country's early stages of rapid urban growth, but can lessen as the experience of rapid urbanization becomes the context for an awakened environmental sensitivity. The growing debates on urban sustainability arose in part from a realization of the deleterious environmental impacts of rapid urbanization, but also from a reconceptualization of an urbanization more in line with ecological realities and long-term sustainability. Cities create parks and ecological reserves as well as parking lots and factories. Rapid and unplanned urban growth can overwhelm and destroy ecosystems, but sensitive and sustainable urbanization can protect and nurture. The process of urbanization raises anew the issue of sustainability.

We can picture a three-stage model of rapid urbanization. In the first stage the increasing population negatively impacts the local environment, often overwhelming traditional methods of coping with the people–environment relations, such as disposal of sanitation and waste. Issues of economic growth and profit maximization trump environmental concern. Then, in stage two, environmental issues become more prominent due to the increasing risk and occurrence of environmental deterioration as well as the political responses of various stakeholders, such as resident groups, citizens and activists to environmental degradation of air, land and water. In stage three, maintaining the urban environment is seen not as peripheral to issues of economic growth, but central to issues of development, equity and justice.

The link between economic growth and environmental quality is sometimes represented as a bell-shaped curve termed the environmental Kuznets curve (Figure 4.2). The adjective is placed in front of the proper noun because there is another Kuznets curve that depicts a similar relationship between per capita income and inequality. At the early stages of the economic growth when per capita income is still relatively low, the environment worsens as aggregate growth is pursued despite the environmental effects. Then, as incomes increase, more people place priority on environmental quality. Evidence can be found that both contests and confirms this dynamic between income and environment. In terms of counter-argument, for example, levels of carbon emission continue to increase with rising per capita incomes. Moreover, richer regions and countries may export the negative environmental

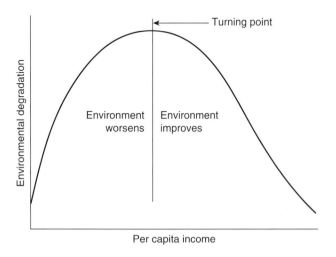

Figure 4.2 The Kuznets environment curve

Source: http://green.wikia.com/wiki/Environmental_Kuznets_curve

consequences of their high growth through shipping pollution waste or off-shoring the more environmentally damaging economic practices. Poorer countries and regions may be able to affect environmental quality much earlier than the curve suggests, as social movements and governments are made acutely aware of the damaging consequences and the false nature of the jobs-versus-environment slogan. Lee Lui, for example, shows how eco-communities in regions across China have promoted environmental improvement despite low per capita income.[13]

Rapid and large-scale urbanization has increased pressure on the ecosystems of sustenance, accelerated global environmental change as well as global climate change and resulted in the development of discourses of sustainability. While rapid urbanization has created major environmental problems, it also created the need and opportunity for more sustainable urbanization.

The rise of big cities

The second trend is that throughout the world, cities have continued to grow larger. In 1800 there were only two cities—London and Beijing—that had more than one million inhabitants: by 1900, there were 13. Today, there are several hundred cities that exceed one million in population and there are more than 35 that have more than five million inhabitants. It is estimated that by 2015 there will be around 400 cities with at least one million inhabitants. The average size of cities has grown dramatically. Table 4.4, for example, highlights selected cities located in the countries previously noted in Table 4.3. What is clearly evident is the rapid rate of growth in these cities. Across the globe cities are getting bigger. One of the more visible aspects of global urbanization has been the rise of megacities, large urban agglomerations with more than ten million inhabitants. They are a recent addition to the

Table 4.4 City growth, 1950–2011: city population (millions)

	1950	2011
Kabul	0.12	3.09
Rio de Janeiro	2.95	11.96
Shanghai	4.30	20.20
Nairobi	0.13	3.36

Source: Population Division of the Department of Economic and Social Affairs of the United Nations Secretariat, *World Population Prospects: The 2010 Revision* and *World Urbanization Prospects: The 2011 Revision.* http://esa.un.org/unpd/wup/index.html (accessed May 12, 2012).

urban scene. In 1950 only New York and Tokyo had populations of more than ten million. Today, there are 26 cities with at least ten million people. Lagos is one example. In 1950 the Nigerian city had a population of only 320,000. By 1965 it surpassed one million and in 2002 it became the first sub-Saharan African megacity when it topped ten million. With an annual population growth rate of 9 percent, it is one of the fastest-growing cities in the world. By 2025 it may have as many as 20 million residents. Table 4.5 shows the megacities in 2012. There are differences in city population totals in this table compared to UN sources because of differences in the definition of metropolitan regions.

Table 4.5 Megacities 2012

Rank	Megacity	Population (millions)
1	Tokyo	34.3
2	Guangzhou	25.2
3	Seoul	25.1
4	Shanghai	24.8
5	Delhi	23.3
6	Mumbai	23.0
7	Mexico City	22.9
8	New York City	22.0
9	São Paulo	20.9
10	Manila	20.3
11	Jakarta	18.9
12	Los Angeles	18.1
13	Karachi	17.0
14	Osaka	16.7
15	Kolkata	16.6
16	Cairo	15.3
17	Buenos Aires	14.8
18	Moscow	14.8
19	Dhaka	14.0
20	Beijing	13.9
21	Tehran	13.1
22	Istanbul	13.0
23	London	12.5
24	Rio de Janeiro	12.5
25	Lagos	12.1
26	Paris	10.1

Source: derived from http://www.citypopulation.de/world/Agglome-rations.html.(accessed May 2012).

Megacities are not just cities of a large size, they are a new distinctive spatial form of social organization that radically transforms the city–nature relationship. Their sheer size exerts a large and heavy environmental footprint. Megacities can impose a heavy environmental toll. Continual city growth generates tremendous rural to urban land-use changes and associated ecosystem transformations. The increasing population also puts extra pressure on the biophysical systems that provide land, air and water.

We can identify both direct and indirect environmental consequences attributed to the magnitude of megacities. For example, megacities are so large, the volume of pollutants is very high, and millions of residents are at risk. Many megacities are located on the coast, including Tokyo, Shanghai, Jakarta, Manila, Mumbai, Karachi, Istanbul, New York, Buenos Aires, Rio de Janeiro and Lagos. Coastal megacities have special concerns such as use conflicts in coastal areas, as well as problems such as coastal erosion, salt water intrusion, fresh water shortage and the depletion of fishery resources.[14] For example, rapid demographic and economic growth in Jakarta has created water pollution, coastal erosion, mangrove destruction and the intrusion of salt water into fresh water supply. In Mumbai, highly polluting industries, including chemicals, fertilizers, iron and steel and petrochemicals, have released semi-treated or untreated waste material into the coastal waters, degrading the beaches and impacting the tourism industry. In Buenos Aires, untreated sewage has been dumped into the River Plate and ultimately into the seas, creating serious coastal environmental problems.

Air pollution can be particularly severe in megacities. Motor vehicle traffic is a significant source of air pollution in all of the megacities; in nearly half of the world's megacities it is the single most important source. Emissions for power generation are also a problem. In China, Beijing and Shanghai are dealing with high levels of sulfur pollution that stem from the use of coal as a major energy source. China is now home to 16 of the world's 20 most air-polluted cities, and demand for fossil fuels (particularly coal) has only increased. However, it is not a constant story of urban growth continually causing environmental deterioration. On a positive note, some megacities have made vast improvements in air pollution in both the developed and developing world. Despite rapid growth, Mexico City has improved air quality. In 1992 air pollution was severe, responsible for 1,000 deaths per year and 35 times as many hospitalizations. Since then, public transport has been promoted, including a free bicycle loan program, polluting industries relocated and automobile exhaust systems improved. The results are dramatic. Lead in the air is reduced by 90 percent, ozone levels have dropped by 75 percent and suspended particulate matter has been reduced by almost

70 percent. Rapid and polluting urbanization can set the context for environment improvements. In many of the very large cities such as Mexico City, the increasing perception and recognition of environmental impacts prompts new and more sustainable policy responses. The low-emissions metrobus system introduced in 2005 annually eliminates 80,000 tons of carbon monoxide. The battle to improve air quality continues. On the one hand, the number of cars has doubled to more than 4.5 million since 1992. On the other hand, in order to further combat air pollution, Mexico City now boasts three vertical gardens, giant green walls of 50,000 plants that create shade, absorb pollution and oxygenate the air. The living sculptures are the result of a public–private partnership that embodies the new civic consciousness of creating a greener city.[15] In Seoul, another megacity long afflicted with poor air quality, the city government created more green spaces, cheaper public transport, the greater use of fuel-efficient cars and constant monitoring and policing of emissions from factories, cars and residences. The result is a vast improvement in air quality.

Large cities create environmental problems but also can provide innovative solutions. São Paulo is the economically most important city region in Brazil. Emission restrictions have helped reduce pollution from many industrial sources. The Brazilian government has encouraged alcohol-fueled vehicles and now more than 60 percent of São Paulo's vehicles run on gasohol—a mixture of gasoline and ethanol. This has helped reduce some air pollutants, although alcohol combustion does generate aldehydes.

Megacities consume vast amounts of resources. Manila and Mexico City, for example, consume vast quantities of water and are already dangerously depleting their groundwater supplies. In addition to the environmental impact that cities have directly, as polluters, there are also indirect consequences of megacities. As megacities grow, their peripheries enlarge, consuming agricultural land, forests and wetlands. Dhaka, Bangladesh is forecast to become the fifth largest city in the world by 2015, with a predicted population increase of 8.5 million over the next 15 years. Yet Dhaka is bounded on the west and south by the flood plain of the Burhi Ganga river and on the east by the flood plain of the Balu river. Both areas are flooded up to four months of a year. Land above the flood plain is high-value agricultural land but is rapidly being converted to urban uses as Dhaka expands. Mexico City has expanded its periphery into surrounding areas, and between 1970 and 1996 over 10,000 hectares of land were converted into urban land use. This expansion has reduced the amount of irrigated areas and forest cover, and extended south into a preservation zone with a high level of biodiversity.[16]

BOX 4.1

Urban regions

There have been a number of attempts to model land use around a city.

A model of intensity around the city was first developed as far back as 1826 by a German landowner, Johan Heinrich von Thunen (1783–1850). Von Thunen noted that there was a pattern to land use around a city: in particular, more intensive uses were closer to the city. Von Thunen postulated a general model of land use around a city situated in a flat plain with homogenous fertility and transportation costs. Farmers' costs were based on land costs and transport costs. Since farmers paid less transport costs closer to the city, land costs tended to be higher. Only farmers growing the more intensive crops, with high returns, could afford the land closer to the city. The net result was a concentric ring pattern with more intensive agriculture closer to the city.

The environmental historian William Cronon developed this model at a larger spatial scale in relation to the commodification of nature. In his 1991 book, *Nature's Metropolis*, he examines the relationship between Chicago and its hinterland from 1850 to 1890. He shows how the physical world was turned into a commodified human landscape as grain, lumber and meat production transformed prairies and woodlands into the physical basis for the city's growth and development. Merchants, railway owners and primary producers transformed the "wilderness" into a humanized landscape that was the basis for the city's impressive economic growth. Cronon's work demonstrates that urban economic growth draws heavily on a physical world.

The von Thunen model is now widely used to understand land-use change and ecosystem transformation at the edge of cities. Antje Ahrends and colleagues, for example, use the model in their analysis of forest cover loss around Dar es Salaam in Tanzania.

We can also imagine a model pitched at the global level. At its heart is the concentrated urban center of the global cities. Around this urbanized landscape is a suburban semi-periphery and beyond that a periphery of agricultural, forest and other areas. We can model the world as one giant urban region of dense urban centers, a suburban semi-periphery and rural/wilderness periphery with flows and interconnected relations between all three. The insights of von Thunen can be recast at a global level.

Sources: Ahrends, A., Burgess, N. D., Milledge, S. A., Bulling, M. T., Fisher, B., Smart, J. C., Clarke, G. P., Mhoro, B. E. and Lewis, S. L. (2010) "Predictable waves of sequential forest degradation and biodiversity loss spreading from an African city." *Proceedings of the National Academy of Sciences* 107: 14556–61. Cronon, W. (1991) *Nature's Metropolis: Chicago and the Great West.* New York: W. W. Norton. Von Thunen, J. (1966) *Isolated State.* Oxford: Pergamon (an English edition translated by C. M. Wartenberg, edited with an introduction by P. Hall).

Although megacities have mega-environmental impacts, it is important to note that many smaller cities have environmental degradation problems that are often more severe. According to the Blacksmith Institute, ten of the most polluted places in the world include Chernobyl (Ukraine), Haina (Dominican Republic), Kabwe (Zambia), La Oroya (Peru), Linfen (China), Norlisk (Russia), Dzerzinsk (Russia) and Ranipet (India).[17] Kabwe, the second largest city in Zambia, is home to 250,000 residents, but decades of unregulated mining and smelting operations have poisoned the soils and water sources. Haina in the Dominican Republic has severe lead contamination, the result of battery recycling, a problem common in the developing world. In some cases the cities have populations of only several hundred thousands, yet the concentrations of pollutants pose extreme threats to human health and are a major source of death, illness and long-term environmental damage.

There are two future directions for megacities of the twenty-first century. They could become places where new policies and politics will reorganize these spaces to make them more livable. Or they could become dysfunctional, non-livable settlements that will cause severe local, regional and global environmental degradation.

In some cases rising affluence is associated with reductions in pollution. Yet sometimes increasing affluence brings its own problem. In Delhi, for example, the growth of private vehicles and the decline of the dominance of public transport is leading to greater energy consumption and an increase in emissions. The private transport sector uses between one-third and two-thirds more energy than either bus or rail.[18]

In general, though, as cities become more affluent there is greater regulation of pollution and more pronounced environmental management. We can distinguish between megacities such as New York and Tokyo, with much better quality urban environments, and the poorer and less regulated megacities such as Jakarta and Manila. Cities such as Shanghai and Mexico City are moving closer to New York and Tokyo in terms of improving the quality of the urban environments. One study looked at subsurface pollution in Asian megacities and found that the poorest cities such as Jakarta and Manila had serious surface pollution from nitrates and trace metals, while the more affluent cities such as Seoul and Tokyo could afford the necessary infrastructure for sewage treatment.[19] The wealth of the city as well as its size influences the quality of the urban environment.

Even in rapidly growing megacities, environmental improvement is possible. Shanghai is one of the largest and fastest-growing cities in the world. In 1950 its population was 4.3 million. After increasing to 6.8 million in 1960, it then steadily declined as the Chinese Communist Party followed an anti-urban policy. People were forcibly removed from the cities. Shanghai's population total fell for the next 20 years; then, as the government policies changed to an encouragement of manufacturing in selected cities, Shanghai rapidly grew. From 1990 to 2011 the population almost tripled from 7.82 million to over 20.2 million. It is estimated to increase to over 28 million by 2025 (see Figure 4.3). Despite the enormous rate of growth, most people in the city have access to piped water and other public services such as garbage collection and regular electricity. In the early years of growth there was severe water and air pollution. The emphasis was on economic growth, not on environmental management. The downside of the rapid economic growth was deteriorating air quality and water pollution and the creation of a more toxic environment. The city still has the highest cancer mortality rate in the country. Again, the problems prompted solutions. Water quality was improved by the construction of waste treatment plants and the clean-up of Suzhou Creek, which runs through the heart of the city. Reponses to the air quality involve a greening of the city through the creation of large parks, and the relocation of factories and power plants away from the city. Air and water quality are both steadily improving. Environmental awareness

Figure 4.3 High-rise development in Shanghai, one of the world's biggest and fastest-growing cities

Source: Photo by John Rennie Short

is now an important part of the discourse of the city. In 2010 the city hosted Expo 2010, which had as its main slogan "Better City: Better Life." More than 73 million people visited the exhibition site on a formerly polluted industrial site. The main emphasis was on the notion of a green city: the Expo used zero-emissions vehicles, was powered by solar energy and used captured rainwater as its main water source. The Shanghai Expo provided demonstrations of what a greener and more sustainable city could look like.

It is clear from this brief survey of the urban environments of megacities that they are the sites of both problems and solutions. The very large cities bring into especially sharp focus the complex relations between the city and nature. While they highlight the enormous environmental costs, they also provide a glimpse of exciting environmental improvements.

The development of giant urban regions

The third global urban trend is the creation of giant urban regions. A number of recent studies suggest that large urban regions are the new building blocks of both national and global economies.[20] Scholars have identified globalizing city regions in which most urban and industrial growth is concentrated. Geographer Peter Taylor describes the world economy as structured around an archipelago of global city regions.[21] These city regions are the loci of control and command functions, with a heavy concentration of advanced producer services such as banking, advertising and business services. In the developing world they are also the site of multinational corporation investments and new techniques of manufacturing, as well as centers of service industries. Three giant urban regions have been identified in Asia Pacific: Bangkok (11 million population), Seoul (22 million) and Jakarta (22 million), which have between 25 and 35 percent of all foreign direct investment into their respective countries and constitute between 20 percent and 40 percent of respective national gross domestic product. In China, for example, the three city regions of Beijing, Shanghai and Hong Kong constitute less than 8 percent of the national population, yet they attract close to 75 percent of the foreign investment and produce 73 percent of all exports. China is less a national economy than three large urban economies. In North America 11 megapolitan regions have been defined as clustered networks of metropolitan regions (Figure 4.4). They comprise 67.4 percent of the population, and will be home to approximately three-quarters of all predicted growth in population and construction from 2010 to 2040.[22]

The largest city region in North America is the urbanized northeast seaboard, a region first named by Jean Gottmann as Megalopolis. Megalopolis stretches from just south of Washington, DC through Baltimore, Philadelphia, New York to just north of Boston. It is responsible for 20 percent of the nation's gross domestic product. In 1950 Megalopolis had a population of about 32 million people. By 2010 the population had increased to almost 45 million. This small area of just over 52,000 square miles with only 1.4 percent of the national land surface still contains over 14 percent of the nation's population. Despite the national redistribution of the US population, Megalopolis continues to remain a significant center for the nation's population. The environmental impact of this population increase is enormous: more people driving more cars to more places; more people running dishwashers, flushing toilets and showers; more people in more and ever-bigger houses. Megalopolis is arguably one the most environmentally impacted regions in the US, subject to the constant, mounting stress of a rising population with an ever-growing list of needs and desires.[23]

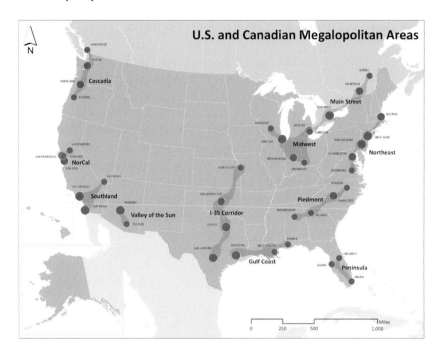

Figure 4.4 US and Canadian megapolitan areas
Source: Lisa Benton-Short

Consider automobiles. Applying the standard estimates of one car for every five people in 1950 and one car for every two people in 2010 yields a total of 6.4 million cars in 1950 and 22.5 million cars in 2010. In the same surface area, the number of cars has almost quadrupled. And this total does not include the buses, cars and trucks passing through the region. There are now, at the very lowest estimate, over 24 million autos burning fuel, releasing exhausts, needing roads and requiring parking spaces. The landscape has been redesigned to give these cars the space and freedom to move throughout the region.

And water usage: in 1950 daily per capita water withdrawals for the US totaled 1,027 gallons, and by 2010 this figure had increased to 1,300 gallons per capita per day. In Megalopolis, not only has the population increased but the per capita daily withdrawal has increased by 30 percent.

A similar picture emerges for municipal solid waste generation. In 2010, 4.4 pounds were generated per person per day. Megalopolis generated approximately *100,000 tons of garbage per day*. Consider the case of New York City, which each day generates approximately 12,000 tons of municipal waste and about that same amount of waste from businesses collected by private companies. Since the closing of the Fresh Kills landfill in Staten Island in 2002, the city now has an elaborate system for the transfer and disposal of the municipal waste. Trash is collected and hauled by 550 trash trucks to transfer stations in the city and New Jersey and then transported for incineration in Newark and dumping in landfills in Pennsylvania, New Jersey and as far south as Virginia. Trash truck journeys have increased congestion and raised the pollution levels along such routes as Canal Street by as much as 17 percent. In some of the receiving areas taxes from landfill companies pay as much as 40 percent of the school budget. Megalopolis is one giant waste-generating, waste-disposal region.

Whatever the measure, it is the same story of increasing population growth in association with increased affluence and spiraling consumption producing a greater environmental footprint and increased strain on the natural systems that sustain and nurture life. As more population crams into the region, an incredible cost is exacted by the environmental transformation that is wrought. Close to 45 million people, with the greatest environmental impact per capita in the history of the world, now live in Megalopolis.

BOX 4.2

Urban networks

Cities around the world, especially large cities, share similar environmental problems. Often the sharing of experience and policy options occurs through visits and the information flowing along more informal networks. There are also cases of city networks specifically designed to foster information sharing and mutual learning. The Cities for Climate Protection, for example, was established in 1991 and now has 650 municipalities from over 30 countries. It is a transnational urban network that aids in the diffusion, adoption and implementation of climate change mitigation programs that promote reductions in greenhouse gases. As more people live in cities the role of these transitional urban networks becomes even more important. In 2005 Ken Livingston, then the Mayor of London, created the C40 initiative that fosters collaboration between large cities seeking to manage climate change. By 2012, C40 included 59 affiliated cities that include Addis Ababa, Bangkok, Copenhagen, New York and Rio. The website is http://www.c40cities.org/c40cities.

While international agreements can flounder as countries seek to maximize their advantages, networks of cities pursuing the same environmental goals can make substantial changes. The C40 cities have the capacity to reduce their carbon output by one billion tons by 2030 by introducing more energy-efficient building codes and lighting regulations. Networks of cities vigorously pursuing shared policies can have as big an impact as international agreements between countries that are often poorly specified and monitored.

There are also national urban networks such as the US Conference of Mayors, which now includes over 1,500 cities that have signed an agreement for urban environmental policies that reduce greenhouse gas emissions, educate the citizenry and lobby state and federal government (http://www.usmayors.org/climateprotection/agreement.htm).

The individual plans for Kansas City, Missouri, and Austin, Texas, for example, can be accessed at these sites: http://www.kcmo.org/idc/groups/citymanager/documents/citymanagersoffice/022729.p; http://www.austintexas.gov/department/austin-climate-protection-program

These global, national and even regional networks allow a sharing of best practices, an opportunity to learn from other cities' experiences as well as powerful platforms for outreach, education and persuasion. As neoliberalism seeks to hollow out the power of the nation-state, these city networks play an increasingly important role as a more democratic and locally based power source that can shape politics and affect the discourse of urban environmental management.

Urban sprawl

Big city regions are now characterized by more dispersed forms of urban development. The steady suburbanization of jobs and residences has extended the urban region further out from central cities. Metropolitan regions can be divided into central city and suburban areas. The most significant feature of the last 60 years is the suburbanization of the population and economic activities. Take the case of the US: in 1950 only 23 percent of the US population was living in suburbs. This figure increased to 46.8 percent in 2010. More people live in metropolitan areas and more of these people are living in extending suburbs. But the process, while very evident in the developed world, is not limited to cities in the developed world. Consider the case of Shanghai. Since 1991, new residential and industrial complexes are developing along the rural–urban fringe, aided and encouraged by local governments who are key stakeholders in the land development system. The end result is a more widely dispersed urban system with land development often leapfrogging across the landscape, prompted more by local governments eager to cash in than by rational metropolitan land-use allocation. The dominant socio-spatial restructuring is now gentrification of the inner city and peripheral expansion as more lower-income households are pushed to the further peripheries. Vast new peripheral developments expand Shanghai's reach far outwards, with networks of inner ringroads and outer motorways, new subway lines, tunnels and bridges and new high-speed trains. Shanghai is now a huge polycentric city connected by fast-moving highways, high-speed trains and contrasting neighborhoods; at the extremes there are exclusive gate communities for the wealthy and enclaves of marginalized migrants.

Urban sprawl is also evident in urban contexts previously marked by high densities. Luca Salvati and colleagues examined sprawl in the Rome metropolitan region of Italy.[24] The previous urban form was characterized by high density, tightly wrapped around a strong urban core, but since the 1980s a more dispersed urban form emerged with more low-density developments located further from the city center. The polycentric dispersed urban expansion triggered land cover changes, especially the disappearance and fragmentation of croplands and woodlands, the destruction of both low- and high-quality agricultural lands and the emergence of a rural, non-farm landscape. This new hybrid landscape is also noted in studies of individual cities, including Athens,[25] Barcelona[26] and Istanbul,[27] as well as in regional surveys of Europe[28] and in measurements of land-use changes in global cities across the world surveys.[29] Australian metro regions are some of the most dispersed and suburbanized in the world (Figure 4.5).

Figure 4.5 Spreading sprawl in Brisbane, Australia. Note how the density declines as the city region spreads beyond the dense urban core

Source: Photo by John Rennie Short

Urban sprawl is an elusive concept. It is embedded in deeper discourses that, for some, tells a story of environmental degradation, social fracturing and loss of community.[30] There are also more recent defenders of sprawl. Robert Breugmann provides a vigorous defense of low-density scattered urban development at the edge of cities. His book, *Sprawl*, is an extended argument that sprawl has always been a part of urban development, and that it reflects the desires of the consumers. The criticisms of sprawl, he believes, reflect the class bias of intellectual elites.[31]

One survey of the social science literature found that sprawl had at least six prevailing definitions:

1 Sprawl is defined by "an example" of typical metropolitan areas such as Los Angeles, Dallas or Atlanta.
2 Sprawl is used as an "aesthetic judgment" about broader and more general patterns of urban development.
3 Sprawl causes "negative economic externalities," including dependency on automobiles, spatial mismatch between jobs and housing, concentration of poor minorities in the inner city and environmental degradation.

4 Sprawl is the effect or cause of an "independent variable," including fragmented local government, exclusionary zoning, or poor planning decisions.

5 Sprawl is defined as "at least one or more existing patterns of development." Some types of patterns are leapfrog, low density, dispersion of employment and housing, and strip mall development.

6 Sprawl is a "process" that gradually occurs over time as regions grow.[32]

David Soule provides a succinct definition of sprawl that builds on three of these six themes when he describes it as a "low density, auto-dependent land development taking place on the edges of urban centers, often 'leapfrogging' away from current denser development nodes."[33] Here, we will use the term sprawl to refer to low-density, car-dependent development with relatively low connectivity. While perhaps easy to use, the concept of sprawl is more difficult to operationalize. Recent, detailed work on measuring sprawl reveals a complex picture. Hal Wolman and colleagues tackle the difficult question of how to define and measure sprawl. They show that the measurement depends on which land area forms the basis of analysis. Even within one country sprawl takes different forms. In the dry Sunbelt of the US, for example, sprawl is of higher density than in the rest of the country because of aridity, public lands and slope constraints.[34]

Sprawl is now associated with a host of negative environmental impacts, including: loss of agriculture in the wake of low-density sprawling development; increased impermeable surfaces that lead to flooding and large discharges of polluted and contaminated water that overwhelm drainage systems and damage ecosystems; the heavy use of vehicular traffic that leads to increased air pollution; and global warming. Let us consider the following in more detail: loss of open space, greater air pollution, higher energy consumption, increased risk of flooding, ecosystem fragmentation and reductions in diversity and species.[35]

In the US, between 1982 and 1997 more than 1.4 million acres were converted each year to developed land. Between 1992 and 2001 the annual rate increased to 2.2 million.[36] The loss of green space is particularly acute in areas of rapid growth. The 3,000-square mile Washington, DC metro region lost 50 percent of its green space between 1986 and 2000. In this short time period the percentage of developed land increased from 12.2 to 17.8 percent. As sprawl continues outwards, more developed land is required. The rapid growth in Montgomery County, just north of Washington, DC, resulted in increased traffic. A new inter-county connector, first proposed in the 1970s, finally opened in 2011. The 18-mile, $2.4 billion road linking Interstates 270 and 95 impacted streams, wetlands and other sensitive ecological areas in its path. Pressure from community and environmental groups resulted in the construction of bridges over the wetlands and floodplains, but the impacts were still large. And in the wake of the road will come more sprawl. Ironically, the existing solutions to sprawl often create the conditions for greater sprawl.

Sprawl is a form of development that is very often too diffuse to support public transport or easy walking. The heavy and in some cases total reliance on private auto transport imposes a heavy environmental price in terms of air pollution, and the increasing dedication of space for roads and parking. The reliance of a built form precariously balanced on one fossil fuel with large and fluctuating costs raises issues of long-term sustainability. Matthew Kahn showed that US suburban households drive 35 percent more and use more than twice as much land as their urban counterparts. However, he also demonstrated that the greater vehicle use did not lead to greater air degradation because of stricter emission controls. In other words, technology and regulations can mitigate against the environmental impacts of sprawl.[37]

Urban sprawl creates more paved surfaces. When 10–15 percent of land surface is paved then increased sediment and chemical pollutants reduces water quality; at 15–20 percent there is markedly reduced oxygen levels in streams; and at 25 percent many organisms die.[38] Even the lowest-density suburban developments have an impervious cover of 16–19 percent. As suburban sprawl spreads outwards there is a seemingly inevitable reduction in water quality and watershed health.

The issue of species depletion and ecosystem fragmentation is more complex. If the land-use change is from a single land use such as green pasture or wheat fields to suburbs then more complex ecosystems may be produced, with public and private gardens providing a wider range of diversity than a simple monoculture. New bird and animal species, for example, adapt to suburban gardens. There is need for more careful assessment of the environmental consequences of land-use changes at the city's edge. If we exchange a field of corn, genetically engineered with yields only sustained artificially high with massive chemical infusions, for suburban garden plots, is this by definition a loss of environmental quality? Much of the debate about suburbanization and urban sprawl has been conducted in the context of an anti-suburban rhetoric. More detailed case studies may provide a more reasoned account.

There is also an emerging body of literature that points to the negative public health effects of suburban sprawl, including a link with increased obesity.

The promotion of a driving lifestyle leads to less physical activity and increases in obesity. Jamie Pearce and Karen Witten describe environments that create higher risk of obesity as "obesogenic."[39] We could make our cities healthier by making them friendlier to walking and cycling rather than making them only convenient for sedentary motorists.

In summary, the environmental impacts of low-density suburbanization include habitat destruction, ecosystem loss and ecosystem fragmentation and degradation of water and air quality. The general conclusion is that sprawl has a heavy environmental impact, and imposes high environmental costs. Even when we compare central cities and suburbs, the differences in environmental impact are substantial. In a detailed study Ed Glaeser demonstrates, for a sample of 48 US metro areas, that carbon emissions were significantly lower for people who live in central cities than for people who live in suburbs.[40] Suburbanites use around 50 percent more energy and produce more carbon dioxide than central city dwellers.

There is also the fundamental issue of the long-term sustainability of sprawl. Low-density suburban sprawl is only possible with relatively cheap fuel and lack of accountancy for the environmental impacts. It is unlikely that the very cheap gasoline that literally lubricated suburbanization will ever return. In the US, for example, the suburbs were built on gas costs of about $27 per barrel (at 2007 prices). In one month in the summer of 2008 they reached over $140. In March 2009, in the middle of a huge recession, the price was $43 a barrel, with OPEC officials suggesting that an ideal price, for them, was $60–70. Prices will remain deflated during recessions but will then tend to rise. There are few large oil reserves left and the price will inevitably rise when the global economy ticks upwards. Where does that leave low-density suburban sprawl that is so reliant on large-scale private car usage? The general answer: in a very precarious position. The long-term sustainability of low-density, energy profligate, heavy ecological footprint sprawl is now a matter of serious consideration.

Guide to further reading

Blais, P. (2010) *Perverse Cities: Hidden Subsidies, Wonky Policy, and Urban Sprawl*. Vancouver: UBC Press.

Bulkeley, H. (2012) *Cities and Climate Change*. London: Routledge

Economy, E. (2010) *The River Runs Black: The Environmental Challenge to China's Future*. 2nd edition. Ithaca, NY: Cornell University Press.

Glaeser, E. L. (2011) *The Triumph of the City*. New York: Routledge.

Hanlon, B., Short, J. R. and Vicino, T. (2010) *Cities and Suburbs*. Routledge: New York.

Jones, G. W. and Douglass, M. (2009) *Mega-Urban Regions in Pacific Asia: Urban Dynamic in a Global Era*. Singapore: Nus Press.

Maciocco, G. (2010) *Fundamental Trends in City Development*. New York: Springer.

Pieterse, E. (2011) "Recasting urban sustainability in the south." *Development* 54: 309–16.

Seto, K.C., Sanchez-Rodriguez, R. and Fragkias, M. (2010) "The new geography of contemporary urbanization and the environment." *Annual Review of Environment and Resources* 35: 167–94.

United Nations (2012) *World Urbanization Prospects, The 2011 Revisions*. http://esa.un.org/unpd/wup/index.htm

Yanarella, E. J. and Levine, R. S. (2011) *The City as Fulcrum of Urban Sustainability*. New York: Anthem Press.

5 The postindustrial city

In the wake of deindustrialization of many cities in the West there is a reorientation in the city–nature relationship with the physical environment that takes a variety of forms, including the reimagining of the city and its environment, the production of postindustrial landscapes and the cleaning up of toxic sites. In this chapter we will discuss each of these elements.

Deindustrialization resulted from the growing efficiency of industrial production that prompted a massive shedding of jobs as well as a shift in corporate investment away from the older industrial cities of North American and Europe toward the cheaper labor in cities in the developing world. Starting around the mid-1970s and continuing still, there was a global shift in manufacturing which resulted in the rise of new industrial cities in places such as China, and the decline of industrial cities in Europe and North America. As an example we will examine the case of older industrial cities in the US.

US cities in the last 40 years saw a shift from a manufacturing to a service economy. Even perennially successful cities such as New York witnessed massive industrial job loss. Between 2002 and 2010, for example, New York City lost close to 50 percent of its manufacturing jobs. However, cities such as New York, with a wide and more varied job base, were able to move from industrial to postindustrial more easily and quickly than cities with a heavy and single reliance on manufacturing. In cities such as Pittsburgh, Syracuse, Buffalo, Akron, Cleveland and Detroit, when companies fired or relocated workers, closed factories and moved out of the region or country there were fewer alternatives. These cities were transformed from "industrial" to "Rustbelt," and at worst from vibrant manufacturing centers to ghost towns of despair. Even growth cities such as Los Angeles and San Francisco struggled to cope with the social and economic consequences of a decline in manufacturing-based employment. The loss of manufacturing employment marked a critical shift in the North American economy. Michael Moore's

1989 documentary *Roger and Me* chronicled the massive job losses and factory closings in Flint, Michigan, home to General Motors. General Motors laid off 40,000 people in Flint between 1980 and 1989, a figure comprising 50 percent of Flint's GM workforce and one of the largest layoffs in American history. Jobs at GM fell from 76,000 in 1978 to 62,000 in 1986, to 19,000 in 2002. By 2012 there were less than 3,000 GM employees in the city that was the birthplace of the company. Flint continues to face hard times. In 2012 Flint's unemployment rate was about 13 percent, much higher than the national average. Compared to the state of Michigan, the city has 2.5 times the number of people below the poverty line and only half the median household income. For many industrial cities, high unemployment rates continue to impact local economies. In traditional industrial cities political and economic leaders have tried to generate strategies to replace lost jobs and investment by attracting other economic sectors such as services or tourism.

Many US cities also became economically and demographically decentralized during this era. Decentralization occurs when city centers lose either population or jobs. A serious result is a decline in the tax base for the city, which limits city funding of social services, education and infrastructure. A downward spiral may occur in which population loss leads to reductions in tax revenue that leads to reduced public services that in turn drives even more people to leave the city. City residents and jobs leave the center of the city for the suburbs or other metropolitan and even non-metropolitan areas. The cities of the industrial Midwest were very hard hit. In 1950, the population of municipal Cleveland was 900,000, but by 2010 it had fallen to 396,815. Detroit, once home to both GM and Ford and dubbed "Motor City" and "Motown," was home to 1.8 million residents in 1950. By 2010 its population had declined to less than 714,000. Former industrial giants have experienced a dramatic reversal of fortune in the wake of deindustrialization. In Flint and Detroit, as in many other cities, the downsizing of the economy and loss of population resulted in an urban fiscal crisis as tax revenues failed to keep up with expenditures. The city was placed under a state-appointed financial emergency manager in 2002–4, and again in 2011–12 after it was declared a "local government financial emergency," to wield even more cuts to the city's budget.

In Detroit, as high-paying manufacturing jobs became scarce and unemployment high, many residents lost their homes or apartments to foreclosure or eviction. Detroit's woes have continued. In 2010 it was estimated that about one-third of the city, some 40 square miles, is vacant. The joke is that the only expanding business in Detroit is demolition and that it is the first US city reverting back to prairie. Redevelopment has been a buzzword since the

BOX 5.1

Shrinking cities and urban farming

In many older industrial cities the previous patterns of growth are now replaced by a trajectory of population loss and employment contraction. The term "shrinking cities" is often used to describe the experience of cities such as Detroit, Liverpool and Leipzig. One of the problems of shrinking cities is the declining tax base, while still having fixed infrastructure and service costs leading to an urban fiscal crisis and/or a crisis in service provision. In the case of Detroit the city owns one-third of the total land area because of foreclosures and housing abandonment. It costs $360 million to service these areas with public services, while these abandoned areas do not generate any tax revenue. In Youngstown, Ohio the city lost almost 20 percent of its population between 2000 and 2010 and was forced to abandon streets, close down infrastructure and consolidate population in a smaller number of neighborhoods.

There are environmental issues associated with shrinking cities. On the one hand, there are a range of possible problems: abandoned houses and sites can become the breeding ground for pests and disease, the lack of investment can lead to a deterioration in environmental management and neighborhood upkeep. Parks and green spaces can become starved of maintenance and improvement funds. On the other hand, shrinking cities involve tears in the urban fabric that can be filled imaginatively with green spaces, urban gardens, open spaces and a return to spaces of greater biodiversity. Urban wildscapes have emerged from derelict, abandoned and marginal urban spaces. In some ways shrinking cities allow an opportunity to see what non-growth cities look like and what a more sustainable city could be, one less based on the assumption of constant growth. Traditional redevelopment strategies of encouraging growth seem unrealistic, which leaves room for imaginative conceptions of more green and sustainable cities.

Perhaps the most dramatic land-use change is the rise of urban farming. There are the small garden plots utilized by households and community groups. These can provide a valuable source of local, nutritious and cheap food. There are also the larger, more commercial agricultural operations. In Detroit the Hantz Farms group is buying up land to create large commercial urban farms. Abandoned houses are bought and demolished and trees and crops are planted. With 40 square miles of vacant land, the city has a vital ingredient for farming—a lot of land. There are plans for organic farms, an urban farming research center and demonstration sites for how abandoned urban land can become productive agricultural land. Urbanization involved the transformation of rural land uses into urban land; shrinking cities are at the reverse tipping point of turning urban back to agricultural.

Sources: Colasanti, K. J. A., Hamm, M. W. and Litjens, C. M. (2012) "The city as an agricultural powerhouse? Perspectives on expanding urban agriculture from Detroit, Michigan." *Urban Geography* 33: 348–69. Haase, D. (2008) "Urban ecology of shrinking cities: an unrecognized opportunity." *Nature and Culture* 3: 1–8. Jorgensen, A. and Keenan, R. (eds.) (2011) *Urban Wildscapes*. New York and London: Routledge. The website of an international study group on shrinking cities: http://www.shrinkingcities.org.

1990s, but redevelopment strategies have garnered mixed results. In the mid-1990s, three casinos opened up in the downtown area. In 2000 Comerica Park replaced historic Tiger Stadium as the home of the Detroit Tigers, and in 2002 the NFL Detroit Lions returned to a new downtown stadium, Ford Field. The 2004 opening of "The Compuware" gave downtown Detroit its first significant new office building in a decade. The city hosted the 2005 Major League Baseball All-Star Game and Super Bowl XL in 2006, both of which prompted more improvements to the downtown area. Currently, Detroit is constructing a riverfront promenade park similar to the one directly across the Detroit River in Windsor, Ontario, replacing acres of train tracks and some abandoned buildings with several miles of uninterrupted parkland. Yet the rebuilding of infrastructure has not necessarily improved economic growth. Detroit remains one of the nation's poorest cities. In 2010 more than one-third of residents were below the poverty line. Abandoned housing ranks as one of the city's most persistent problems. In 2010 a total of 78,000 housing units were vacant or abandoned; of that number 55,000 were in foreclosure. Detroit, already hit hard by abandoned housing in the 1980s and 1990s, is one of the US cities hit hardest by the 2008 foreclosure crisis.

Cleveland has also struggled with the legacy of deindustrialization to reinvent itself in the more competitive global economy. Initiatives to rebuild Cleveland have replicated the formula that many cities have employed: new museums, sports stadiums, convention centers; the renovation of old industrial warehouse districts for housing and retail; and waterfront development. Pundits dubbed these efforts "the Cleveland Comeback." One of the most successful projects is the Rock and Roll Hall of Fame and Museum, which opened to the public in 1995. The building, located on the shore of Lake Erie, was crucial in the redevelopment of Cleveland's waterfront area. New downtown sports stadiums for the professional teams have also aided revitalization. The Gateway sports complex cost $360 million and included an open-air stadium for baseball, and an indoor arena for basketball. Currently, the city is redeveloping the waterfront along both Lake Erie and the Cuyahoga River as a destination for tourists and locals alike. The city has also become a regional and national player in health services by capitalizing on the wealth of educational and medical facilities in the region to produce economic growth. Both the Cleveland Clinic and University Hospitals have announced billions of dollars in investment in new facilities, such as a new heart center for the Clinic, as well as a cancer center and new pediatric hospital. Despite these efforts, some experts claim the Cleveland Comeback has stalled. Cleveland suffered one of the largest proportional population losses in the country: the city shrank by 17.8 percent between 2000 and 2010. In addition, many of Cleveland's inner suburbs continue to decline and overall urban growth remains negligible. The case studies of Detroit and Cleveland show that cities continue to experience mixed results in efforts to realign and reinvigorate their economies.

Reimagining the city

The city of Syracuse, New York is a good example of the reimagining of the industrial city. This city of around 145,000 people in upstate New York has an industrial history based in salt production and later in a range of manufacturing and metal-based production. A major effect was the pollution of the local environment, especially centrally located Onondaga Lake. The city celebrated its industrial base with images of factories and salt fields. In 1972 the mayor of the city organized a design competition to replace the 100-year-old seal. There was community resistance, and it was only in 1986 that another mayor was able to introduce a new city logo. This logo represents a clean lake and an urban skyline with not a factory chimney to be seen (see Figures 5.1 and 5.2).

Industrial cities in the developed world have a difficult time in an era of world competition and the global shift of industry toward much lower cost centers. To be seen as industrial is to be associated with the old, the polluted and the out-of-date. Cities such as Manchester in the UK, Syracuse, Pittsburgh, and Milwaukee in the US, and Wollongong in Australia all have been (re)presented in more attractive packages that emphasize the new rather than the old, the fashionable postmodern rather than the merely modern, the postindustrial rather than the industrial, consumption rather than production, spectacle and fun rather than pollution and work.

The rebranding of formerly industrial cities is now a global phenomenon. Take the case of Wollongong, an industrial city on the New South Wales coast of Australia. The urban economy was dominated by a massive steelworks.

Figure 5.1 Syracuse logo, circa 1848–1987. This logo celebrates the industrial city

Source: Photo by John Rennie Short

Figure 5.2 Syracuse logo, 1987–present. This image represents the postindustrial city

Source: Photo by John Rennie Short

The steelworks shed 15,000 jobs from the early 1980s to the mid-1990s. The rate of job loss was only one major strand in negative imagery associated with the city in the national imaginary and amongst investors. City leaders decided to rebrand the city in the public imagination. An image campaign was built around the idea of "innovation, creativity and excellence." An important part of the campaign involved the steelworks planting half a million trees on the site while the council found funds to clean up the beaches and construct bicycle paths. The greening of the city is now an integral part of a city's shedding its hard industrial image for a softer postindustrial imagery.[1]

Creating the postindustrial urban landscape: the example of waterfront redevelopment

The process of deindustrialization sometimes allows opportunities for urban redevelopment as factories are abandoned and new geographies of production and circulation leave old docks and railway lines economically redundant.

Beginning in the 1960s, continuing in the 1970s and accelerating in the 1980s, waterfront redevelopment became a widespread process in the production of new urban landscapes. The processes of deindustrialization left many of these cities with abandoned warehouses and buildings and unused port facilities on their waterfronts. The advent of port containerization meant many older port facilities were inadequate for the new technology and became obsolete. As a result, urban waterfront redevelopment projects were among the most prominent examples of urban renewal in the 1970s and 1980s.[2] Cities in the US, Canada, Europe, Australia, New Zealand and Japan have transformed their waterfronts into vibrant, public spaces that attract locals and tourists. The waterfronts of Boston, Pittsburgh, Toronto and Vancouver have become the new festival spaces, filled with sports stadiums, restaurants and hotels. Even smaller cities such as Austin, Buffalo, Charleston, Cleveland, Ottawa, Savannah, Syracuse and Victoria have transformed their harbors, lakes or riverfronts. Similar trends are evident throughout the world. In London, the Docklands were transformed into a vast complex of multi-use spaces that include office buildings, shops, museums and residences. This project stretched from the city through the East End, some 5,500 acres. Figure 5.3 shows construction along the waterfront in the 1980s and Figure 5.4 shows a completed residential project.

Despite the substantial costs, waterfront transformations represent a dramatic story of urban rebirth—economically and environmentally. In addition

Figure 5.3 Construction at London Docklands
Source: Photo by John Rennie Short

Figure 5.4 Residential housing at London Docklands
Source: Photo by John Rennie Short

to new hotels and an aquarium, Boston's waterfront has been transformed by eliminating the old elevated portion of Interstate 93. The freeway was demolished and moved underground as part of the well-known "Big Dig." The resulting open space has become the Rose Kennedy Greenway, a linear park. Waterfront redevelopment has restored the centers of cities to economic, social and ecological health. Baltimore's Inner Harbor is often cited as a model US waterfront redevelopment project. It has become the city's gathering place: home to the national aquarium, two sports stadiums, hotels, restaurants, museums, high-rise condominiums and hotels (Figure 5.5). In Barcelona the urban renewal associated with the 1992 Olympic Games involved the opening up of the old docks to a harbor waterfront with a pedestrianized walkway. Until the mid-1980s, the city had turned its back to the sea, leaving only warehouses and docks at the water's edge. After the makeover the harbor front became a congenial place for leisure consumption rather than for production and storage. Waterfront development is not limited to cities in the rich world. In Beijing, waterfront development has occurred alongside rapid economic growth (see Figure 5.6).

Such large-scale development as in the Docklands or Baltimore is not without its costs. The reconstruction of Baltimore's Inner Harbor cost $2.9 billion. But there are also social costs. The diversion of funds to Baltimore's Inner

Figure 5.5 Baltimore's Inner Harbor
Source: Photo by John Rennie Short

Figure 5.6 Waterfront development in Beijing
Source: Photo by Michale A. Judd

Harbor contrasts with the city's poor public school system and the perceived decline in many public services. In some cases waterfront development is part of a valorization of selected parts of the urban landscape that allow for the further enrichment of real-estate interests, sometimes at the expense of social welfare programs. While Baltimore's Inner Harbor flourishes, many inner-city neighborhoods continue to experience high crime rates, population loss and housing abandonment.

Cleaning up the toxic environment

Industrial manufacturing processes in urban areas often generate hazardous waste. For much of the twentieth century, hazardous wastes were largely unregulated. During the 1970s and 1980s many cities of North America and Europe saw a significant decline of industrial manufacturing. Many industrial manufacturers closed, reduced their workforce or relocated to countries where labor was less expensive or environmental regulations were more lax, abandoning their factories and leaving behind a legacy of pollution and environmental contamination.

In the US, the 1980 Comprehensive Environmental Response Compensation and Liability Act (CERCLA), more commonly known as the Superfund, established a multi-billion dollar federal trust fund (Superfund) to pay for emergency measures as well as clean-up of sites for which responsible parties could not be identified. Superfund sites are found everywhere, including rural areas. Many Superfund sites are in and around urban areas, where there has been a long history of water or land pollution. A full list of current Superfund sites can be found at http://www.epa.gov/superfund/sites/index. htm. The Superfund legislation set an important precedent: polluters were responsible for clean-up, and in cases where polluters were no longer in business or there was no identifiable responsible party, the federal government would assume the costs of clean-up. In addition, 1986 amendments to Superfund established a Community Right to Know Act that requires industry to report publicly their use and releases of more than 100 toxic chemicals. That information must be made available at County Boards of Health as well as public libraries. This amendment not only provides incentives for pollution reduction, it adds the element of the fear of public disclosure for those companies not in compliance with regulatory standards.

The EPA Superfund website gives background history to each of the listed sites. Consider just one, the 19-acre Central Chemical site in Hagerstown, Maryland. From the 1930s to the 1980s it was the site of a chemical plant

BOX 5.2

Love Canal

In the last decade of the nineteenth century a developer named William T. Love had a plan to dig a six-mile canal to link the upper and lower Niagara Rivers in upstate New York. The plan floundered after only part of the canal was dug. The land was sold at auction in 1920 and became a chemical dump. The city of Niagara dumped waste. So did the US Army and private companies. Between 1942 and 1953 the Hooker Chemical Company dumped a giant cocktail of 21,800 tons of 82 different chemicals, including dioxins and benzyl chlorides. Much of it was in steel drums. In the 1950s the company sold some of the land to the school board for a dollar, who in turn sold some to a housing developer. The construction of a storm sewer and a new road to service the development disturbed the hidden toxic waste. The steel drums corroded and leaked their poison into the local soil and groundwater. Toxic sludge seeped into the basements of local houses. By the 1970s there were mounting health problems with abnormally high birth defects, seizures, infections and miscarriages. After intense agitation by the local community, with a local resident, Lois Gibbs, playing a leading role, the site was declared a health hazard by New York State. Love Canal was evacuated and property owners were compensated. President Carter declared it a national disaster and the clean-up of the site prefigured and foreshadowed the Superfund legislation of 1980. The searing experience of Love Canal and the distressing scenes of sick children living close to a toxic waste site helped propel the Act's passage through Congress.

Sources: Gensburg, L. J., Pantea, C., Fitzgerald, E., Stark, A. and Kim, N. (2009) "Mortality among former Love Canal residents." *Environmental Health Perspectives* 117: 209–16. Gibbs, L. and Levine, M. (1982) *Love Canal: My Story*. Albany, NY: SUNY Press.

that blended agricultural pesticides and fertilizers. Waste was buried on the site. Chemical contaminants including arsenic, lead, benzene, aldrin, chlordane, dieldrin, DDT and methoxychlor were found in the soil and groundwater. In 1997 the EPA reached an agreement with the owners of the site and a fence was constructed to keep people away from the contamination. In 2005 all the old buildings were demolished, and by 2011 a remediation plan was agreed upon. What is revealing from the story is the cavalier dumping of hazardous chemical wastes up until the 1980s, the high level of toxicity almost 30 years after pollution ended and how long and drawn out the process is as problems are identified and agreed remediation plans slowly emerge from the interaction between public and private interests.

Another site is the General Electric (GE) Co. Spokane Shop in Spokane, Washington. From 1961 to 1988 GE repaired and cleaned transformers and electric equipment on the site. Polychlorinated biphenyls (PCBs) contaminated the soil and groundwater. The site lays above an aquifer that is the water source for 200,000 people living within a three-mile radius of the site. In 1990 the state demolished all the buildings and removed the surface pollutants. By 1999 remediation had turned the most polluted sludge into a solid, while less polluted soil was removed. The site remains highly contaminated and groundwater monitoring continues. While the problems are identified and the more mobile sludge is fixed in place, it is still a toxic waste site sitting on the region's aquifer.

The EPA has identified a National Priority List (NPL) of some 30,000 sites they consider to be the most immediate hazards and to which they prioritize funds for clean up.[3] Sites are placed on the NPL through a process that begins with a preliminary site inspection that identifies sites that pose a threat to human health and the environment. Any person or organization can initiate this preliminary inspection. The EPA then employs a hazardous ranking system based on three categories: (1) if the site has or could release hazardous substances; (2) the toxicity and amount of the waste; and (3) the amount of people and sensitive environments affected by the release. Scores are calculated for the four pathways of groundwater migration, surface water migration, soil exposure and air migration. The highest-rated sites are placed on the NPL. States can also nominate one top-priority site regardless of score.

In 2012 the Southern Avenue Industrial Area in South Gate in Los Angeles County in California was added to the list. The site is located in a mixed industrial–residential area in a minority low-income community. In 1972 hot-melt adhesive tape for the carpet industry was made at the site. County authorities cited the facility for the improper use of trichloroethylene (TCE) and improper disposal of oils and solvents. In 2002 the EPA found high concentrations of TCE in the groundwater. The property owner and operator refused an agreement and, so far, nothing has been done. This case highlights the limits to effective action in the face of inaction and non-compliance by private owners and operators. This contrast with the Spokane example cited above, where a giant multinational company, with deep pockets and a keen eye on public relations, readily and easily found the money to remediate the site. Small companies and cash-strapped property owners simply may not have the resources to clean up polluted sites. The listing identifies hazardous sites and enables full-scale investigations of the contaminated soil and drinking water sources, but it does not secure action or guarantee remediation. It is the first step on a long and expensive road to environmental clean-up.

Brownfields

Despite Superfund intentions, by the mid-1990s many derelict site areas had not been cleaned up or were a low priority on the Superfund list. A new term emerged to describe these contaminated areas: brownfields. Although brownfields can be found anywhere, many are concentrated in cities that hosted manufacturing activities. Brownfields share numerous characteristics outlined in Table 5.1. A widely accepted definition of brownfields, provided by the US EPA, is "real property, the expansion, redevelopment, or reuse of which may be complicated by the presence or potential presence of a hazardous substance, pollutant, or contaminant."[4]

Table 5.1 Characteristics of brownfields

Industrial heritage
Integral part of the urban structure
Real or perceived contamination problems
High local unemployment
Decline of municipal revenues
Adverse effects on urban life
Need for intervention from outside to be brought back to beneficial use
Offer development without encroaching on existing green spaces

Source: Adapted from US Environmental Protection Agency.

The EPA estimates that there are more than 450,000 brownfields in the US. Brownfields, unmediated (not cleaned up) can have health impacts that include cancer, mortality, reproductive effects and chronic disease morbidity. In order to promote the clean-up and redevelopment of these urban sites, the EPA has developed the Brownfields Initiative, a successor of sorts to the Superfund. It has encouraged many states and local governments to establish programs that provide tax credits and other financial incentives such as low-interest loans to attract private investment. There are grants to provide funding for brownfield inventories, environmental assessments and community outreach. There are also loans that provide funds to clean up these sites and job training grants that provide training for residents of brownfield communities. Many loans and grants prioritize brownfields in low-income areas or economically depressed areas.

Some brownfields are large and require extensive rehabilitation and decontamination work, and the costs are considerable. Prospective buyers and developers of contaminated properties are often leery of becoming responsible for clean-up and many of the parties who caused the original contamination may be long gone. As a result, federal and state governments have attempted to reduce the barriers to redevelopment and to financially support projects through grants, low-interest loans and other subsidies. The US EPA created a grant program to help cities redevelop brownfields. Over 1995–2004 the EPA invested over $700 million in brownfields, providing grants to states, local governments and nonprofits.[5] The grants are used to help plan the decontamination and the redevelopment of these spaces. In addition, the EPA has tried to scale back environmental requirements that would be perceived as too onerous for development. It may not be necessary to have the same clean-up requirements at a parcel slated to be an industrial reuse area compared to a children's daycare facility (Figure 5.7). State and federal governments have tried to tailor clean-up requirements to meet the expected future use of the properties.

In 2003 the town of Babylon in Long Island, New York received a $200,000 grant to assess a 0.68-acre site of three adjacent properties that were used for parking, storage and manufacturing. The town was able to consolidate and clean up the property, demolish existing buildings and make the site ready for a new community center. Environmental remediation was an important part of an urban renewal and an improvement in community facilities.

One example of a successful public–private brownfield venture is the Highland Marine Terminal in the heart of southeast Baltimore's industrial port district.[6] For nearly a century, the site was a copper-processing plant.

Figure 5.7 In Baltimore, developers redesigned a brownfield, the old Proctor and Gamble factory. The redevelopment is now called "Tide Point" and consists of five main buildings, each named for products once produced on the site: Joy, Cascade, Ivory, Dawn and Tide. The old factories have been transformed into offices, a daycare facility and a restaurant

Source: Photo by Lisa Benton-Short

The site was contaminated with lead, arsenic, beryllium, copper, nickel, silver and selenium. As one of the first of Baltimore's brownfield projects, plans were to turn the site into warehouse space in a part of the city that had insufficient warehouse space. The redevelopment deal consisted of a state-subsidized loan, loan guarantees and $40,000 in grants from the city and state. The state of Maryland also put in another $1 million in low-interest loans. The developer razed some 100,000 square feet of buildings and cleaned up the site. The project built or rehabilitated more than 700,000 square feet of warehouse space and seven acres of outside storage, all of which is currently fully leased. Developers were quick to state that without the brownfield program, environmental issues might have deterred them from undertaking the project. Several decades of successful clean-ups and redevelopment projects show that given incentives, developers find it profitable to absorb the risks of purchasing, cleaning and reusing contaminated land.

Cities in Europe are also affected by brownfields. The amount of land defined as brownfields is difficult to assess. In Germany there may be as many as 362,000 sites covering about 128,000 hectares; in France approximately

200,000; in the UK some 105,000 sites; and in Belgium some 50,000 sites covering at least 9,000 hectares.[7] Environmental contamination not only has an adverse effect on the environment, but is also considered a significant barrier to the economic and social redevelopment of the city. In addition there is growing evidence to suggest that many urban brownfields are situated in areas with higher concentrations of minority populations and households below the poverty level, raising important questions about environmental justice.

In countries in Central and Eastern Europe the scale of urban pollution is often greater than in Western Europe because of the emphasis on economic growth at the expense of environmental concerns during the Socialist era from the 1940s to 1989. Cities in the former East Germany have higher levels of urban contamination than cities in the former West Germany. However, the wealth and commitment of Germany to environmental improvement allows a strong response. In Leipzig, for example, the coal-mining area just outside the city was transformed into a theme park, Belantis, which opened in 2005 and draws around half a million visitors each year. There is also an ambitious plan to create 17 interlinked lakes. By 2015 it is hoped that the former industrial-mining site will be the home to tourist-attracting marinas, beaches, meadows, forests and lakes.

Large brownfield sites are a common feature of cities in the former Eastern Bloc countries. Krakow in Poland has 700 hectares of brownfields within the city perimeter. Remediation started relatively late, only beginning in many cities after the fall of Communism in 1989. The Czech Republic has an estimated 10,000 brownfield sites, many of them agricultural and military as well as industrial and urban. The first industrial brownfield program in the Czech Republic only began in 2002.[8] Renewal programs are made problematic by the uncertain nature of land ownership and the high level of "public" land where the polluting industries have long since disappeared into bankruptcy, closure and privatization.

Where there is a functioning land market, brownfields are redeveloped in a variety of ways: for industrial reuse, commercial or residential uses, and also as green spaces such as parks, playgrounds, trails and greenways. The projects can be modest—the reuse of a single isolated property—or more ambitious, such as the revitalization of an entire distressed neighborhood. In the US, there has been a tendency to redevelop brownfields into spaces that have immediate economic benefits and cities have focused more on commercial or residential uses. Pauline Deutz looked at what she termed eco-industrial parks in North America and found that they were often used as a form of place promotion, with economic development issues often trumping ecological objectives.[9]

In European and Canadian cities, however, there has been more emphasis on converting derelict brownfields into green spaces. For example, between 1988 and 1993 19 percent of brownfields in the UK were converted into green spaces.[10] In Derbyshire, the former Markham colliery (coal-mining facility) is being transformed into Markham Vale, a 200-acre business–industrial park planted with willow trees. The willow trees will be harvested and used as a renewable energy resource to heat boilers for nearby commercial buildings.

Christopher De Sousa has examined the transformation of several brownfields into green spaces in the city of Toronto.[11] He found that many of the larger brownfield-to-green space projects involved sites that were near or within existing parklands or floodplain areas. The projects reintroduced native trees, shrubs, wildflowers and plants that enhanced the ecological integrity of the area and improved flood protection, stormwater control and offered increased recreational opportunities for many under-serviced communities. De Sousa notes, however, that all the projects were carried out by the public sector, as private interests were deterred by the perceived costs and lack of benefits. De Sousa also looked at the effect of brownfield development on land values. He found that the positive spillover led to a net increase of house prices of nearly 12 percent in Milwaukee and 3 percent in Minneapolis.[12]

In general, brownfield redevelopment is a marked improvement in city–nature relations. On the positive side, brownfield redevelopment is a form of land recycling that can restore and regenerate formerly derelict and toxic urban spaces. Brownfield initiatives have been integrated with community economic redevelopment and job creation. The EPA estimates that each dollar spent on the program leverages another $18 and in total in the US generates over 76,000 new jobs. The program also improves health and safety issues, a vital part of neighborhood restoration. The redevelopment of brownfields is also an effective reuse of urban space to counter suburban sprawl. In formerly industrial cities, reintegrating brownfields into urban space can promote sustainable development. On the negative side, the process can be regressive, in that low-income communities, affected worst by the pollution, do not benefit the most.[13] There is also an emerging issue of the environmental costs of remediation of brownfield sites. Remedial technologies produce waste streams, and do not always remove contaminants. Brownfield policy is often dominated by a chemical quick-fix. A number of researchers have drawn attention to more simple remediation techniques used carefully over the longer term. The emphasis on the chemical quick-fix, a process driven and dominated by short-term commercial interests, can sometimes produce new environmental problems.[14]

"Brownfield" is just one color-coded urban category. Other, newer terms include redfields (see Box 5.3), which are sites in financial distress and grayfields, a liminal category between a greenfield and a brownfield (see Figure 5.8). Peter Newton has examined grayfields, turning his attention to many of the inner suburbs of older industrial cities in Australia.[15] These aging suburbs provide a challenge but also an opportunity to channel urban growth into existing communities and promote long-term sustainability.

BOX 5.3

From redfields to greenfields

In many cities abandoned office buildings and factories, dead or dying shopping malls and foreclosed commercial properties are a financial drain. The city must still provide services such as fire and police for these places, yet no longer receives taxes on these properties. For these reasons, the city is "in the red" on these properties.

Organized by Georgia Tech University and the City Parks Alliance, The *Red Fields to Green Fields* concept seeks to acquire financially distressed properties, properties "in the red," and convert them into green space: public parks, for example. *Red Fields to Green Fields* seeks to turn non-productive assets into well-managed green space for neighborhoods and cities. The recent financial crisis that began in 2008 has helped promote the Red Fields project, as numerous commercial properties have gone into foreclosure, or are abandoned.

The *Red Fields to Green Fields* alliance proposes to create a federally funded Land Bank Fund with an initial investment of $200 billion. The bank would offer zero-interest loans, and would issue loans to public–private partnerships to buy distressed properties and remove buildings. Part of the land would be turned into a park; the rest would be redeveloped later to retire the loans.

Proponents of this approach see it as a win–win situation. Parks improve the economy, environment and health of a city, helping to foster more livable, healthier communities. At the same time, redeveloping these abandoned and deteriorating areas helps to avert the imminent failure of many banks and businesses by removing bad loans from the banks' books. In addition, some economic gains could also include the creation of new jobs in park design, construction and maintenance. Proponents also argue that the redevelopment of redfields into greenfields increases the value of businesses and residences near finished parks.

Several cities have been involved with this program, including: Atlanta, Cleveland, Denver, Miami, Philadelphia and Wilmington. For example, Cleveland is a city with plenty of surplus commercial real estate and some 17,000 vacant, foreclosed homes. The Cleveland team developed a plan that envisions a downtown public square that reclaims roads and other paved surfaces to create a signature park with an extended towpath trail and a state-of-the-art rowing facility. With a $1 billion investment, they propose to take 1,850 acres of real estate off the market, create 120 miles of connected greenway, restore the air and water, create 8,000 jobs and develop an attractive waterfront that could draw new businesses to the city.

Another proposal comes from Atlanta. Georgia leads the nation in bank failures, and Atlanta is one of the most overbuilt real-estate markets in the country. A preliminary study suggested that:

- The Atlanta region has 24 million square feet of vacant office space and has vacant lots selling for 25 cents on the dollar.
- The metropolitan region could gain 6,000 acres of park space and 780 miles of trails by acquiring selected properties.
- The parks initiative could generate $20 billion in economic development, 175,000 temporary construction jobs and 100,000 permanent jobs.

- That transforming these properties into parks could
  - put over 100,000 people back to work;
  - remove over 22,000 acres of non-performing real estate from outside the perimeter.
  - inject liquidity into the southeast US banking system to fuel residential and small business economic development.

Whether the idea of Red Fields to Green Fields will be successful depends upon many factors, but it does provide another example of the ways in which creative people can see the city as the solution to problems, not simply as the cause of problems.

Source: The Red Fields to Green Fields Organization at: http://rftgf.org/joomla.

Figure 5.8 The grayfield suburbs of Australia: Adelaide, South Australia
Source: Photo by John Rennie Short

Guide to further reading

Adams, D., De Sousa, C. and Tiesdell, S. (2010) "Brownfield development: a comparison of North American and British approaches." *Urban Studies* 47: 75–104.

BenDor, T. K., Metcalf, S. S. and Paich, M. (2011) "The dynamics of brownfield redevelopment." *Sustainability* 3: 914–36.

Desfor, G., Laidley, J. and Schubert, D. (eds.) (2010) *Transforming Urban Waterfronts: Fixity and Flow*. London: Routledge.

Dixon, T., Otsuka, N. and Abe, H. (2011) "Critical success factors in urban brownfield regeneration: an analysis of 'hardcore' sites in Manchester and Osaka during the economic recession (2009–10)." *Environment and Planning A* 43: 961–80.

EPA websites: http://www.epa.gov/superfund/sites/npl; http://www.epa.gov/brownfields/basic_info.htm.

Hollander, J. B. and Nemeth, J. (2011) "The bounds of smart decline: a foundational theory for planning shrinking cities." *Housing Policy Debate* 21: 349–67.

Power, A., Ploger, J. and Winkler, A. (2010) *Phoenix Cities: The Fall and Rise of Great Industrial Cities*. Bristol: Policy Press.

6 The developing city

Compared to the easy identification of the industrial and postindustrial city, the city in the developing world is a more complex phenomenon. For example, while the cities in the developing world are sites of industrial production, there is a rich variety including rapidly growing industrial cities in places such as Vietnam; some industrial areas such as Bangkok metro region that are moving up the value-added chain toward more advanced industrial production; and some, like Shanghai, shifting toward a postindustrial emphasis on services.

The easy and common identification of developing cities with intractable environmental problems also ignores the complicated reality that, despite facing many environmental issues, there are numerous examples of innovative solutions.

In this chapter we will explore the diversity by considering four types of cities in the developing world: megacity, industrial city, greening city and slum city. Many actual cities have elements of some or all of these types.

Megacity

We have already commented on the rise of megacities. In Chapter 4 we noted that a significant global trend was the development of megacities, especially in the developing world. The size and rate of growth of megacities is a significant feature of urban development in the developing world. In 1980, the city of Shenzhen in China, for example, had a population of only 332,900. But as official restrictions on rural to urban mobility were lifted and rapid industrialization was promoted, more rural migrants moved to the city. Economic growth reached a staggering 30 percent in annual increases. By 2012 its official population was just under ten million, with some estimates as high as 14 million. The impact on the ecosystem was dramatic. Fresh water supplies, for example, were quickly depleted by demand and contamination. One of the fastest-growing megacities in the world is Lagos in Nigeria, with a population estimated as close to 16 million. People migrate to the city from throughout West Africa in search of jobs and better opportunities. More than 10,000 people arrive in the city each week, helping to increase the population by around 600,000 each year. More than one in four people in the city live in slums that are spread throughout the urban region, ranging from permanent buildings on illegal sites to makeshift shelters beside rubbish dumps. The city has myriad environmental problems, including inadequate fresh water supply, flooding, sewerage management and environmental contamination of air, water and soil. The mounting refuse poses major problems for public health, with increases in diseases such as typhoid and salmonellae typhosa.[1]

The ecological issues facing megacities are immense. They range from the problem of fresh water supply of Shenzhen, the persistent flooding of Dhaka, to the air pollution of Beijing. There are, however, significant differences between the megacities of the developing world in their ability to cope with these issues. It is apparent that in some of the middle-income range of countries, a combination of an increasingly strong regulatory framework and greater public and private affluence is creating vast improvements. Air quality in Chinese and Brazilian megacities is poor compared to cities in the developed world, but it is steadily improving. Mexico City, long derided for its poor air quality, has made significant improvements. In other megacities the situation has deteriorated. In Kolkata, India for example, air quality has declined as a result of rapid industrialization, higher levels of energy consumption and increasing traffic, especially as more private cars and motorcycles clog the streets.[2] The environments of megacities are increasingly differentiated. Consider the health risks related to air pollution: São Paulo has a low risk of chronic pulmonary disease while the highest risk are found in Dhaka and Karachi. The difference was 50 hospital admissions per year in São Paulo compared to 2,100 for Dhaka and Karachi.[3]

There are now substantial differences between the qualities of the urban environment in megacities in the developing world. Fast-growing cities in Brazil, China and India initially showing a marked deterioration as rapid industrialization and growth of energy consumption overwhelmed urban ecosystems. Then, as economic growth continued, there was a greater acceptance of the need for environmental regulation and improvement. The steady improvement in Seoul's air quality parallels the steady rise from a lower-income to higher-income economy (Figure 6.1). Environmental improvement is made easier when countries move into the middle- and upper-middle-income bracket of countries. It is also made easier when strong

Figure 6.1 Seoul, one of the world's megacities
Source: Photo by John Rennie Short

regulatory frameworks can be quickly imposed and monitored. The strong state systems of China, South Korea and Singapore, for example, are more able and willing to impose environmental regulations. In other countries such as India, which lack a strong central state, the implementation of stricter environmental controls can be longer and less compelling. The more dysfunctional nature of Nigerian governance is partly responsible for the chaos that is Lagos, where it is not only the huge population upsurge that causes the environmental problems, but also the poor physical planning, inadequate enforcement of existing laws, inadequate funding and lack of proper coordination among agencies.[4]

Similarities remain, especially in terms of vulnerability to climate hazards. Lagos, like many large coastal cities, is vulnerable to sea-level rise associated with climate change. Ocean surge and coastal inundation are likely to increase endemic flooding and seawater contamination of fresh water and ecosystems.[5] Problems are heightened by continued urbanization of areas at risk from flooding and a lack of investment in coastal sea defenses. Alex de Sherbinin and colleagues looked at the vulnerability of three coastal cities: Mumbai, Rio de Janeiro and Shanghai. Given the expected increase of at least over 1°C and a sea-level rise of 50 cm by 2050, their analysis reveals a stress bundle of population increases and flooding that will make all three cities very vulnerable to global climate change unless long-term adaption measures are adopted.[6] This message does not apply only to the cities of the developing world, and this realization again reminds us of the interconnecting and shared fate of cities across the surface of the globe.

The industrial city

In the wake of a global shift in manufacturing employment, new industrial cities have developed in developing countries. We can follow the wave of industrialization cities from Japan and Korea in the 1960s and 1980s through to coastal China in the current era. These new industrial cities often emerge in an intense period of rapid economic and population growth coupled with few environmental regulations or little enforcement of environmental regulations. The result is unprecedented levels of air, land and water pollution. In many ways, the new industrial city resembles the industrial city of the nineteenth century, as unprecedented levels of urban pollution are generated.

In an event that replicated the worst excesses of the nineteenth-century industrial city, an explosion at a pesticide plant in Bhopal, India operated by

---

BOX 6.1

Cubatão and "the valley of death"

In the 1950s government planners in Brazil designated Cubatão as the center of the nation's nascent oil refinery industry. It was an ideal location, in a valley beside a river, close to the coast that allowed the petroleum and steel industries to import raw materials and ship out finished products. Large state corporations like COSPIA (steel) and the oil monopoly Petrobras established a gigantic refinery and nearby chemical and fertilizer plants. Private corporations soon followed. The city became heavily concentrated with industrial manufacturing, with few environmental controls. By 1985 Cubatão accounted for 3 percent of Brazil's GDP. Lack of enforcement resulted in thousands of tons of pollutants daily. By the early 1980s the city recorded the highest infant mortality rates in Brazil and over one-third of the population suffered from pneumonia, tuberculosis or emphysema. Two alarming developments occurred in the late 1970s and early 1990s. First, dozens of babies were born without brains, although researchers were never able to prove that the birth defects were caused by pollution. One in three new-born babies failed to reach their first birthday. Second, the city experienced several mudslides down the mountains, which had been denuded. Cubatão became known as "the valley of death" and "the most polluted city on Earth" and was even the subject of an ironic pop song called "Honeymoon in Cubatao." By 1983, the state government demanded industries start implementing pollution control; many responded, but others were notoriously lax.

In the last 20 years air quality has improved significantly, and respiratory ailments are half of what they were in 1984. Storage ponds that used to contain toxic waste have been cleaned out and turned into small lakes; the once barren hillsides have reforested. And although Cubatão is still one of the most polluted areas in the state, many of the pollutants are within World Health Organization recommendations. Although things have improved a lot, it is impossible to clean completely the soil and underground water. Government officials and industrialists now hold Cubatão up as a model of environmental recovery, but environmental groups and scientists see the city as a danger zone where contaminated air, soil and water are impacting people's lives. In 2010 Cubatão was ranked number seven on the "Nine Most Horrible Places in the World" list, perhaps more a testament to the enduring power of its legacy than to today's reality. Although industrial pollution is more controlled than it once was, high sewage levels continue to contaminate the estuary.

The lessons of Cubatão highlight two important points. First, at the same time developed countries were instituting air pollution reforms, developing cities were experiencing rapid growth with little legislative or institutional measures in place. Second, although the global community recognized pollution as early as 1972 when the United Nations established its Environment Programme, pollution regulation in developing countries tended to occur a decade or two behind the reforms in the industrialized North.

Sources: "Nine Most Horrible Places in the World." http://news.xinhuanet.com/english2010/culture/2010-06/20/c_13359106_7.htm (accessed July 2012). Campos, V., Fracacio, R., Fraceto, L. F. and Rosa, A. H. (2012) "Fecal sterols in estuarine sediments as markers of sewage contamination in the Cubatão area, São Paolo, Brazil." *Aquatic Geochemistry* 18 (5): 433–43.

United Carbide Corporation occurred on December 2–3, 1984. The explosion released 40 tons of methyl isocyanate from one of the storage tanks. The toxic gas permeated slums bordering the plant. The resulting leakage of lethal gas caused the deaths of close to 4,000 people and close to 120,000 people still suffer from the ailments. Faulty and insufficient safety systems, poor maintenance and use of unskilled technicians for specialized work were blamed for the accident.[7] The incident confirmed the worst-case scenarios of early industrialization: foreign companies, weak regulatory systems and little concern for the health and welfare of workers and nearby residents. It is probably the world's greatest industrial disaster.

Waves of industrialization have swept quickly over the developing world, moving to ever-cheaper labor areas, transforming them into places of higher paying jobs than existed for agricultural workers. Cities in Japan, then South Korea, then China and Vietnam became successively centers of industrial production and sites of a new middle-income group. The rise of China as a new global economic force has been unrivaled in recent years. GDP has quadrupled and per capita incomes have risen threefold in many cities.

Since 1979, China has implemented economic reforms that transformed selected cities, designating "free-market zones" that encouraged foreign investment. These zones include the cities of Shanghai, Beijing and Guangzhou. Two decades of unparalleled economic growth, swelling urban populations and often unchecked emissions from automobiles, factories, domestic heating, cooking and refuse burning have made China's cities prone to air pollution.

Consider the capital city of Beijing, which has been rapidly transformed in the past 20 years. Foreign investment focused on export processing, retail and insurance has generated new levels of affluence; in addition, the government has attracted industry through development zones that concentrate the location of certain types of economic functions. For example, to the southeast of Beijing, the smaller city of Tianjin has become a major international port city; to the east Tangshan is now a major center of heavy industry and coal mining. Forming a ring around Beijing is a series of major industrial areas, including textile mills, iron and steelworks, machine shops, chemical plants and factories manufacturing heavy machinery and electronic equipment. The city downtown has large commercial/financial districts, busy shopping areas and thousands of new stores and restaurants. Growing per capita income has generated a construction boom of residences and commercial properties. But at the same time there has been unparalleled environmental pollution. A combination of coal-fired power generation, expanding car ownership and polluting industries create poor air quality in cities. Beijing is listed as one of the ten worst cities in the world for both air and water pollution. As in most Chinese cities, soft coal is the predominant form of energy. The heavy reliance on coal means enormous amounts of sulfur and nitrogen oxides are released. Respiratory diseases have become one of the country's biggest health risks. Almost half a million Chinese die every year prematurely because of air pollution, especially as a result of the large amount of particulate matter in the air. In 2010 Beijing recorded a particulate matter content that exceeded World Health Organization standards by over 600 percent. Two-thirds of Chinese cities fail to meet Chinese government standards for air quality.

All households in Shanghai have access to piped water, electricity and other services. But, as with Beijing, Shanghai relies on soft coal as a source of fuel for both industrial energy and residential heating. It too suffers from significant air pollution. In addition, some four million cubic meters of untreated human waste enters the Huangpu River each day. Currently, the metropolitan region is home to a significant heavy industry sector, machinery manufacturing, textiles and steel. The toll of unrestrained economic growth in tandem with unimplemented environmental regulation has created cities in desperate need of pollution control. When *Time* magazine published a list of the most polluted cities in the world it was no surprise that the top two, Linfen and Tianying, were in China.[8]

In many ways the recent rise of the new industrial cities resembles the coke-towns of the industrial revolution of the eighteenth and nineteenth century. Pollution has blotted out the sun; sanitation systems cannot keep pace with population growth and increasing consumer demand. These new global industrial cities are in need of the same sort of sweeping and comprehensive air, land and water reform measures as those of the previous era.

Environmental factors also place limits on economic growth. Of China's 600 cities, 400 of them have water shortages and approximately 100 have severe water shortages. This puts an environmental brake on continued economic growth. One solution, involving piping water from the south of the country to the north can only be implemented with heavy costs, involving environmental disruptions and the forced relocation of over 300,000 people. Break-neck economic growth with few environmental controls creates not only environmental pollution but also ecological bottlenecks to sustained growth.

The ecological problems of the industrial city in the developing world are also producing a counter response. A number of factors are at work. First, there is recognition of the costs of industrial pollution. One study estimated that the cost to China of pollution was between 2 and 10 percent of GDP.[9] Environmental degradation exacts a cost equivalent to 9 percent of GDP. Poor water quality costs India 6 percent of national income. The effects of pollution on human health can become a political issue when citizens have more information on the levels and effects of pollution. Second, there is the important image issue, especially for ambitious countries like China, of national prestige compromised by poor environmental standards. It can become a matter of national pride. Consider the enormous stride made by the Chinese government to improve air quality in Beijing just before the 2008 Olympic Games. When the city became a stage for the globally important event, the city's poor air quality became a source of national embarrassment.

Chinese cities are shifting emphasis away from a sole reliance on economic growth as the metric of success, toward a more explicit commitment to environmental improvement and urban sustainability. As well as steadily increasing environmental standards, there are also model experiments (Figure 6.2). The Sino-Singapore Tianjin Eco-city, for example, is a collaboration between the governments of China and Singapore to build a green city. Located 40 km

Figure 6.2 A green urban park in the middle of Shanghai, a megacity with environmental problems but also a site of vast improvements in environmental quality

Source: Photo by John Rennie Short

from the existing city of Tianjin, the Eco-city site is non-arable land close to a wastewater pond. Carbon emissions will be reduced by the mass transit system, and waste is to be recycled and reused. More than 50 percent of the water supply will come from recycled water and desalinized water. It is the modern incarnation of Howard's garden city. The Eco-city is scheduled for start up in 2012 and to be completed, with a population of around 350,000, by 2020.[10]

China is a special case. The government is able to organize and devote enormous resource to environmental improvement. Other developing countries, and especially those further down the income scale with smaller, weaker economies and governments, lack the resources or the willingness to make major improvements.

From the 1970s there was a global shift in manufacturing production from the developed to the developing world. There were still new and expanding centers of industrial innovation, what Peter Hall and Manuel Castells describe as technopoles of the world, but the center of gravity shifted toward the developing world, where de-skilled manufacturing processes, business-friendly governments, cheap labor and low transport costs created the new industrial city.[11] By the second decade of the twenty-first century the process has become more complex. As manufacturing becomes even more automated, the more industrial jobs can return to the cities of the developed world. There was also an upward shift in urban economies of the developing world, what Peter Daniels and colleagues describe as the shift from industrial restructuring to the cultural turn.[12] There are now many industrial centers in the developing world with high value-added processes where the impact of design and innovation is central. Vu Hoang Nam and colleagues looked at industrial clusters of iron and steel production in Vietnam. Even in this older style of industry they found the importance of upgrading product lines, marketing and management.[13] In the new industrial city of the developing world, the simple manufacturing of cheap goods is being replaced by more sophisticated production techniques. And one consequence is a shift away from the extensive pollution of the earlier round of industrialization toward more modern techniques where waste management and pollution control are a more integral part of the industrial process. As production matures from the earlier stage of simple technology transfer—often prompted by the difference in environmental standards, as polluting industries escaped more regulated to less regulated systems—there is the real possibility that the levels of pollution will decline due to improvements in production and growing regulation. The rapid growth of the garment industry in Bangladesh provides just one example of this shift from simple transfer to growing maturation with associated environmental improvements.

There is also the suburbanization of industry in the developing city. The older central cities' factories are being closed as industry shifts to new greenfield sites. In Bangkok, for example, by the mid-1990s manufacturers were threatened by lower-cost producers in Indonesia, Bangladesh and China. Industry in Bangkok moved up the value-added chain with industries such as appliance and consumer electronics now located in peri-urban industrial areas to the north and east of the city, in some case up to 120 miles from downtown Bangkok. Here, as in many developing cities, the latest round of industrial expansion is occurring in purpose-built estates in industrial suburbs. In the process the central cities are deindustrializing and in some cases, such as Bangkok and Shanghai, are evolving into centers of producer services. There is the closure of older, more polluting city plants as production shifts to more modern and potentially less polluting sites and practices. There is also, however, the resultant land-use change and ecosystem transformation in formerly rural areas as the expanding peri-urban fringe continues to move outward.

The greening city

There is a popular view that sees a world divided into the cities of the rich world with environmental regulations and cities of the developing world with less regulation and much poorer and deteriorating urban environments. The green/non-green city is often used to overlay other perceived differences between developed and developing societies. However, the simple distinction of green, developed city/ non-green, developing city ignores the ways in which many cities in the developing world are becoming laboratories for new forms of urban sustainability. Here we come to an interesting paradox that the cities of the developing world have much to teach the cities of the developed world. Take the case of recycling. The fact of poverty in developing cities means that the waste stream is not simply something to be exported, buried or burned; it is a valuable source of income to marginal groups— people who recycle and reuse. The pickers at the waste dumps provide an important lesson to the rest of us about the possibilities of salvaging something valuable from trash and garbage. In the main dump of Lagos, for example, more than 5,000 people work at picking over the garbage. In the process the waste stream is recycled and income and employment is afforded to people on the margins.

Returning to the case of Shenzhen, one plan is to use wastewater to irrigate rooftop gardens in the slum area of Chengzhongcun.[14] This plan utilizes the wastewater to promote algae fuel production on the roofs of the poor neighborhoods, combining ecology with the economics of providing an income source for poor people. The city of Langfan, another of China's eco-city projects, is also planning to use byproducts from coal plants, the main source of electricity, to grow algae for biofuel production. The rapid urbanization of countries such as China poses enormous threats and danger, but they also provide the opportunity for practicing and imagining sustainable forms of urban living.

Sometimes the crisis of developing cities can provide a model for the richer cities as they face a more uncertain future and less predictable supply of cheap oil. Emma Piercy and colleagues looked at Cuban cities' response to the Special Period, the name given to the era immediately after 1989 when Soviet aid was cut off abruptly and imports of energy and food were drastically reduced. The response to more stringent energy supplies created a model of what a past peak oil city could look like.[15] Cities in developed economies can learn from the crises and experiences of the developing world, where scarcity and limits to resources have

BOX 6.2

Stimulating demand for better environmental quality

"Households in Gurgaon, India, that were told that their drinking water was 'dirty' were 11 percentage points more likely to begin some form of home purification in the next seven weeks than households that received no such information.

A water test, which costs less than $0.50 per sample, is available off the shelf from many nongovernmental organizations in Delhi and simple enough for households to use themselves.

The study shows that the impact of a water test on the probability of purification is about 25 times that of an additional year of schooling and more than two-thirds that of a move from one wealth quartile to the next. This result suggests that public programs that focus on disseminating health information are cost effective and relatively easy to implement in low-income countries. Such efforts can stimulate demand for better environmental quality through political expression or increased willingness to pay for improvement of environmental services."

World Bank (2008) *Poverty and the Environment; Understanding Linkages at the Household Level.* Washington, DC: World Bank, p. 36.

created difficulty but also innovative models of possible routes to urban sustainability.

There are also the model cases that continue to provoke wider interest. Curitiba in Brazil, one of the best-known examples of urban sustainability in a developing city, is often cited and has proved an inspiration—especially for its mass transit system—for other cities, including Bogotá, Guatemala City, Jakarta and Kuala Lumpur. We note its commitment to sustainability in Chapter 16. Curitiba's historic trend of sustainability is impressive, but new challenges are appearing as the world continues to globalize. During the recent period of growing economic activity, Curitiba's city planners have managed to keep the urban impact on the environment very low. Despite having 1.78 million residents in 2010, the city has been quite successful at integrating human and natural activity, and has avoided many of the congestion, pollution and lack of open space problems that plague other cities. In 2010 the city was awarded the Globe Sustainability award for its sustainable urban development. Yet Jeroen Klink and Rosana Denaldi qualify Curitiba's environmental success by pointing out how globalization's restructuring pressures are re-arranging the local regime that helped the city achieve its current high level of sustainability. They suggest that the ability of Curitiba's local government to guide development may be waning due to the competition and power of the private market.[16]

Slum city

The term slum (also called shantytown, informal housing and squatter housing) refers to unplanned, illegal, informal housing.[17] Initially the term referred to rundown parts of cities across the world, but in recent years it is commonly used to refer to the informal settlements in the cities of the developing world. The housing is considered illegal because the occupiers hold no title to the land, do not pay taxes and have constructed some form of shelter that does not meet building code. Because these are illegal structures, they often lack government provisions such as sewage and sanitation infrastructure or services such as clean water (Figure 6.3).

The most recent global survey of slums was published in 2003 by the United Nations.[18] The survey estimated that around one billion people lived in slums. The figure is estimated to rise to two billion by 2030 as people migrate to the cities and urban population growth continues apace.

Slums arise due to the inability of formal markets and public authorities to provide enough affordable and accessible housing. They grow in size because of continuing rural to urban migration and natural increase. In one sense they represent endemic poverty. More than 50 percent of the population of Lagos, for example, live below the poverty line; in Monrovia, Liberia it is closer to 60 percent. The World Health Organization estimated that in 2000 more than 600 million urban dwellers in developing cities had no access to clean water, sanitation or drainage. While this figure is disturbing, the figures for rural dwellers are much worse. In Nigeria, for example, around 75 percent of urban dwellers have access to good drinking water; the figure for rural dwellers is a staggeringly low 50 percent. While things may be bad for people in the slums, in many cases they represent an improvement on conditions in rural areas. That is why people move in their millions from the countryside

Figure 6.3 Slum housing beside luxury living in Saigon, Vietnam
Source: Photo by Becky Barton

to the city—they are moving to improve their living conditions and gain better access to employment, services and for the possibility of a brighter future for their children.

Some slums go through a process from temporary accommodation built on the most marginal sites to fully formed urban neighborhoods. Consider the case of one neighborhood in Mexico City, a magnet for rural migrants for the past 60 years. In the 1950s some rural migrants moved to Ciudad Netzahualcóyotl, an area on the outskirts of Mexico City, where they built their own homes on appropriated land using whatever materials they could find. Over time more people came from rural parts of the country, the place expanded and more permanent buildings were constructed. With size and permanence came greater political leverage to demand better services and legal status. The area was designated a municipality in 1963 allowing the formal provision of public services such as potable water, pavements, sewerage and electric lighting. By the 1980s public buildings such as hospitals and schools were being built. In 1995 the area exceeded one million residents. Temporary shelter for recent migrants became a slum and then an integral part of the city.

BOX 6.3

The informal sector in Accra, Ghana

"The geography of informality has five features. First, there is the development of large slum/squatter areas ... that have emerged to provide affordable accommodation for migrants and other urbanites. These areas provide a reserve of cheap labor for sectors such as food and informal construction. Second, there are areas where housing conditions have severely deteriorated.... More crowding in these areas coupled with residents being tighter squeezed economically has resulted in a different type of building boom: adding rooms/kiosks and shops incrementally for renting. Third, there are kiosks and semipermanent front and backyard workshops that dot upscale residential areas.... Fourth, there are temporary workshops, primarily serving the residential construction industry.... Fifth, there are microenterprises in services and production that have sprung up everywhere, and cluster heavily along major thoroughfares, in road reservations, and idle parcels of land."

Grant, R. (2009) *Globalizing City: The Urban and Economic Transformation of Accra, Ghana.* Syracuse, NY: Syracuse University Press, pp. 115–16.

Those who live in slums, especially slums that fail to achieve the size, permanence or political power of Ciudad Netzahualcóyotl, face insecure land tenure, exposure to hazards and often lack political voice.

There are different types of slums. In Abidjan, in Cote d'Ivoire, slum dwellers represent one-fifth of the city's population. There are three types of slums. Zoe Bruno contains buildings of permanent material and basic infrastructure and is only different from the formal areas of the city by the illegal land occupation. In Blingue the buildings are made of non-permanent materials and the area has low levels of infrastructure. In the worst areas, such as Alliodan, makeshift buildings have no infrastructure. There is a marked difference between different types of slums. A distinction is sometimes made between early self-help housing, which has communal provision of services, and unauthorized housing illegally occupying land. The worst slums often have to occupy the most hazardous sites, on steep slopes, on areas vulnerable to flooding, landslide and other environmental and social hazards. The poorest urban dwellers are exposed to a myriad of environmental and social problems that include:

- lack of infrastructure providing water, sewerage, electricity, trash collection;
- disease-causing agents (pathogens) in air, food, water or soil that impact human health;
- pollutants in air, food and water that impact human health in both the short term and long term;
- congestion;
- physical hazards such as accidental fires, floods, mudslides or landslides.

In the rest of this chapter we will examine these items in more detail.

The lack of infrastructure is widespread in cities in the developing world. Consider the following statistics: in Bangkok, 33 percent of the population does not have access to a clean water supply; in Kolkata, five million are without clean drinking water. In Khartoum, 90 percent of the population has no municipal sewerage system; in many cities in India more than one-third of the population have no latrines and must rely on buckets to remove human waste. In Bogotá, more than 2,500 tons of trash goes uncollected each day. On July 11, 2000, the collapse of a rubbish dump in Payatas, Manila killed 218 people living in slums at the bottom of the site and left another 300 people missing, trapped under the rotting garbage. A United Nations report noted: "The tragedy of their burial underneath the trash of a world city, off its edge and in the darkness of night, symbolizes the invisible, daily plight of innumerable poor people."[19]

It is estimated that hundreds of millions of urban dwellers do not have piped water supplies and thus have no alterative but to use contaminated water, or water whose quality is not guaranteed. In Libreville, Gabon, for example, only 60 percent of the population have access to clean water. This is not to say they have no access to water; they do. In some slums there are public stand posts or public fountains from which residents can fill buckets and other containers. However, there is a significant amount of time needed to obtain water, and often long distances must be traveled to collect it. In addition, there are water vendors (usually private firms) who sell water to the poor, but they often charge 5–10 times as much as the rates for water delivered by a public water system. Many of the urban poor cannot afford private water sources, and hence have an inadequate supply. In a fascinating study,

Kirsten Hackenbroch and Shahadat Hossain describe the struggles of slum dwellers in gaining access to public space and water supply in Dhaka. They describe the unequal interaction between slum dwellers and local political leaders as "the organized encroachment of the powerful."[20] The urban commons that could provide a starting point for poverty alleviation instead is a site for the projection of power and the embodiment of unequal power.

Many slum residents have to use whatever water they can find. Often it is contaminated. Waterborne diseases take a tremendous toll on human health. Diarrheal diseases affect an estimated 700 million people each year and account for most water-related infant and child deaths. A high proportion of the slum residents have intestinal worms that cause severe pain and malnutrition. In slums across the world diseases such as hepatitis and tuberculosis flourish. And cholera, typhoid and dysentery, long eliminated in the developed world, continue to wreak havoc. Among slum residents neonatal deaths are two times higher, mortality from respiratory disease six times higher and mortality from septicemia eight times higher than among the middle-class or wealthy in that same city. Poverty often means that infants and children do not always receive their vaccines for measles, whooping cough and diphtheria, exposing these children to diseases that more wealthy populations no longer contend with. An infant is 40–50 times more likely to die in a slum than in a city in North America or Europe.

The lack of social services, such as schools, impacts educational attainment. In Delhi 75 percent of all men and 90 percent of women living in slums are illiterate; 40,000 slum children work as laborers, 30,000 work in teashops and 20,000 in auto repair.

A particular type of pollutant unique to populations living in slums is indoor air pollution. One of the major sources of indoor air pollution comes from indoor smoke, a result of cooking over open wood or dung fires. Most slums lack fans or exhaust systems, and because they often have no electricity or gas, cooking takes place over open fires. The health impacts of consistent exposure to smoke and fumes have been underestimated and understudied, but we do know that burning coal, wood or other biomass fuels can cause serious respiratory and eye problems. Research has shown that concentrations of total suspended particulates are 10–100 times higher in indoor dwellings. Chronic effects of exposure include inflammation of the respiratory tract, which in turn increases vulnerability to acute respiratory infections such as asthma, bronchitis and pneumonia. Women are often heavily exposed because they spend several hours each day at the stove; infants and young children may also be heavily exposed since they remain close by their mothers.

BOX 6.4

Changes in a Bangkok slum, Klong Thoey

"To qualify for a government school, a child has to have a birth certificate, but to get such a document the parents have to live in a registered house; the problem, of course, was that none of the houses in slums were, or could be registered, since that would imply legal ownership of the land.

The great majority of Klong Thoey's families accepted this as merely another fact of slum life, along with crime, the lack of sanitation and the constant threat of eviction. Prateep's mother was an exception. She was determined to give her daughter at least the rudiments of an education, and to do so she was willing to somehow scrape together the necessary money to send her to a cheap private school on the fringe of the slum. By the time she was 15, she had saved enough to enroll in an evening school for adults, where she managed to complete six years of study in only two and a half.... Finally, certificate in hand, she turned one small room of the family shack into a sort of daycare center for young children of working parents, some of whom paid her a baht a day for the service....

The Port Authority informed Prateep and some two thousand of her neighbors that the area in which they lived—officially designated as Block 12—was required for 'development' and that they would have to leave.... But even when squatters have inhabited an area for upward of 30 years, as many in Klong Thoey had, they have no legal rights in Thailand. No one paid much attention until a reporter from the *Bangkok Post* happened to hear about it, saw the possible makings of a good story....

Cast in the role of the villain, the Port Authority compromised. The residents of Block 12 would still have to move, but they could relocate to some empty land not yet needed, further back in the slum. The move turned into a show of unprecedented slum solidarity. The three-hundred-odd families involved selected representatives to divide the new area into housing sites, and when the time came they helped one another take down the old shacks and reassemble them. Half an acre was reserved on which to build a real school for the young teacher, now regarded as a leader."

Warren, W. (2002) *Bangkok*. London: Reaktion Books, pp. 134–6.

The exposure of infants and young children to indoor air pollution, combined with malnutrition (often common), leads to a greater prevalence of chronic bronchitis.

Another indoor/outdoor pollutant slum dwellers must deal with is the removal and safe disposal of excreta and wastewater. While some cities have concentrated on improving the provision of clean water, fewer have dealt with issues of sanitation. Many of the waterborne diseases, however, are excreta-related, such as schistosomiasis, hookworms and tapeworms. Slum dwellers dispose of excreta in pit latrines, bucket latrines or "flying toilets." Flying toilets refer to defecation that has been placed in plastic bags, which are then thrown into ditches, gullies, streams, canals and rivers, where the human waste remains untreated and may further contaminate water supplies down-river. Figure 6.4 depicts a "flying toilet" in Mahare, a slum in Nairobi, Kenya.

High density and congestion are hallmarks of slums. In the city of Kolkata, there are some 1,500 slums. The average dwelling measures 6 feet by 8 feet, yet houses 6–8 people. Such high population density facilitates the spread of contact diseases such as influenza, meningitis and tuberculosis, as well as food-borne diseases. The frequency of contact, the density of the population and the concentration and proximity of susceptible people in an urban population promotes the transmission of the infective organisms.

The urban design of slums exacerbates risks. The dense settlements mean that narrow paths or streets (usually unpaved) restrict vehicular access. As a result, ambulances, fire-fighters and police often have trouble accessing areas. There is also the lack of recreational space or safe places for children to play. We note the particular problem of fires in slums in Chapter 8.

The urban explosion of the past 50 years has forced expansion onto new spaces on more vulnerable sites such as steep hillsides, flood plains or in

Figure 6.4 "Flying Toilets" in an open sewer in a Nairobi slum
Source: Photo by David Rain

areas with unstable soil conditions (Figure 6.5). Many slum structures are often erected on such marginal lands because standard, legal housing long ago claimed the best, most secure land in the city. The poor are often forced to settle on land subject to higher risks. Slum homes perched precariously on hillsides are often swept away with heavy rains, which also flood slums located on floodplains. Heavy rain contributed to the landslides and the flooding, but the "real causes" are due to the fact that low-income groups could find no land site that was safe and to the failure of government to ensure a safer site or to take measures to make existing sites safer.

Slums are very vulnerable to disasters such as landslides, fire and erosion. Those living in informal settlements have a higher probability of being killed or injured in an accidental fire, earthquake or flood, or during a storm, than people in the formal housing sector of a particular city. Fires are common in slum dwellings because the building materials tend to be combustible and most of the cooking takes place over open stoves or fires. The lack of electricity means people heat and light their dwellings with candles or kerosene lamps.

Figure 6.5 Informal housing on precarious hillside in Roseau, Dominica
Source: Photo by John Rennie Short

A study of injuries in slums in Karachi found that most of the injuries were due to falls, burns and cuts.[21] Most of the burns were suffered by women who were cooking and by young children who accidentally burned or scalded themselves.

The impacts of slum life disproportionably affect women and children. In most slums, infants, children and women bear the heaviest burden of air and water pollution and vulnerability to environmental hazards.

Slums can provide a platform for rural–urban migrants to the city, but at some cost. The dichotomy between slums of hope and slums of despair summarizes the benefits and costs. Slums provide easy access to the benefits of the city, to those such as recent rural migrants with minimal income. The community of the slums may provide a basis for mutual support—slum residents often come from the same region or province—a tight social network that can provide mutual support and leads and connections to economic opportunities. Yet the environmental costs can create a web of multiple deprivation that traps the poorest. The slum conditions can exacerbate the risk to human health and the vulnerability of these residents to other problems. For example, the lack of basic infrastructure increases the likelihood of pathogens spreading diseases. The lack of water services means that fires are more difficult to contain and put out, and the lack of paved roads poses difficulties for emergency and rescue services. Many of the urban poor in developing cities have multiple deprivations. A recent study of slum dwellers in Dhaka, for example, highlighted the problem of mental health for those with the poorest sanitation and housing quality and the highest flood risk.[22] Mental health problems are just some of the many unrecorded costs of slum dwelling, especially for the most marginal women.

The lack of legal ownership makes slum dwellers particularly vulnerable to government clearances. On May 19, 2005 the government of Zimbabwe, under the dictatorial control of Robert Mugabe, launched a new urban policy. It was called operation Murambatsvina, literally translated as "Drive Out the Rubbish." As part of a nationwide campaign to beautify the cities, destroy the black market, reduce crime and undermine the support base of their political opponents, the government authorized the bulldozing of squatter settlements that fringed the city's larger cities. The squatter settlements had long been the main source of opposition to Mugabe's rule. The campaign destroyed the homes of 700,000 people and wiped out the informal economy that provides livelihoods for 40 percent of the population. Almost 2.4 million poor people were faced with increased economic hardship. In July 2012 the Lagos state government in Nigeria moved to evict residents of Makoko, a slum neighborhood of around 200,000. Many of the dwelling rest on stilts over a lagoon. The residents were given 72 hours to vacate before men starting chopping down the dwellings. The official letter notes that the slums are an environmental nuisance, impede waterfront development and undermine the megacity status of Lagos. Officials felt that a 2010 BBC television documentary program, *Welcome to Lagos*, which told the story of Makoko's residents, presented the country and the city in a negative light. In fact, the documentary was a celebration of the ingenuity and vibrancy of slum life in the city. While it noted problems of flooding, irregular power supply and poor environmental conditions, it was lyrical in its invocation of the resiliency and vitality of marginal groups living in slums on the economic margins. Rather than a depiction of gloom and darkness it highlighted the incredible dynamism and energetic entrepreneurship of the slum dwellers.

Conclusions

Rapid urban growth in the developing world need not produce either the heavily polluted industrial city or the particularly vulnerable slums—some cities such as Curitiba in Brazil have mitigated environmental degradation. And there have been efforts to help improve the social and environmental quality of the urban poor. For example, in Belo Horizonte, Brazil, a PROFAVELA project helped slum residents obtain land tenure, and thus connection to municipal service networks. Some slums in Lima, Peru, have organized trash pick-up teams, which pedal tricycle-like carts along fixed routes in the slums.

One characteristic of cities in the developing world is the resiliency of the inhabitants. Even the middle-income groups in the developing world often have to face urban difficulty of irregular power, uncertain employment, inadequate public services and poor environmental conditions. But it is the poor and the marginal that have to scrimp and save, hustle and bustle in the face of a formal market that cannot provide jobs and housing. It is left to the citizenry themselves to find their jobs and make their accommodation. This is not to romanticize the slums of the world, but to note that people are crafting a life of ambition and hope, often in appalling conditions.

Guide to further reading

Daniels, P. W., Ho, K. C. and Hutton, T. A. (2012) *New Economic Spaces in Asian Cities*. New York: Routledge.

Davis, M. (2006) *Planet of Slums*. London: Verso.

Hackenbroch, K. and Hossain, S. (2012) "The organized encroachment of the powerful: everyday practices of public space water supply in Dhaka, Bangladesh." *Planning Theory and Practice* 13 (3): DOI:10.1080/14649357. 2012.694265

Hsing, Y. (2009). *The Great Urban Transformation: Politics of Land and Property in China*. New York: Oxford University Press.

Neuwirth, R. (2005) *Shadow Cities: A Billion Squatters; A New Urban World*. London: Routledge.

Power, M. (2006, December) "The magic mountain: trickle-down economics in a Philippine garbage dump." *Harper's Magazine*, pp. 57–68.

McGee, T. G., Lin, G. C. S., Marton, A. M., Wang, M. Y. L. and Wu, J. (2007) *China's Urban Space*. London: Routledge.

Mitlin, D. and Satterthwaite, D. (eds.) (2004) *Empowering Squatter Citizen: Local Government, Civil Society and Urban Poverty Reduction*. London: Earthscan.

Roy, A. (2011) "Slumdog cities: rethinking subaltern urbanism." *International Journal of Urban and Regional Research* 35: 223–38.

The website for UN-Habitat is: http://www.unhabitat.org/categories.asp?catid=9.

Part III
Urban physical systems

7 Urban sites

We can make a distinction between location and site. Location refers to relative space, the space of connections, hierarchies, economic transactions and social relations. It is a space abstracted from territory. Site, in contrast, refers to the absolute space that a city occupies. In much of recent urban geography and indeed of urban studies in general, the abstract space of location is the more dominant theme, an intellectual trajectory that tends to ignore, marginalize or simply forget the importance of absolute space to understanding cities.

But there is also a large and growing body of work that looks explicitly at the place of cities. These studies range in theoretical orientation from the landscape analysis school to political ecology to critical social theory.

Rebecca Solnit explores the city as a distinct and unique place in her highly imaginative atlas of San Francisco.[1] The chapters in her book, *Infinite City*, show how different inhabitants experience the city. The city is depicted as a mosaic of interleaving, overlapping and co-existing spaces: the green space saved by women activists; butterfly habitats and salmon migrations; as well as cinematic spaces and maps of neighborhoods. The various sites, sights and cites of the city are revealed by looking with different eyes, at different scales of analyses at different times. The work is a collaboration between artists, cartographers and writers. The 22 maps reveal something of the richness of the city as a particular site composed of overlapping places and shared, multiple and differing experiences.

In direct contrast, Luis Bettencourt and Geoffrey West examine cities in relation to their relative population size.[2] With each doubling of their population size US cities, at least according to their calculations, are, on average, 15 percent more innovative, 15 percent wealthier and 15 percent more productive. Cities for these two scientists are sites of superscaling rather than linear scaling. This study contrasts with the more emotionally evocative

Solnit approach: Bettencourt and West emphasize metrics rather then experiences and focus on the city as a site along a non-linear scale rather than the city as site of emotion, memory and desire. The Solnit atlas and the Bettencourt and West statistical study represent two poles on the continuum of approaches to understanding the site of individual cities.

New Orleans, for example, is the subject of a number of works. The cultural geographer Pierce Lewis looks at the making of the urban landscape, while Craig Colten, from a political ecology perspective, tells of how the city was shaped in the attempt to conquer nature.[3] Reyner Banham also used the physical geography of a city in his evocative and classic conceptualization of Los Angeles. He outlines the place and role of what he terms surfurbia, foothills, plains of id and autopia in his part-social, part-physical model of the city.[4] In his book, *Paris: Capital of Modernity*, especially in part 2, Materializations, David Harvey adopts a more critical perspective in examining the relationships between an urbanized nature and emerging social relations.[5] And in his study of Mumbai, Arjun Appadurai paints a detailed picture of the urban landscape and its connection with cash and capital. He invokes a sense of an "immense landscape of street-level traffic" and shows how urban cleansing was invoked in ethnic sectarianism.[6] While these and other studies tend to focus on the local, the ideographic and the unique, collectively they remind us of the role of place in the making of the cities, how the urban transformations of place affects both physical and social relationships. Cities occupy sites and this occupancy creates constraints and opportunities, provides the site of material culture as well as the imaginings of urban dreams and fantasies. The site of cities is the basis for economic transactions, the place of social relations and the setting for the production of urban images.

The specific site of cities is also an important element in *bioregionalism*, a concept that can be defined as a sensitivity to regional variations in climate, culture and environment. There is a homogenizing character to modern consumer capitalism with similar lifestyles, technologies and urban forms promoted across very different ecoregions. The result can be inappropriate designs for houses, neighborhoods and entire cities. Take garden design: the classic English garden is suitable for the cool, damp climate of England, but is not the most ecologically sensitive design for gardens in hot, dry Australia or the continental US. And yet English gardens figured largely in the garden history of Australian and US cities until comparatively recently. There is now a clearer understanding that urban sustainability is intimately connected to situating the city more appropriately in the space of opportunities and constraints provided by the local ecosystem, locating the city more to its place in the world rather than simply as a node in the homogenizing consumer space of contemporary global capitalism. When we have similar housing, neighborhoods and cities across the world, they may link into a shared consumer culture or global capitalist market, but they neither reflect nor embody their local environments. One of the significant strands of urban sustainability is about making cities more responsive to their local environments, their variations in temperature, their rainfall levels, their altitude, latitude, longitude and biotic zone. Planting indigenous vegetation, encouraging local species, removing invasive species and building to the context of local ecosystems all connect cities to their immediate environment more snugly. Urban bioregionalism highlights the constraints and opportunities afforded by the particular location and the specific site of individual cities. Buying food from local farmers, for example, cuts down on transport costs, encourages the local economy, enhances food freshness and promotes better nutrition. While much of the urban economy is about connecting with wider markets, much of the health of a city, broadly defined, is a function of the strength and resilience of its local ecological connection. There is now the development of ecovillages, which are communities concerned with environmental sensitivity and urban sustainability. They range from the radical to the corporate as savvy builders and developers adopt sustainability and green themes in their design and especially in their marketing. Across the world urban developments are increasingly referencing their local environments. No longer the serial urbanization of before, urban planning and design is now more imbued with a bioregional awareness, a sensitivity to the local environment and the forging of a much closer connection between cities and their particular sites.

Stephen Graham and Lucy Hewitt argue that much urban research focuses on the horizontal rather than the vertical. They suggest more emphasis on urban verticality and in looking at cities in three dimensions.[7] More specifically, William Meyer, for example, highlights the role of altitude in the early development of the US city. In some cities the high ground was avoided because it was difficult to pump water to or to get horse-drawn fire services to, so it was more of a fire hazard. The early forms of public transport also found steep hills hard going. Later, as altitude was more easily overcome, the high ground became the site of more elite residences away from the pollution and dirt of the lower ground.[8] In Los Angeles, as Banham noted, there is a clear link between altitude and socio-economic status, with the Hollywood Hills and the plains of Crenshaw providing a polar example.

Altitude also plays a more general role in the (literal) life of a city. Mexico City, for example, is surrounded by endemic dengue fever, but the city has remained relatively free of the disease because of its altitude of 2,485 meters. The fever is contracted through the bite of an infected Andes mosquito. This type of mosquito does not survive at higher altitudes and so dengue fever is more prevalent at lower altitudes. In Venezuela, for example, most of the cities are at a much lower altitude and dengue fever poses a greater threat.[9]

Adding the third dimension provides a greater understanding of cities. But more than just an understanding of altitude is needed to obtain a multi-dimensional view of cities. We have to understand the environmental milieu that surrounds, encapsulates and structures a city. Consider just one type of environmental context, rivers. Many cities are situated beside rivers; they grew up beside specific rivers and the history of the city is, in large part, a tale of its unfolding and changing relationships with the river, as mediated through the prisms of social and economic power. To understand Glasgow, London, Florence, Cairo, Phnom Penh, Dublin or Kolkata it is necessary to understand the Clyde, the Thames, the Arno, the Nile, the Mekong, the Liffey and the Hooghly. The evolution of London, for example, can only be understood through its changing relationship with the Thames; and not only for specific former dockside areas such as Canary Wharf, but also to explain the social geography of areas north and south of the river.

Tricia Cusack explores the role of the Hudson, Seine, Shannon, Thames and Volga rivers in the creation of national and urban identities. The rivers were vital elements in the representation and understanding of place, nation and cultural identity.[10] In the case of the River Thames, for example, nineteenth-century realities of urban pollution clashed with the images of patriotism and pageantry that had been long invoked. The state of the Thames was thus seen as a national scandal. Moreover, the embankment of the river, a major feat of civil engineering, was undertaken to provide a fitting appearance to a national and imperial capital. And even in cities where the river does not immediately spring to mind or is barely visible in the landscape, the urban–river connection is still there. Blake Gumprecht tells the story of the Los Angeles River. Before western invasion the river system supported a dense network of Native American settlements. Later it proved the main reason for the location of a new city. The river provided drinking water and irrigation for farms. Orange groves were watered with river water. As urbanization extended along the riverbank the seasonal floods caused more and more damage. A series of winter storm floods in the late nineteenth and early twentieth centuries led to calls for greater river management. The river was channeled into 51 miles of concrete culverts. Now it is a completely urban river occasionally running through an entirely human-made channel system, little more than a small trickle of water flowing through a wide concrete scar through the city. Gumprecht's book is subtitled *The Life, Death and Possible Rebirth*. He suggests that the exhumation of the river and the greening of its riverbank could become important goals of ecological restoration, environmental improvement and urban renewal.[11] Around the world, cities in developed countries are reassessing and reevaluating their river connections, often moving from a narrow economic resource dependency and dumping ground mentality to a sense of a more sustainable recreational mixed-use set of options. We can tell much about a city from the nexus of its river connections.

In the rest of this chapter we will approach this broad theme of city sites with a series of theorized case studies of particular types of city–environment relations. It is important to remember that the environment structures but does not mechanically determine a city's development.

On the beach

One of the most iconic images in the photographic history of Australia is Max Dupain's 1937 "Sunbaker." It shows the head and shoulders of a young man on the beach. His head lays on his forearms, the short shadows suggest an overhead summer sun. Max Dupain (1911–92) was one of Australia's most celebrated photographers. He photographed many Sydney scenes, including streetscapes, beaches and buildings, all in a distinctly modern style. By the 1970s his images were recognized as capturing the essence of Australia. But among his thousands of photographs, it is the black-and-white one of a man sunning on a beach that has become the most widely recognized. The icon of the man on the beach entered the national imaginary as well as international associations of Australia and Sydney.

The earliest settlers who inhabited the area where the city of Sydney now stands were the Eora people. The coming of the First Fleet in 1788 rudely disrupted this coastal people's world of fishing, farming and hunter-gathering. Botany Bay and then Sydney Cove became a far-flung outpost of an imperial Britain. The city grew to become the largest in the nation and by the late twentieth century was the global gateway of Australia.

Sydney grew from the Inner Harbour in all directions, both inland and along the coast. The result was an incorporation of beachscapes into the urban fabric (Figure 7.1). There are 38 beaches along the sea coast that range in size from the expansive 5 km of Cronulla to the pocket-sized 70 m of Clovelly. They are strung out along the coast from Palm Beach in the north to south of Botany Bay. There are also 32 beaches inside the Harbour.

Figure 7.1 A Sydney beach
Source: Photo by John Rennie Short

The ocean beaches are particularly striking because they have beautiful golden sands, clean ocean water, high swells that provide photogenic waves and a geology of sandstone headlands that surround the beaches in a graceful crescent arc. The Harbour and ocean beaches are an integral part of the urban fabric, adding light and space and a closer connection to the sea. Sydney's reputation of a free and easy attitude may in part be due to its less confining urban space. In contrast to the sense of confinement in built-up inland cities, Sydney's expansive relationship to the ocean and the open water creates a feeling of space and openness.

The beaches are also an integral element of life in the city. The English emigrant Henry Gilbert Smith turned the fishing village of Manly into an ocean resort in 1857. Bathing was initially regulated. A law that allowed bathing only after eight in the evening and six in the morning was only relaxed in the early twentieth century. The editor of a local newspaper first publicly defied the ban in 1902 and police refused to prosecute. People soon came to the beach to swim and sunbathe and hang out. Swimming clubs, life-saving clubs and later surfing clubs provided the social binding that connected people in what could truly be called an urban beach culture. The male life-saver became an image of Australian masculinity, surfing helped to define a youth culture, topless sunbathing was an important element in women's liberation in Australia and in the articulation of changing sexual mores. Sydney was a city where the body on the beach was an important element of the urban experience. Social changes were reflected and refracted through the prism of this sensual beach experience.

BOX 7.1

Sydney and the beach

"The beach is Australia's true democracy ... we Australians did not derive our freedom from bewigged Georgian founding fathers and their tablets of good intentions ... we have found our freedom by taking our clothes off and doing nothing of significance, and by over the years refining and elevating this state of idleness to a culture now regarded highly in the world's most fashionable places.... Whatever racists and Jeremiahs may say, Australia, a society with a deeply racist past, has absorbed dozens of diverse cultures peacefully. The beach and the way of life it represents are central to this. Today the sons and daughters of these people are often the majority on Bondi Beach, where lifesavers have Italian, Greek and Turkish names and board rides are Vietnamese ... the beach is theirs now."

Pilger, J. (1989) *A Secret Country.* London: Jonathan Cape, pp. 10–12.

There were so many beaches that there was marked differentiation. The northern beaches, located in the suburbs, such as Curl Curl were considered more sedate than the more youthful and cosmopolitan Bondi. Around the larger, more popular beaches restaurants and hotels, bars and shops soon developed. The beach and the coastline also shaped the social geography. In Sydney the rich tend to live in the inner city, along the waterways and coast; the poor live further inland, more distant from the beach. The inland western suburbs are more blue-collar than the inner-city Double Bay.

The beach has become a more contested place. In December 2005 civil unrest broke out along the beaches of Cronulla, south of Botany Bay. The general context was not only the changed world of post 9/11, but also the particular Australian experience of a terrorist bombing in Bali in 2002 that killed 88 Australians. Anti-Islamic feeling was high. The more local context was the location of Cronulla. It is one of the few beaches accessible by rail and hence more accessible to the poorer residents of the western suburbs. The suburb itself is one of the more Anglo-Celtic suburbs of the city, less impacted by the large foreign-born immigration of the city. Because of its easy accessibility from the western suburbs, Cronulla beach had become a favored beach for poor immigrant youths. Lebanese youths had attacked two lifeguards the week before. Anecdotal evidence from local informants suggests that the conflict had been caused by competition for the attention of young women on the beach. On Sunday December 11, 2005 a mob of over 5,000 white men marched along the beach attacking people they thought were Lebanese. The mob had been organized by right-wing and neo-Nazi groups eager to show that the beaches were only for "real" Australians.

The beach has also played a part in the selling of Sydney. The beach and bridge, the sun and Opera House are part of the internationally marketed and promoted images of the city. The success of the city in competition with Melbourne in the race to become Australia's primary global gateway can in part be put down to the more recognizable urban iconography and the role that the tempting beach culture has played in the successful international marketing of the city. Sydney has an ensemble of some of the most internationally recognized urban images, the Harbor Bridge, Opera House and Bondi Beach. The hedonistic pleasures of the beach beckon people from near and far.

Cities in the desert

Mecca and Las Vegas: two unlikely cities to be paired together; one a destination point for religious pilgrims, the other for gambling and entertainment. But both cities are located in a desert and this location has helped to shape the cities. The idea of a pilgrimage, whether for the soul or the body, is enhanced by the sense of a literal or metaphorical journey through the desert to an oasis.

Mecca is situated in the Sirat Mountains 45 miles from the coast of the Red Sea. The city is dry and hot. Rainfall is less than five inches per year, although winter flash floods can sweep through the usually dry streambeds of the Wadi Ibrahim. Temperatures are high and in the summer can reach over 100°F. The city has long been an oasis, wells capture underground water sources and one well, Zam Zam, was considered a gift from God to Abraham and his wife Hagar. An oasis in the middle of harsh desert was interpreted as divine intervention. By the time of the Romans and the Byzantines the city was an important trading center and a center of pilgrimage. Pilgrims came to visit the Kaaba, a 50-foot high cubical building, supposedly built by Abraham and his son Ishmael over 4,000 years BCE, housing a black stone that was said to have fallen from God at the time of Adam. The city was the birthplace of

Figure 7.2   Pilgrims in Mecca

Source: http;//www.worldcity~photos.org/SaudiArabia/SAU-mecca-webshothai=jj20012.jpg

the prophet Muhammad around 570. He was forced to flee the city in 622 but returned eight years later, destroyed the pagan idols and declared the city the center of Muslim faith. The city is the most sacred site of Islam and a pivotal point in the life of Muslims. Devout Muslims pray five times each day in the direction of the city and are supposed to make a pilgrimage, the hajj, to the city at least once in their lifetime (Figure 7.2).

Islam has 1.6 billion adherents and is the fastest growing religion in the world. With the growth of the believers has come the growth of the hajj. In 1950 approximately 250,000 people made a pilgrimage to the holy city; by 2012 the number had increased to over 2.8 million. In the simplest form of the hajj, pilgrims make their way to the city in the Islamic month of Dhu al-Hijjah, the men wearing only a simple plain garment composed of two white sheets. Recreating Hagar's desperate search for water they walk seven times between the two hills of Safa and Mawah and then circle the Kaaba seven times counterclockwise, the first four at a quick pace, the last three more leisurely. The longer hajj also involves travel to Mian and a return to Mecca.

Catering to pilgrims has been Mecca's main industry for over 1,300 years. In the past pilgrims would converge on the city across the desert on foot, donkey or camel, reinforcing the sense of supplication, a journey through the wilderness heightening the sense of a blessed arrival at a holy site. Today, many fly into the international airport at Jeddah, travel by car and bus and then walk to the giant mosque that encircles the Kaaba. The Saudi Arabian government has spent vast sums improving access to the city and providing pilgrims with transport, food, accommodation and health facilities. It is estimated the Saudi government spent $14 billion in the past 25 years. They built the world's largest abattoir to provide meat, constructed a water factory that provides 50 million bags of fresh water and every year they transport half a million people by Saudi Air to Jeddah. The revenues from oil have allowed improved facilities and access to the city. Crowd stampedes in 1990, 1994 and 1998 also prompted changes to the spatial layout, facilitating easier movement by the vast crowds, and extra money is spent on security and surveillance to hold in check the threat of violence and terrorism by religious fundamentalists and dissident groups.

Traditionally, water in Mecca came from surrounding wadis and connecting tunnels that ranged over 20 miles in the south and 60 miles in the north. Now the government, rich with oil revenues, can make water and freely give it away.

BOX 7.2

Cooling desert cities

Issues of global warming are often presented as topics suitable only for international forums and the coordination of national policies around the world. Increasingly, however, as some national governments, especially the US, resist global regimes, regions, states and even cities are getting more involved. In 2006, 22 of the world's largest cities made a combined pledge to reduce greenhouse gases by implementing policies and sharing technology. In the US the more progressive states have set emissions targets for greenhouse gases. Some require power plants to generate a portion of their electricity from renewable resources and 11 states have followed California's lead in adopting stricter vehicle standards for greenhouse gases than the federal standard.

Even small cities can play a part. Keeping cool in the desert involves smart design as well as sophisticated technology. It also involves a range of innovative programs. Alice Springs is situated in the middle of Australia. The semi-arid climate produces hot, dry summers and warm winters. A local initiative called desertSmart Cool Mob, has created a program to reduce electricity, transport and water use, minimize waste and encourage recycling. More than 500 households have signed up to achieve a more sustainable lifestyle. The positive environmental outcomes are pursued by the provision of energy-saving information to households, access to a range of discounts and workshops aimed at retrofitting older housing and building more environmentally sensitive, newer housing.

Source: http://desertsmartcoolmob.org.

Las Vegas is also a desert city but one that hides its dry location with extravagant displays of water in the tourist areas and the luxury hotel districts. The city grew up in the modern era as a remote frontier post, a stepping-off point on a journey to somewhere else. In 1905 it was a railroad town, but because of its location, further growth was difficult to imagine. The town is in the middle of the Mohave Desert, far from population centers, with precipitation less than five inches per year and summer temperatures that can soar well above 100°F. The building of the Hoover Dam, only 30 miles from Las Vegas, initially named the Boulder Dam, was a significant event in the development of Las Vegas. It began in 1929 and brought in thousands of federal workers. While there were only 5,000 jobs, almost 42,000 people came in the job-starved Depression years of 1931 looking for work. The state legislature, realizing an economic opportunity, legalized gambling. The dam, eventually completed in 1936, laid the basis for plentiful water and cheap energy. The federal government further helped the growth of the city by building a base for training pilots, later named Nellis Air Force Base, that housed 10,000 people. The federal government laid the basis of the economic development of the city, but it was mob figure Bugsy Siegel who saw its full potential. His *Flamingo Hotel* opened in 1946 and inaugurated a new gambling era. Casinos funded and operated by organized gambling made Las Vegas a destination point for more people. In 1950 the permanent population was still only 24,624. There was a reaction to the organized crime.

In 1955 the state of Nevada passed legislation that required all shareholders to be licensed. It was designed to keep out known mobsters. However, it enshrined rather than replaced organized crime. In an era of corporate capitalism the large financial institutions would have to license every one of their hundreds of thousands of shareholders. This requirement was only overturned by the 1967 Corporate Gaming Act that paved the way for corporate investment in the city. Since the early 1990s, the city has changed as mega resorts have replaced casinos. The 1989 opening of the Mirage marked a movement away from a model of money being made from casino gambling to more profits being generated from entertainment. While some casinos still exist, such as the Golden Nugget, Four Queens and the Horseshoe, others such as the Dunes, Sands, Aladdin and Hacienda were demolished and replaced with hotel and entertainment spectaculars such as the Bellagio, Venetian, Paris and New York. Even the name changes from the Dunes and the Sands to the Venetian and New York represent a change from a casual reference to the desert to a postmodern simulacrum of places from around the world. The themed hotels and casinos suggest a fantasy world (Figure 7.3).

Figure 7.3 Las Vegas fountains display the city's fantasy of extravagant disregard for water

Source: Photo by Joe Dymond

Las Vegas is a desert city, with only average rainfall of less than five inches per year but with the highest per capita water consumption in the US. By 2011 the city had grown to over 583,000 and water consumption was just over 400 gallons per person. The city gets most of its water from Lake Mead, a reservoir of the Hoover Dam. Since 2000 the water level in the Lake has dropped 100 feet, resulting in a loss of five trillion gallons of water. There are conservation plans that have resulted in a reduction in water demand by restricting car washing and lawn watering during peak times. There is also a very successful rebate program that aims to lessen the square footage of turf use. The program offers $2 per square foot of turf that is removed, replaced by dry land, using low-water and native plants. Las Vegas' water is managed by the Southern Nevada Water Agency (SNWA), which has proposed the development of a 327-mile pipeline that would pump groundwater from an aquifer in eastern Nevada, pulling 175,000 acre-feet/year to serve Las Vegas. The total cost of the project is estimated to be around $15 billion.

Water plays a crucial role in the fantasy presentation of the city, including the spectacular presentation of hotel architecture such as the winding Grand Canal of the *Venetian*, the extravagant water displays of choreographed water jets to the glistening water surfaces of the swimming pools and the lush green fairways of well-watered golf courses. Las Vegas defiantly negates its desert setting in a fantasy of urban spectacularization for its 35 million annual visitors.

In Mecca the desert is used to reinforce a sense of purity, sublimation and redemption through difficult pilgrimage, its austerity perhaps echoed in the simple and fundamental tenets of Islam and the fundamentalist interpretation of Saudi Wahhabism. Strong rules for people living in a harsh environment. In Las Vegas it is the negation of the dryness of the desert that is a persistent theme. The waterfalls, pools and lush greenery all signal not so much a defeat of nature but indifference. The air-conditioned rooms not so much defy the dry heat but ignore it. The desert is significant only in its silence.

And yet there are similarities between the two cities in the desert. There is the same sense of pilgrimage, a journey to an oasis. The same promise of a life-changing experience. Located in the most inhospitable places each city offers something to the traveler: riches and redemption, spiritual connection or easy money, enlightenment or entertainment. The desert location informs the identity and functioning of the city both in its presence and its absence.

City on the delta

The city of New Orleans owes its origins, economic rationale and a possible source of its destruction to the Mississippi. Situated close to the mouth of the world's third largest river system, this location provides both the *raison d'etre* and possible annihilation for the city. The city's history, geography and future is intimately bound up with its position on the delta.

New Orleans is less a result of a careful assessment of environmental suitability and more the result of geopolitical power struggles. It began as a French city. Early French incursions into North America were along the St. Lawrence and then down the Mississippi. After the French founded Quebec in 1602 they began to move along the inland riverways and waterways, lured in part by the prospect of controlling the fur trade. Traders, Jesuit priests and functionaries traveled along the Mississippi river in search of pelts, converts and allies. They sold liquor, metal tools and blankets in return for fur, religious conversion and political alliances. In 1682 Rene-Robert, Cavalier de La Salle, leader of 40 French and Indians, entered the great river from the Illinois River and moved south, paddling downstream. By April they had reached the mouth of the river. On April 9 they planted a cross and raised the French standard to formally claim possession of the entire river basin.

A merchant company was given title to Louisiana by the French government in 1717. The government gave the company, known as the Mississippi Company though formally entitled the Company of the West, a trade monopoly for 25 years and a directive to establish 6,000 free settlers and 3,000 slaves. This was a common practice of imperial power to retain overall control of overseas territory but to offload the costs and everyday management of colonization to a merchant company. The city of New Orleans was founded by the merchant company in 1718. The attempt to build a trading city close to the mouth of the river encountered the watery geography of a giant delta, half marsh, half mud, a floating spongy raft of shifting vegetation. The city was located 120 miles from where the river flows into the Gulf of Mexico, at a bend in the river close to Lake Pontchartrain, a site that enabled the portage of goods from the lake to the city. It was easier to ship goods to the lake and transport them to the city than to sail up the ever-shifting Mississippi river.

The city was sited on a relatively high piece of land where French traders had already been encamped close to a Native American trail. The city's origins are thus part Native American, part French, and the joint ancestry is apparent in names. The city was named after the Duc of Orleans who was regent and ruler of France in 1718. The French wanted to call the river the St. Louis, but the Native American name of Mississippi persists to this day. The French legacy is apparent in the French names along the lower Mississippi, as well as the long-lot land-use patterns. Under the French, land holdings were divided up as long strips with narrow river frontages.

The city was an outpost of the French empire, part of a global network of colonial possessions that stretched from the Americas to Africa and Asia. French power in North America was concentrated along the St. Lawrence. New Orleans was at the outermost edges, at the end of a long river journey south across lakes and rivers and down the Mississippi. Other French towns along this river highway included Quebec, Montreal, Detroit, St. Louis and Baton Rouge. The city was also part of wider French interests along the Gulf coast. Other cities in this Gulf cluster included Mobile and Biloxi.

Large land grants were awarded and several thousand colonists were brought from France, Canada, Germany and Switzerland. Native Americans resisted the incursions into their land, but the French military defeated them and subsequently shipped many of them off to the French colony of St. Dominique as slaves. On the lower Mississippi, the colony soon evolved as a plantation economy growing sugar, rice, indigo and tobacco. Labor was scarce so slaves were imported from Africa to work in the fields. At the center of this new colony was New Orleans.

Cartesian order—the imposition of the grid—was introduced to the North American wilderness. New Orleans was laid out as a rectangular grid of 44 blocks, 11 blocks running alongside the river and four away from the river. The city was enclosed in fortifications. At the center block immediately facing the river was the open square of Place d'Armes surrounded by government and religious buildings. This street pattern remains a distinctive feature of the French Quarter, or *Vieux Carré*, of contemporary New Orleans.

The city grew slowly at first. A 1764 map of the city shows that one-third of the blocks were empty. And even by the end of the eighteenth century not all the land in the city boundaries was occupied. It took some time before reality filled in the promise of the grid. Buildings were constructed initially out of wood, to be replaced as time and wealth accumulated by more substantial and permanent brick buildings often stuccoed in white or yellow. Houses were built on pillars with large verandahs that wrapped around the entire floor; a design that maximized air circulation in an oppressive climate. The low water table created difficulties in construction, and the river threatened the city with regular flooding.

By 1825 the fortifications were pulled down, creating opportunities for the construction of broad boulevards. Canal Street, North Rampart Street and Esplanade Avenue are sited along the wide locations of the early fortifications. As the city grew, the river long-lots of plantations were subdivided into gridiron strips. The open square, broad streets and the grid became distinctive features of the city's morphology. The city grew first along the river and then north toward the lake. Initially development was restricted to the higher ground above sea level because of the fear of flooding, but improvements in pumping technology in the early twentieth century encouraged more development in the lower-lying areas of the city.

In 1763 the city and the wider province of Louisiana changed hands. In the wake of France's defeat in the Seven Years War (the French and Indian Wars as they applied to North America) the city and territory changed from French to Spanish possessions. It made little real difference. New Orleans was always at the end of a very long tentacle of French government interest and involvement. The Spaniards proved efficient administrators. They rebuilt the city after the destructive fires of 1788. The buildings of the so-called French Quarter are more Spanish design than French. But Spanish power in the region was short-lived. The territory was ceded back to the French in 1801 and then two years later it was included in the Louisiana Purchase. New Orleans became an American city. It also became more strategically located. As settlers pushed west to the Mississippi and beyond, New Orleans was now at the mouth of a giant economically productive river basin. Commodities from the interior were shipped down the river through the port of New Orleans. Slaves, goods and commodities were also shipped up river.

Steam power shrunk the distance within the giant basin while economic growth increased the traffic along the giant river system. New Orleans was a pivotal hub in this new economic geography. In the early nineteenth century the city grew from around 10,000 in 1800 to 102,193 in 1840, when it became the third largest city in the nation, behind only New York and Baltimore in population size. From 1830 to 1860 it was one of the largest half-dozen cities in the entire country, the single most important export center of the whole economy with the largest slave market in the nation. The city was at its high economic point on the eve of the Civil War when technological change and new economic geographies propelled the city to a prime position. Subsequent technological changes and further transformations in economic geography undercut its strategic location. As rail replaced river traffic the city was no longer such a vital hub. As the economic center of gravity shifted ever westward and as the national economy transformed from agricultural to more manufacturing based, New Orleans lost its importance. Railroad cities such

BOX 7.3

New Orleans and the Mississippi

"As one looks upon the Mississippi, which curves and winds round New Orleans, as it does in every part of its course, and from which the title of the Crescent City is derived, we look in vain for the reasons, which prompted the choice of the site. So far as the river is concerned, the city might have been a hundred miles higher or fifty lower … the ground is so low that the drainage runs away from the river. Immediately behind the city is a low swamp, which generates fever and disease, and which is the secret of the unhealthy condition of the place. On such low ground the city is built."

Kingsford, W. (1858) *Impressions of the West and South, During Six Weeks' Holiday*. Toronto: Armour, p. 54.

"As sediment slides down the continental slope and the river is prevented from building a proper lobe—as the delta subsides and is not replenished—erosion eats into the coastal marshes and quantities of Louisiana steadily disappear. The net loss is over fifty square miles per year.… A mile of marsh will reduce a coastal storm-surge wave by about one inch. Where fifty miles of marsh are gone, fifty inches of additional water will inevitably surge. The Corps [of Army Engineers] has been obliged to deal with this fact by completing a ring of levees around New Orleans, thus creating New Avignon, a walled medieval city accessed by an interstate that jumps over the walls."

McPhee, J. (1989) *The Control of Nature*. New York: Farrar Straus Giroux, pp. 62–6.

"The average flow of 600,000 cubic feet per second that courses by Baton Rouge and New Orleans had, in the minds of authorities an almost limitless capacity to assimilate discharges to the point that they were harmless. Not only was the river a bottomless 'sink,' the state touted its huge quantities of fresh water to

> prospective industries that built plants along the waterway....
> While contesting pollution elsewhere in the state, Louisiana
> authorities took virtually no enforcement actions along the
> lower Mississippi until the 1960s.... From the great fish kill
> in the mid-1960s to the cancer scare in the mid-1970s the
> lower Mississippi was a focal point for public debate over
> water quality."
>
> Colten, C. E. (2000) "Too much of a good thing: industrial
> pollution in the lower Mississippi River," in *Transforming New
> Orleans and Its Environs*, C. E. Colten (ed.). Pittsburgh:
> University of Pittsburgh Press, pp. 141, 159.

as Chicago grew to prominence and industrial cities such as Cleveland, Buffalo and Pittsburgh all eclipsed the crescent city. The population continued to increase, but its relative position was slipping. In 1900 the city's population had increased to 287,104, but the twentieth century saw the long, slow relative decline of the city as it moved from twelfth largest city in 1900 to sixteenth in 1950, twenty-fourth in 1990 and thirty-first in 2000. After Hurricane Katrina it fell to fifty-first in 2010. New Orleans lost influence and national significance. Factories were locating elsewhere, waves of economic growth passed it by. The city remained deeply divided along lines of class and race. New Orleans seemed to miss out on the creation of a substantial middle class. The city languished in relative decline.

The river that gave the city economic prominence also threatened its destruction from two sources. The first is flooding. The history of New Orleans is one of floods, responses to floods and measures to avoid floods. When the snow melts in the upper regions of the vast river basin and rainfall is heavier than usual, then the river can flood its banks. The city was first constructed only months after a major flood. Since then, floods have been an integral part of the city's history. The city flooded in 1731 and 1752. In 1816 the city was flooded for a month. The flood of 1828 prompted the mandating of taxes to pay for levees. In 1849 the city was flooded for 48 days. The floods of 1850 prompted the federal government to pay for levees. The river had major floods in 1882, 1884, 1890, 1891, 1898, 1903, 1912, 1913, 1922 and 1927.

When 20 inches fell on the city in 1995 the heavy rain caused the city to flood.

One response was to build levees to keep the water in check. Since the first four foot levees were built in 1722, levee construction has been the predominant response to the threat of flooding. Keeping the river inside high channels was the solution. One influential theory in the nineteenth century argued that levees would not only keep floods at bay, but by channeling the water in a narrow area would increase flow and hence scour out the riverbed bottom. The bankruptcy of the policy was revealed by the great flood of 1927, when more than one million people were flooded from their homes. Thereafter spillways, which would release the floodwaters in a controlled way, supplemented levees.

Urbanization created a greater risk of flooding. As the amount of impermeable surface increased in both the city and throughout the entire Mississippi river basin, so increased the risk of flooding. Every new parking lot and residential development reduces the gradual absorption properties of the more permeable surfaces and increases the likelihood of flooding.

Because the river keeps depositing sediment, river levels rise and so, to be effective, levees have to be raised. In order to be safer levees have to be increased in size, which means more people are thus located below the river in the event of flooding. The Mississippi now flows through New Orleans as an elevated water highway. Rain in the city has to be pumped out and this has led to a steady sinking of the city below sea level. The more the city sank the more pumping was necessary and the higher the levees needed to be. Within the city residential development followed the space-packing process common to other private market societies. The rich get the best sites with lower-income groups getting less desirable places to live. In New Orleans the most powerful groups had commandeered most of the high ground. There was a connection between altitude and class. The poorest in the city live in neighborhoods like the Ninth Ward, which was well below sea level and most vulnerable to floods. The higher elevation Central Business District and the more upmarket Garden District and French Quarter escaped the worst flooding. The lower-lying neighborhoods of poor, black people in rental accommodation were the most severely affected by the post-Katrina flooding.

Levees need continual maintenance and monitoring. Yet levee maintenance lacks political support until it fails. It was levee failure that led to the catastrophic flooding after Hurricane Katrina in 2005 that killed over 1,000 people and devastated the city.

The second problem for New Orleans is the dynamic quality of the Mississippi. It is best likened to a giant hose. As water squirts out, the hose moves from side to side. Over the long term the river has shifted its position as it deposits sediment and creates new river courses. The river is continually changing course because of the changing topography and the gravitation pull of searching for the quickest way to the sea. Left to themselves the giant loops of a meandering river would constantly fantail across a very broad area. The channelization of the river keeps things in place, but at a cost. Three hundred miles upstream from the mouth of the river the US Army Corps of Engineers had to build giant dams to keep the water flowing down the Mississippi rather than into the Atchafalaya, which has a sharper gradient and quicker route to the sea. Without the human-engineered system the river would divert to the west and south and New Orleans would no longer be a river city. The Old River Control Project was begun in 1954 to keep the water split between the Mississippi and the Atchafalaya at 70/30. Conservative estimates put the project final cost at $1 billion, covering the construction of earthen dams and intricate flood control systems. The control of nature is achieved at tremendous cost and constant vigilance.

Dhaka: another city on a delta

The city of Dhaka in Bangladesh is also situated on a delta, right in the heart of the Ganges–Brahmaputra deltaic region. It sits on a low-lying plain criss-crossed by many streams and rivers in the middle of a vast river system that meanders and bends, and breaks up into different channels as it approaches the ocean. It is a watery city, only around 20 feet above sea level. It sits in the track of annual monsoons that bring moisture-laden winds from the Arabian Sea and the Bay of Bengal. From June until September the monsoon down-pours drench the city. Since 2000 the average monsoon rainfall has been almost 79 inches (2,000 mm). In 2010 over 13 inches of rain fell in just one 24-hour period.

The city became an important node in global trading as English, French and Dutch merchants set up trading stations. It emerged as an important administrative city in the British colonial control of the Indian subconti-nent and developed after independence, first from the British and then as a breakaway state from West Pakistan, as the capital and largest city of Bangladesh.

There is one strain of urban environmental writing that locates cities in the developing world very much in a deterministic model. They have problems,

BOX 7.4

Dhaka: contrasting city lives

"There is a stark contrast between the rich and poor.... The inequality aspect in the city is brought out vividly by the latest model imported cars on the roads and crowded shopping plazas and malls filled with costly imported goods, on the one hand, and by the ever increasing number of disabled beggars and vagrants on the pavement.... The inhuman conditions under which the poorest live in the slums and on pavements under open skies have few parallels. Such dwellings can be found on ditch embankments and on the edges of lakes, rivers and sewers and beside railways tracks.... With uncollected and rotting garbage all over, cave-like thatched dwellings on the roadside, railway lines flanked by squatter settlements, brackish water bodies, streets with gaping holes and no street lights, open drains and sewerage lines with malodorous spillovers often running onto the thoroughfares, stagnant ponds at prominent locations filled with water hyacinth and waste, most of Dhaka city indeed offers a most dismal spectacle. During the monsoons the misery of most of the city residents, and particularly those living in the slums, assumes an indescribable state. On the other hand the oases of affluence, with luxurious buildings, brand new cars and lifestyle involving conspicuous consumption provide a sharp contrast to the general picture of poverty prevailing in the city."

Siddiqui, K., Ahmed, J., Siddique, K., Huq, S., Hossain, A., Nazimud-Doula, S. and Rezawana, N. (2010) *Social Formation in Dhaka, 1985–2005*. Farnham: Ashgate, pp. 14–15.

so the argument runs, primarily because of the environmental context. Their social problems are read off from their environmental context. The underlying assumption is that cities in the developed world can transcend their environmental constraints while cities in the poor parts of the world are always constrained by theirs. A city located in the middle of a deltaic plain, right in the center of the southwest monsoon winds, is not in itself a problem. The river system lays down layers of alluvial soil, ideal for a productive agriculture. The monsoon rains bring life and lushness. The coming of the monsoon was traditionally welcomed as a sign of the ending of summer's oppressively high heat, as well as the bringing of water vital to make things like rice, jute and sugarcane grow. The first rains of the monsoon season are celebrated with recitations, dances and songs. The rains revive plants and crops parched by summer heat and provide moisture to soils sun-baked by the high summer temperatures. Fragrant flowers such as jasmine begin to blossom and fill the air with the scent of a green renewal. Guava and pine-apple fill out to juicy succulence.

It is not the site of the city in itself, but a number of other processes that have made the location less a source of economic potential and more a hazard. Two in particular can be noted. The first is that climate and environmental change increase the problems of the city. The rise of global temperatures increases cyclonic activity that generates more storms and increased rainfall. Situated so close to sea level, the entire country is vulnerable to climate change. Increasing sea levels and rising temperatures have increased the intensity of monsoonal flooding and storm damage. Increased snowmelt in the Himalayas, the source of the rivers, increases storm run-off and the possibility of flooding. Dwellers in the land of braided streams and muddy lowlands are particularly vulnerable.

The second, and related, process is the large influx of rural migrants to the city. More than 400,000 people from rural Bangladesh relocate to Dhaka each year. Many of them are environmental refugees driven off the land by rising sea levels, flooding and storm damage that have taken away their land and livelihood. Even in good years most fields in the vast delta flood once each year. The increasing rainfall and storm run-off flooding now washes away fields and farms. Bangladesh is on the front line of global climate change and Dhaka is in the eye of the storm. The impacts of climate change prompt many of those living in rural Bangladesh, a largely low-lying country in the vast delta, to move to the city, reinforcing the congestion in Dhaka. As the country's primary city, Dhaka attracts all the people from the countryside displaced by rising sea levels, severe storms, flooding and economic marginalization. The sustained rate of growth has overwhelmed the ability of the formal economy to provide jobs or the housing market to provide accommodation. The results are slums located throughout the city. This unplanned self-built housing is, on the one hand, a remarkable testament to human ingenuity in the face of immense hurdles. But, on the other hand, this also has a major impact on the urban ecology. Wetlands have been destroyed, green spaces built over, groundwater polluted. From 1960 to 2008 half of the wetlands and one-third of the water bodies in and around the city were lost to urbanization. These had acted as sponges to soak up storm water. Now, the monsoon rains bring not only relief and moisture, they also bring urban flooding. The urban hydrological system is overwhelmed by the rate and scale of urbanization.

Dhaka is one of the faster growing megacities in the world. In 1950 its population was only 336,000. By 1980 the figure had risen to 3.2 million and in 2011 it was 15.3 million. The city doubled in size from 1990 to 2005. The 2025 estimate is for a city population of 20 million. More than half live in slums, self-made settlements widely distributed in the marginal spaces in the city, on empty sites, green places and flood plains. The rapid growth, especially of informal slum areas, has made the city more vulnerable to flooding because of the reduction in green space, the increase in impermeable surfaces, waterlogging and settlement on flood-prone areas. The very poor often have nowhere else to go but to sites that are particularly vulnerable to flooding.

The flooding of Dhaka is now a regular occurrence. Disastrous flooding occurred in the city in 1998, 2004 and 2011. In 2004 flooding marooned 2.3 million people as all the roads were under water and businesses were shut down for almost a month. Waterborne disease affected more than half a million people. In July 2011, 1.3 inches (33 mm) of rain fell on the city in a very brief shower on the morning of July 19. Since the drainage system can only remove half an inch (12 mm) of water an hour, the result was widespread flooding. People had to go to work through dirty water as traffic jams also clogged the city; rickshaw drivers could raise their rates. It is no surprise that Dhaka ranks, according to the *Economist*, as the world's second least livable city, ranking only after Harare.

Guide to further reading

The city–site nexus is a major theme of the new urban environmental history. See articles in the journal *Environmental History*, as well as the 2012 special issue on "History of urban environmental imprint" in the journal *Regional Environmental Change* and the 2011 special issue on "Methods and contents in landscape histories" in the journal *Landscape Research*.

Melosi, M. V. (2010) "Humans, cities and nature: how do cities fit in the material world." *Journal of Urban History* 36: 3–21.

On bioregionalism and cities:

Berg, P. (2009) *Envisioning Sustainability*. San Francisco, CA: Subculture Books.

Gabor, Z. (2013) *The No-growth Imperative; Creating Sustainable Communities Under Ecological Limits to Growth*. New York: Routledge.

Thayer, R. (2003) *LifePlace: Bioregional Thought and Practice*. Berkeley and Los Angeles, CA: University of California Press.

The website of the global ecovillage network: http://gen.ecovillage.org.

There are many books on Sydney:

Connell, J. (ed.) (2000) *Sydney: The Emergence of a World City*. South Melbourne: Oxford University Press.

Falconner, D. (2010) *Sydney*. Sydney: UNSW Press.

Noble, G. (ed.) (2009) *Lines in the Sand: The Cronulla Riots, Multiculturalism and National Belonging*. Sydney: Institute of Criminology.

Spearritt, P. (2000) *Sydney's Century*. Sydney: UNSW Press.

For a more recent update of environmental plans and policies, see the website of *Sydney 2030*, which is the city's plan to make the city more sustainable; it also gives an annual benchmark: http://www.sydney2030.com.au.

On Las Vegas:

Ferrari, M. and Ives, S. (2005) *Las Vegas: An Unconventional History*. New York: Bullfinch.

Land, M. and Land, B. (2004) *A Short History of Las Vegas*. Reno, NV: University of Nevada Press.

Roman, J. (2011) *Chronicles of Old Las Vegas: Exposing Sin City's High-Stakes History*. New York: Muesyon.

Schumacher, G. (2010) *Sun, Sin and Suburbia*: *An Essential History of Modern Las Vegas*. Las Vegas, NV: Stephen Press.

Mecca as a pilgrimage site is covered by:

Hammoudi, A. and Ghazaleh, P. (2006) *A Season in Mecca: Narrative of a Pilgrimage*. New York: Hill and Wang.

Wolfe, M. (ed.) (1999) *One Thousand Roads to Mecca: Ten Centuries of Travelers Writing About the Muslim Pilgrimage*. New York: Grove.

On New Orleans, see:

Colten, C. (2004) *An Unnatural Metropolis: Wresting New Orleans from Nature*. Baton Rouge, LA: LSU Press.

Gotham, K. F. (2007) *Authentic New Orleans: Tourism, Culture, and Race in the Big Easy*. New York: New York University Press.

Lewis, P. F. (2003) *New Orleans: The Making of an Urban Landscape*. 2nd edition. Santa Fe, NM: Center for American Places.

Powell, L. N. (2012) *The Accidental City: Improvising New Orleans*. Cambridge, MA: Harvard University Press.

On Dhaka, see:

Byomkesh, T., Nakagosshi, N. and Dewan, A. M. (2012) "Urbanization and green space dynamics in Greater Dhaka, Bangladesh." *Landscape and Ecological Engineering* 8: 45–58.

Hossain, A. M. M. and Rahman, S. (2011) "Hydrography of Dhaka City catchment and impact of urbanization on water flows: a review." *Asian Journal of Water, Environment and Pollution* 8: 27–36.

Islam, N. (2005) *Dhaka Now: Contemporary Urban Development*. Dhaka: Bangladesh Geographical Society.

Siddiqui, K., Ahmed, J., Siddique, K., Huq, S., Hossain, A., Nazimud-Doula, S. and Rezawana, N. (2010) *Social Formation in Dhaka, 1985–2005*. Farnham: Ashgate.

Sultana, M. S., Islam, G. M. T. and Islam, Z. (2009) "Pre- and post-urban wetland area in Dhaka City, Bangladesh: a remote sensing and GIS analysis." *Journal of Water Resources and Protection* 1: 414–21.

8 Hazards and disasters

Cities are places where the threat of hazards and the prospect of disaster are always a possibility. We are reminded of this whenever major disasters are televised and searing images are broadcast around the world, whether it be the drowning of New Orleans after Hurricane Katrina, the tsunami waves that swept across the littoral of the Indian Ocean, the destruction of Port au Prince after the 2010 earthquake or the tsunami that battered Japan's Pacific coast in 2011. Throughout history cities are places where disasters have occurred.

Some definitions are perhaps in order: *hazards* are extreme events such as severe weather, earthquakes and tsunami. Environmental hazards include floods, windstorms, landslides and many others. There are also more social hazards such as fire and nuclear plant meltdown. The distinction between the two is problematic. In the case of the 2011 tsunami in Japan, the environmental hazard of an earthquake-induced tsunami triggered the spread of nuclear radiation fallout, a social hazard. The distinction between the two is elastic and the categorization is porous, as environmental hazards merge into social hazards. *Risk* is the probability of an extreme event; it can also be more closely defined as the product of a person's exposure to hazards and their ability to anticipate, respond and recover from a hazard. *Vulnerability* is a measure of the exposure to hazards. Two types of vulnerability can be identified: locational and social. There is a geography to vulnerability. Some cities are more vulnerable to certain hazards than others because of their location—Mexico City is more vulnerable to volcanic eruption than New York City. Table 8.1 notes ten of the largest cities and their associated hazards. But vulnerability to hazards is also socially constructed. Poorly planned urban growth, deforestation and poor medical provision are just some of the factors that increase the chance that hazards become disasters. Many cities, and particularly poorer citizens in these cities, are vulnerable to environmental hazards that can turn into disasters. This relational use of

Table 8.1 Cities and hazards

City	Population (2010)	Hazards
Tokyo	36.67 million	Earthquake, storms, tornado, storm surge
Delhi	22.16 million	Earthquake, storms, flood
São Paolo	20.26 million	Storms, flood
Mumbai	20.04 million	Earthquake, storms, flood, storm surge
Mexico City	19.46 million	Earthquake, volcano, storms
New York	19.43 million	Earthquake, storms, storm surge
Shanghai	16.58 million	Earthquake, storms, flood
Kolkata	15.55 million	Earthquake, storms, tornado, flood, storm surges
Buenos Aires	13.07 million	Storms, flood
Jakarta	9.21 million	Earthquake, flood

Source: United Nations Department of Economic and Social Affairs (2010) "Urban agglomerations, 2010." www. unpopulation.org (accessed March 13, 2012).

vulnerability highlights the role that socio-economic status, wealth and power play in the experience of hazards. Vulnerability to hazards and exposure to risk varies across the social spectrum, with the poor most at risk and most vulnerable. A number of scholars employ ideas of marginalization and the production of unequal risk, which both reflects and embodies socio-economic differentiation.[1] A *disaster* is the negative impact of a hazard. Disaster results from vulnerability to hazards. The risks are often greatest in cities because there is a greater concentration of both people and hazards, an increasing number of hazards, and the possibility of hazards interacting with each other—as in floods causing disease outbreaks or earthquakes causing fires. Disasters are especially prevalent in cities where emergency and response mechanisms are often too little too late.

As the title of one edited book about the 2005 Hurricane Katrina proclaims, *There is No Such Thing as a Natural Disaster*.[2] It seems more appropriate to use the term environmental disaster rather than natural disaster, as there is nothing "natural" about the effects of a disaster; rather, they are first and foremost social. Disasters reveal our social fault lines and make our political structures clearly visible. There is growing literature on the relationship between cities, hazards and disasters that raises issues of vulnerable cities, at-risk urban dwellers, the social differences in how urban disasters are experienced and the resiliency of cities to bounce back from a disaster.[3]

In the 1980s 177 million people were affected by disasters; by 2002 this had increased to almost 270 million due to population growth, rapid urbanization, environmental degradation and climate change. Ninety-eight percent of those affected live in low-income countries. In 2005, 430 environmental disasters killed almost 90,000 people, the vast majority citizens of low-income countries. Disasters differentially impact the poorest in the cities of the developing world and they play a part in sustaining poverty and inequality.[4] Some 373 disasters killed over 296,800 people in 2010, affecting nearly 208 million others and costing nearly $110 billion.[5] Each decade in the US, the cost in constant dollars of damage to property doubles. The cost of the 2011 disaster in Japan is estimated to be around $300 billion. Table 8.2 lists the major urban disasters of recent years.

Urban environmental disasters show, in the starkest terms, the vulnerability of cities. Analyzing the connections between hazards and disasters and outcomes in cities highlights an important link in nature–society relations. For example, flooding, caused by increased rainfall due to global warming, that impacts slums is one of the many connections between environmental issues such as global warming and social justice. Cities are the nexus of interaction between environmental and social issues.

More attention is now focused on the vulnerability of cities to environmental and social disasters. There is a rereading of history to excavate the role of

Table 8.2 Recent major urban disasters

City	Date	Hazard type	Reported deaths
Sendai Ichihara, Fukushima, Onagawa	2011	Earthquake, tsunami	5,178
Port au Prince	2010	Earthquake	222,570
Beichuan, Miazhu, Juyuan, Chengdu	2008	Earthquake	87,476
Yangon, Myanmar	2008	Tropical cyclone	138,366
Yogyakarta	2006	Earthquake	5,778
Muzaffarabad, Pakistan	2005	Earthquake	73,338
New Orleans	2005	Storm surge	1,833
Banda Aceh	2004	Tsunami	167,000
Bam, Iran	2003	Earthquake	26,300
Paris	2003	Heatwave	14,800

Source: "World Disasters Report 2010, OCHA." http://www.guardian.co.uk/global-development/datablog/2011/mar/18/world-disasters-earthquake-data#data (accessed May 12, 2012).

disasters in the evolution of cities as well as an examination of current levels of risk, preparedness and responses. In this more anxious age there is a reexamination of cities as sites of disasters. In the wake of 9/11, tsunamis, Hurricane Katrina and earthquakes in cities around the world there is a more marked sense of the vulnerability of life and the precarious nature of urban civilization (Figure 8.1). This perspective has shaped urban studies focused on urban disasters.

For example, in his book *Ecology of Fear*, Mike Davis examines some of the disasters that continually threaten the Los Angeles metropolitan region. From the storms that sweep across the LA basin from the Pacific, to the wildfires of summer and the landslides that push expensive houses into the valley or sea below, Davis presents a picture of a city in an ecological disaster zone. Although exaggerated, Davis does point to the often-fragile presence cities have on the surface of the earth.[6]

Mark Pelling looks at cities in the Caribbean as sites of disasters. With reference to Bridgetown, Georgetown and Santo Domingo he highlights the issue of social vulnerability. He also examines the resiliency of cities that rebuild after a disaster strikes.[7]

Figure 8.1 Earthquake damage in Port au Prince, Haiti

Source: Photo Marco Dormino/The United Nations: http://en.wikipedia.org/wiki/File:Haiti_Earthquake_building_damage.jpg

In the remainder of this chapter we will discuss the major urban disaster categories of fires, floods and earthquakes, then consider one case study in more detail, examine the vulnerability of cities to social hazards such as oil dependency, and finally discuss the notion of resilient cities.

Environmental hazards

Fire

The Great Fire of London began on Sunday morning, September 2, 1666. The fire started in Pudding Lane and swept through the city for five days until the wind died down and the use of gunpowder explosions created firebreaks. The fire was not the first. Previous fires in 1133 and 1212 also caused extensive damage. The city, full of timber-built, high-density buildings and the extensive use of candles and wood and coal fires for heat and light was a conflagration waiting to happen.

The Great Fire caused major damage: 13,200 houses were destroyed, as were 87 churches and many fine public buildings. Over 200,000 people were displaced. The official death toll was only six, but many more are presumed to have died, especially from smoke inhalation. The fire swept across 436 acres, burning almost all the buildings inside the medieval walls.[8]

The Great Fire also led to the planning and rebuilding of London. In the fire's aftermath new building codes were introduced. Wood buildings were replaced with brick and stone buildings. Laws restricted dwellings to two floors in lanes and three in the larger streets, and mandated wider and straighter streets to eliminate congestion and to prevent the quick spread of fire in the future. Reforms enacted primarily in response to the great fire would result in better sanitation and help control future disease outbreaks.[9] New laws also required owners to obtain insurance and the new insurance companies quickly realized that employing men to put out fires could minimize their losses. Private fire-fighting companies were established, the forerunner of public fire protection.

The architect Christopher Wren planned much of the reconstruction of London and built 51 new churches, including his masterpiece, St. Paul's Cathedral. From the ashes of the Great Fire a new London was constructed, a more modern London in terms of building design, street layout, sanitary levels and the provision of collective services such as insurance and fire protection.

Throughout history fires have been an urban hazard. Cities composed of wooden structures with widespread use of fire as a power source are an ideal recipe for fire. Small-scale fires were regular occurrences. Epic fires were rarer. A small fire in Chicago in the fall of 1871 quickly turned into a widespread conflagration. It began around 9 p.m. on Sunday, October 8, 1871 and lasted until the morning of October 10, when a light rain helped to douse the remaining flames. The fire killed 300 people, left 100,000 homeless and destroyed most of the downtown.

Yet the fire, although a traumatic event, barely seemed to halt the growth of the city. Within five years the city was rebuilt and in 1891 hosted the Columbian Exhibition, which showcased the City Beautiful Movement and celebrated the rebuilt city.[10]

BOX 8.1

Old books, recurring themes

The recent focus on urban environmental disasters is only the latest manifestation of a long fascination. In 1913 areas in the Midwest of the US were hit by a series of tornadoes, floods, storms and blizzards. To match the near apocalyptic events Logan Marshall wrote a marvelously overwritten book, *Our National Calamity of Fire, Flood and Tornado*. The elaborate style makes it a product of its time but raises issues that still resonate down the years:

> Man is still a plaything of Nature. He boasts loudly of conquering it; the earth gives a little shiver and his cities collapse like a house of cards.... He imprisons the waters behind a dam and fetters the current of the rivers with bridges; they bestir themselves and the fetters snap, his towns are washed away and thousands of dead bodies float down the angry torrents.... He burrows into the skin of the earth for treasure, and a thousand men find a living grave.

Logan, M. (1913) *The True Story of Our National Calamity of Fire, Flood and Tornado*. Lima, OH: Webb Book and Bible Co, p. 11.

Once a fire has been extinguished, then, if the city is large enough and its economy is buoyant enough, the city can be rebuilt. The Great Fire destroyed medieval London but the rebuilding laid the basis for modern London. Chicago survived the great fire of 1871 to come back bigger and more confident than ever. Another response is the institutional rearrangements that mitigate a repeat of the fire hazard. Insurance and fire companies are created, building codes are enforced to both plan for and negate repetition of the tragedy.

In some cases fires can initiate major legislation and new policy. At the beginning of the twentieth century New York City experienced rapid industrial growth. Around the garment district many factories employed female immigrant labor in poor working conditions. A fire that started on March 25, 1911 in the Triangle Shirtwaist factory on the eighth floor of a building killed 146 people. It was particularly horrific as the mostly young women were trapped behind locked doors, their plight visible and audible from the streets below. The next day more than 100,000 mourners passed through the temporary morgue that was set up. The official response to the widespread public revulsion included the indictment of the factory owners and the establishment of a Factory Investigating Commission that over the next three years enacted 36 new safety laws for the city. In the wake of public reaction to the fire came new forms of regulation and safety concerns that were soon imitated by other cities throughout the country. The reaction to spectacular disasters is often a pivotal point in urban public policy.

Disasters can also influence public opinion and public behavior. In 1987 a discarded match ignited litter and grease underneath a wooden escalator at King's Cross in London's Underground. The fire quickly spread and 31 people were killed. In response, there were policy changes—all wooden escalators were replaced and automatic sprinklers and heat detectors were installed in escalators. But there was also a major change in public attitudes. A no-smoking policy, introduced two years earlier but seldom acted upon, was more strictly enforced after the fire. In stations as well as in trains, smoking became a socially unacceptable thing to do in the Underground.

The risk of fires is brought under control as stone, brick and concrete replace wood, and as cleaner fuel sources provide more light and power. But there is always the danger of fires from industrial sources. Chemical fires are a hazard in certain industrial areas as well as in the transportation of these materials. On June 1, 2006 an explosion in a chemical plant in Teesside in northern England, heard up to 20 miles away, caused a fire involving a deadly mixture of hydrogen, nitrogen and ammonia. Two people were injured, roads were sealed and local residents were warned to keep windows and doors closed. The risk of such chemical fires is always present. The effects are mitigated by land-use planning that segregates hazardous facilities well away from population centers.

Another source of fires is caused by urban/suburban growth into rural and wilderness areas which are susceptible to fire. Bushfires, brush fires and forest fires have long been a feature of ecosystems. Fires caused by lightning are an environmental fact of many arid and semi-arid and dry ecosystems. These fires are natural in the sense that they may occur naturally, but they are social in that their effects are exacerbated by urban growth that has pushed settlement into fire hazard areas. Cities in places such as California and much of Australia are so affected.

BOX 8.2

Wildfires in Los Angeles

"Why is the American West burning again, and what might we do about it? There is a short answer to the first part: The American West has large wild-land fires because its extensive wild lands are prone to burning. Planning policy is much harder and requires us to consider fire history.

Natural fire regimes beat to the rhythm of cyclic wetting and drying: it must first be wet enough to grow combustibles and then dry enough to get them ready to burn. Wet forests therefore normally burn during dry spells, deserts after rains. Fire also demands a spark, and under wholly natural circumstances, this means dry lightning. The eastern United States has wet lightning, which normally accompanies dousing rain; only in Florida do thunderstorm days and lightning-kindled fires routinely overlap. The West has dry lightning — and that is why, with or without people, significant fractions of the American West will burn.

The fire 'problem' resides, apparently, in the West. Why? The obvious reason is that the place is intrinsically fire-prone.

The deeper reason comes from the second force: it is that the American West experienced what a historian might call an 'imperial' narrative. In the 19th century, state-sponsored conservation policies encountered a landscape that had become largely emptied because the indigenous peoples had been driven off by disease-driven demographic collapses, wars, and forced relocations. It thus became possible, during that historical vacuum, for the young federal government to establish 'public' lands that would exclude agricultural settlement. In doing so, it created a habitat for free-burning fire....

The problem that has grabbed public and political attention is the spectacle of burning houses—the problem the agencies call the 'wild land–urban interface fire.' These fires might better be called 'intermix fires.' They occur in lands whose use has become scrambled into an ecological omelet, involving abandoned agricultural land as well as public preserves. Their existence and the hazards they pose are simply the result of unmanaged growth: the untrammeled growth of natural vegetation and the uncontained growth of our increasingly far-flung suburbs. The wild and the urban have become the matter and antimatter of the American landscape. When they collide, we should not be surprised by the occasional explosion."

Pyne, S. J. (2001) "The Fires this time and next." *Science* 294 (5544): 1005.

Oakland Hills of Oakland California was badly damaged by a fire in 1991. The neighborhood of timber-frame houses sat in an area of dry chaparral. A fire that started from a garbage fire quickly spread in the tinder-like environment; the blaze killed 25 people and destroyed 2,843 houses. Much of coastal California is semi-arid and susceptible to fire. These areas become urban hazards as suburban growth snakes its way into the less urbanized, more vegetated—and hence more at risk—environments. At the end of summer, with dry vegetation and high Santa Ana winds, southern California is particularly vulnerable. In late October 2003 fires scorched parts of San Diego County in California as two wildfires killed 16 people and destroyed 2,427 structures.[11]

Australia is a dry continent and the fire damage between 1969 and 1999 was estimated at an annual cost of A\$77 million. Australia's most devastating bushfires occur in eucalyptus forests in the suburban fringes of major cities. In the hot, arid summers eucalyptus trees produce a flammable gas that is easily ignited and quickly spreads. A bushfire is a series of minor explosions as the gas catches fire. The trees are adapted to fire: their bark and leaves can withstand high temperature and after a bushfire the vegetation will return. However, as suburbanization has shifted more people out toward the bush, these fires have more devastating effects on humans and their property. In some cases arsonists intentionally light fires. But "natural" bushfires continue to occur. In January 2003 fires caused by lightning strikes burned the national capital of Canberra. The dry vegetation and high winds provided the perfect recipe for a major fire that burned for almost a week in the surrounding bush before entering the suburbs on January 18. Four people were killed and 816 houses destroyed. An official inquiry suggested a greater use of controlled burning to minimize the build-up of dry vegetation and reduce the possibility of widespread bushfires. In February 2009 a series of ferocious bushfires spread across the state of Victoria. High temperatures and strong winds and possibly arson spread the fires. More than one million animals were killed, almost half a million hectares were burned, 2,000 houses were destroyed and 173 people lost their lives, notably in the suburban swathe of towns northeast of Melbourne. A long drought created tinder-dry conditions and high winds spread the fires. Because of their damaging and deadly impact the event is referred to as the Black Saturday bushfires.

As suburbs spread into drier environments the risk of fire increases. In the longer term controlled burning may provide prevention. But perhaps there is a need to reconsider the siting of residential areas in fire-risk areas; this poses a more problematic political conundrum with long-term fire safety considerations often outweighed by development interests and the forces of housing development and urban growth. In the wake of the Black Saturday bushfires an expert panel recommended that the state government ban new housing in the highest fire-risk areas.[12] The urge to rebuild needs to be balanced by the wisdom of rebuilding in areas highly vulnerable to fire.

Fire is also a major hazard in the slum dwellings of developing cities. High-density dwellings, often built of flammable, combustible materials, make fire a constant threat for slum dwellers across the developing world. A fire beginning in a slum community quickly gets out of control and affects large areas. The slum area of Baseco in Manila regularly suffers from fires; a 2002 blaze left 15,000 residents homeless. A devastating one in 2004 claimed the houses of 20,000 households. And a fire in 2010 left

4,000 homeless. Tightly packed housing, prolific use of open fires and stoves and often-inadequate fire protection services lead to frequent fires. In the slums of Nairobi, Kenya the high density combined with informal housing built of wood and corrugated tin makes fire an ever-present danger. In addition to poorly constructed houses, many of which have access to electricity illegally through unsafe wiring, extremely narrow and crowded roadways severely hamper the ability of firefighters to get to blazes. Thousands of people are killed and made homeless by slum fires in Nairobi every year.[13] In addition, many believe that developers intentionally set the fires at times in order to clear the land for permanent dwellings with higher rent. In September 2011, an oil pipeline running through the middle of a Nairobi slum caught fire and exploded, killing over 100 people in the explosion and subsequent fire.

Slum fires can also play a role in urban change. Loh Kah Seng tells the story of rapid urbanization in Singapore. In the 1950s the margins of the city were surrounded by low-income, informal, wooden settlements called *kampongs*. Fires were a frequent occurrence. The response by authorities helped to create a strong state presence, while the volunteer fire service aided popular political mobilization. As Seng notes,

> the increasingly well-organised rehousing operations following the fires progressively integrated families en masse into the social fabric of the nation-state. Their swift relocation to emergency public housing meant that the families now could obtain their accommodation only on the terms of the state. At a strategic level, the fire site became a vital springboard for the authorities to effect the clearance of nearby kampongs.[14]

Floods

Many cities are close to water. Flooding has long been a hazard of such locations, whether it be inundations by the sea or river floods. The historian John Barry tells the tale of the great Mississippi flood of 1927, which inundated 27,000 square miles and flooded the homes of nearly one million people. The flood was in part caused by steady rains through the entire river basin for months beforehand. The river flooded for 1,000 miles from Cairo, Illinois to New Orleans, Louisiana.[15]

The Mississippi has always flooded. What made the river banks so fertile was the steady deposition of alluvium by the river's changing course as it meandered its way to the sea. The flooding become an "urban" hazard as more people settled and moved to the delta regions attracted in part by the rich soils. Attempts to control the river led to the construction of levees that effectively channeled the river, which in turn made the floods that much more devastating. Barry reveals that the response to the flood had important social and political effects. The immediate response reflected existing social and economic power. The evacuation of many black people in the delta region of Mississippi was delayed because the white plantation owners believed that black sharecroppers, often heavily indebted, would not return. A cabal of New Orleans business leaders purposely dynamited the levees of two parishes down-river of the city to protect their interests. Responses to the flood included not only a change in river management away from the levees-only policy to a greater use of spillways. It also brought about profound socio-economic changes including a recognition that the federal government should play a larger role. And it ushered in a change in power relations. In the immediate aftermath of the 1927 flood, the great migration of blacks from the delta to the cities of the north began and in Louisiana the populist power of Huey Long overturned the old patrician ways.

Flooding is a regular part of many river ecosystems, but can be aggravated by land-use changes associated with rapid urbanization. Seoul, South Korea has endemic flooding problems caused by river overflows from the river Han and its various tributaries in the wake of heavy rains and rapid snow melt. Between 1960 and 1991 131 people died in floods. An analysis of the floods by Kwi-Gon Kim shows that the loss of green spaces led to more flooding. Seoul's rapid growth, from a population of 2.4 million in 1960 to 10.9 million in 1990, meant a significant loss of land in agriculture and forestry and an increase in urban land uses. This conversion of permeable to impermeable surfaces raised the risk of flooding. Kim's analysis shows that the presence of green space explained 31 percent of the variation between the different wards of the city in deaths and building damage.[16]

We can compare the differential experience of environmental risk in general and flooding in particular by looking at cities in the same river basin yet on opposite sides of a border between a rich and a poor country. Timothy Collins looks at the 2006 floods that impacted the US–Mexico border towns of El Paso (US) and Ciudad Juarez (Mexico). Between July 17 and September 7, 2006 the river Paso del Norte received twice its average annual rainfall. Widespread flooding resulted. Yet because of socio-economic differences the impacts were experienced asymmetrically. He describes the vastly divergent experiences of residents of the informal settlements in Ciudad Juarez compared to the residents of a rich neighborhood in El Paso. While one witnessed widespread damage and limited public response, the other escaped severe damage and received the full weight of public services. This interesting case study reveals how the unequal production of risk mirrors differences in income, wealth and political power.[17]

In some cases international development agencies provide aid to cities in the developing world. A $250 million project funded in part by the World Bank aims to reduce flooding in Bogotá, Colombia. Río Bogotá flows 370 km through central Colombia and through the capital city of Bogotá, which has a population of around 9.5 million. Rapid urban development since 1950, has not only resulted in the deterioration of water quality in the river, but it also caused the channelization of the river and the reduction of wetlands from 50,000 ha to less than 1,000 ha in 2009, both of which make floods more likely and more devastating when they do occur. Low-income neighborhoods along the river are especially prone to flooding. The World Bank's Río Bogotá Environmental Recuperation Project and Flood Control Project incorporates ecological design into flood controls by going beyond the typical deepening and channelizing of the river and raising embankments, by restoring meanders, creating riparian zones and maintaining connections with adjoining wetland areas to provide a measure of flood detention.[18]

In the cities of the developed world, flood hazards, while not eradicated, have been minimized by extensive and expensive flood control measures as well as organized rescue and recovery systems. However, in the poorer cities of the world this is less of an option. The capital city Dhaka, in Bangladesh, is located on a flood plain, with much of the city only several meters above sea level. The city is located on the north bank of Buriganga River in the Ganges delta. It has grown enormously from approximately 335,926 in 1951 to almost 15 million in 2010. Bangladesh is one of the poorest countries in the world, with a gross national income per capita in 2010 of only $610. The equivalent for Switzerland, US, UK and the Netherlands are, respectively, $67,700, $44,999, $35,980 and $46,954. Flooding is a regular occurrence in the city, especially during the monsoon season from July to September, and has become more pronounced as low-lying areas of the city have seen residential and commercial development. In addition, many of the city's poor have constructed shantytowns on steep slopes and hillsides, or in the low-lying parts of the city. Drainage backups and storms regularly lead to inundations of the low-lying parts of the city. When the monsoon rains are particularly intense the rainfall overwhelms the drainage system. Mudslides can also occur, wiping away precariously perched shantytown dwellings. In 1988 a flood covered 78 percent of the city. In September 2004 more than half of the city was flooded, including all the main roads. International aid donors have funded a Dhaka City Flood Control and Drainage Project that relies heavily on Western engineering solutions, including the construction of expensive embankment and pumping stations. The operational costs are a heavy burden for such a poor county. The continual flooding of Dhaka is thus a function of unplanned urban growth into low-lying vulnerable areas, highly populated sub-standard housing that makes the poor vulnerable to floods, mudslides and fires, and the poverty of the country that makes it difficult to afford efficient flood-prevention systems common to richer countries such as the Netherlands.

The effects of monsoon flooding are also impacting the ability of Bombay/ Mumbai to position itself as global city. The city's 150-year-old drainage system was overwhelmed by severe monsoon flooding in 2005 and again in 2006. In 2005 floods killed over 1,000 people. In 2006 exceptionally high monsoon rainfalls incapacitated the city's infrastructure, delaying flights and paralyzing road and train travel. During high tides the city closes the storm drains to prevent seawater from entering the city; even with moderate monsoon rains the city easily floods as the rainwater cannot escape. On June 28, 2012 there was 75.4 mm (almost three inches) of rainfall, which resulted in widespread traffic congestion as subways were flooded and road traffic slowed. The inability to deal with monsoon flooding is dampening the city's ambitions plan to became a hub in the financial network of global commercial centers.

Earthquakes

On November 1, 1755 the city of Lisbon in Portugal was hit by a massive earthquake estimated at between 8.5 and 9.0 on the Richter scale. The epicenter was located in the Atlantic Ocean and the city was swept by three tsunamis caused by the underwater tectonic movement. Soon fires also engulfed the city. This major disaster destroyed most of the city and killed almost 90,000 people in a city that had a population of 275,000. The city was rebuilt but never fully reclaimed its former pivotal position in the global urban network. The decline of Portugal as a major imperial force was also part of the story, but the Lisbon earthquake stands as testimony to the destructive power of earthquakes and the possibility of consequent urban decline.

San Francisco tells another story, one of resilience. On the morning of April 18, 1906 the city was shaken by an earthquake that registered 8.25 on the Richter scale and destroyed 25,000 buildings. Gas mains broke open and power lines were downed. Fires raged through the city for three days. More than 700 people were killed and 250,000 were made homeless. The city rose

BOX 8.3

Scales of disaster

Three of the main disaster types—hurricanes, tornadoes and earthquakes—have scales that allow measurement and comparison.

The Saffir–Simpson hurricane wind scale is a five-fold categorization based on wind speed. It was developed in 1971 by Herbert Saffir and Bob Simpson. The scale also gives indications of possible storm damage.

Category	Wind speed (mph)	Storm surge (feet)	Damage
One	74–95	4–5	Little damage to secure buildings
Two	96–110	6–8	Damage to vegetation, some flooding
Three	111–129	9–12	Flooding possible up to 15 miles inland, damage to property, evacuation of low-lying residences
Four	130–156	13–18	Extensive property damage, massive evacuation of up to six miles inland
Five	≥157	>18	Massive evacuation up to ten miles inland

The Enhanced Fujita scale is a six-fold categorization of tornadoes based on estimated wind speed and assessed damage. It was originally devised by Ted Fujita in 1971 and revised in 2007. The new enhanced scale better takes into account the materials of a structure when assessing damage.

Category	Wind speed (mph)	Damage
EF0	40–72	Little damage to secure buildings
EF1	73–112	Minor damage
EF2	113–157	Roofs torn off
EF3	158–207	Walls collapsing
EF4	208–260	Buildings blown down
EF5	>261	Buildings blown away

Charles F. Richter devised the scale that bears his name in 1934. It measures the amplitude of the largest seismic wave. Values are recorded from one to ten in a logarithmic progression. Each number increase on the scale represents 31.6 times more energy released.

Magnitude	TNT equivalent (metric tons)	Damage
1.0	0.00048	Not felt
2.0	0.015	Not felt
3.0	0.48	Sometimes felt, no damage
4.0	15	Indoor items may shake, little damage
5.0	475	Slight damage
6.0	15,023	Damage in more densely populated areas
7.0	475,063	Serious damage
8.0	15,022,833	Major earthquake with severe damage
9.0	475,063,712	Major earthquake with devastating damage
10.0	15,000,000,000	Widespread destructive damage, never recorded

from the destruction relatively quickly, rebuilding within a few years. However, the threat of earthquakes remains as the city sits on the San Andreas Fault, an unstable plate boundary. In 2006, to coincide with the 100-year anniversary, researchers simulated the effects of a similar-strength quake. At worst the scenario predicted 3,500 people killed, 130,000 buildings damaged and 700,000 people made homeless. The 1906 earthquake affected a city of 400,000; today there are more than seven million people in the entire Bay area. The scary truth is that the issue is not *if* the big one happens but *when*?[19]

Rebuilding a city after an earthquake reveals social and political cleavages. Diane Davis documents the rebuilding of Mexico City after the earthquake struck on September 19, 1985.[20] The quake reached 8.1 on the Richter scale. Another, almost comparable, quake hit the next day. The official death toll, probably a significant underestimate, listed 5,000 killed, 14,000 injured and one million residents temporarily made homeless. Hundreds of thousands were made permanently homeless. The greatest damage was in the city center, the administrative heart of the city. The jobs of 150,000 public service employees were permanently relocated to other parts of the city. Davis' research of the response reveals a mixed story. While there was rebuilding and reconstruction, with repairs to more than 240 public buildings, many of the private buildings, particularly in the private renting sector, remained unrepaired for years, notably those catering to middle- and lower-income groups. While the government wanted to reassure investors and maintain its legitimacy, citizen groups wanted to recover their city and restore dignity and accountability. The effects of the earthquake revealed violations in building standards and evidence of obvious government corruption. The reconstruction of the buildings went hand in hand with a citizens' attempt to rebuild the city as a place of more democratic discourse. Environmental disasters revealed cracks in the political system as well as faults in buildings. In Mexico City the earthquake de-legitimized the government, led to a citizens' grassroots movement and the election of a new, more democratic mayor. As Davis notes, "The reverberations of the earthquake, in short, were deep and long lasting, and they extended far beyond the built environment to the social and political life of the city."[21]

Responses to earthquakes are not always a route to greater democratization, they can also reinforce the power of central authorities. On the morning of July 28, 1976 an earthquake that registered 7.8 on the Richter scale hit the Chinese city of Tangshan. Three-quarters of the industrial buildings were destroyed and almost all the residential buildings. The official death toll is listed at 24,000 but may be twice as high. The low-rise industrial city was almost totally obliterated. Within a decade the city was rebuilt. The 1976 Comprehensive Plan imagined a modern city more resistant to earthquakes. Under the very strong and centralized control of the Chinese Communist Party a more planned city was reconstructed that had increased green space and greater land-use controls. The reconstruction of the city was "an arena for the display of political authority in the communist regime."[22]

Just before five in the afternoon on January 24, 2010 an earthquake measuring 7.0 on the Richter scale hit Haiti's capital city of Port au Prince. The epicenter was located just 16 miles west of the city. The effect was devastating.

BOX 8.4

Megacities and earthquakes

"Megacities are something new on the planet. Earthquakes are something very old. The two are a lethal combination as seen in the recent tragedy in Port-au-Prince.... The next Big One could strike Tokyo, Istanbul, Tehran, Mexico City, New Delhi, Kathmandu or the two metropolises near California's San Andreas Fault, Los Angeles and San Francisco. Or it could devastate Dhaka, Jakarta, Karachi, Manila, Cairo, Osaka, Lima or Bogotá. The list goes on and on.... The next Big One could be on the isthmus of Panama, where Panama City sits just six miles from a major fault that hasn't ruptured in four centuries.... Or the next catastrophe could be in Caracas, Venezuela, where millions of people live in poverty near a boundary of two tectonic plates, including the one that created the fault that broke in Haiti.... Another seismic bull's-eye is Mexico City, which sits on the worst possible soil, a drained lake bed that will intensify seismic waves. The city also is in a basin in the mountains, which essentially traps the seismic waves. The devastating earthquake of 1985, which killed about 10,000 people, was centered hundreds of miles away but managed to ring Mexico City like a bell.... Although New York City is rarely thought of as earthquake country, the region experiences many small tremors that indicate that larger ones are possible. The good news is that a magnitude-6

> earthquake should happen only every 670 years or so.... The bad news is that there is a massive amount of infrastructure built without earthquakes in mind.... Urbanization is a steady process. In the next half-century, the planet will add about 5 billion people and build about 1 billion housing units.... The question is whether those people will live in buildings designed for a sometimes shaky world."
>
> Achenbach, J. (2010, February 23) "The deadly plates under the world's megacities." *The Washington Post*, A1, A9.

More than 230,000 are estimated to have died, many injured and over a million were made homeless. Government was rendered ineffective as many public employees were killed and vital services, including power, water supply and sewage disposal systems were destroyed. Port au Prince is the primate city of the country, housing 3.7 million people, almost half the country's population. Because the earthquake destroyed much of the city it wreaked havoc on the entire nation. Connections with the outside world were also cut. Telephone lines were severed, the seaport was closed for an entire week and the airport was out of commission until a new control center was established. The earthquake was so devastating because of the size of the quake—it destroyed a quarter of a million residences and 30,000 commercial buildings—but also because of the poverty of the country. With the capital rendered inoperative for all intents and purposes there were few internal resources that the country could call upon. International relief took some time to get up and running and the primary emphasis was on the emergency relief of temporarily housing and feeding people. A year later there were still close to one million people living in tents. Much of the aid was poorly coordinated. Many aid officials with vital local connections and crucial local knowledge were killed. Inexperienced aid agencies with little local knowledge took some time to get a handle on the depth, complexity and severity of the problems. Two years after the earthquake, half a million refugees remain homeless. Less than 50 percent of the promised aid has been disbursed. The problems are exacerbated by the underlying poverty, the large number of aid agencies with different agendas and the unwillingness of many of these agencies to trust the Haitian government because of its long record of corruption and cronyism. The US government awarded only 23 out of 1,490 contracts to Haitian companies. While this may have avoided local corruption it meant the local skills and knowledge were also rarely tapped. The case of Port au Prince highlights the problem of a major urban disaster in a poor society with a dysfunctional government, allied to the problems of coordinating diverse relief agencies faced with huge problems. The recovery is limited, with estimates suggesting it will be ten years to return to something close to normal in the capital city. The rebuilding of the city is painfully slow.[23]

The city and disaster: a case study of Hurricane Katrina

We can perhaps bring out the more subtle connections in an urban environmental hazard by considering one case study in some more detail.

On Thursday, August 25, 2005 a tropical storm reached the coast of Florida. As winds reached over 73 mph it was upgraded to hurricane status. Hurricane

Figure 8.2 Hurricane Ivan, 2004. After wreaking havoc in the Caribbean, especially on the island of Grenada, Hurricane Ivan moves toward the mainland US. See also Figure 8.3

Source: http//images.google.com/imgres?imgurl=http://www.nnvl.noaa.gov/hurseas2004/ ivan1945zB-040907-1kg12.jpg&imgrefurl=http:www.nnvl.noaa.gov/cgi-bin/index.cgi%3Fpa ge%3Dproducts%26category%3DYear%25202004%2520Storm%2520Events%26event%3D Hurricane%2520Ivan&h=1199& (public domain)

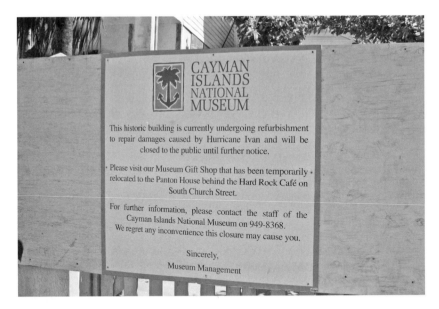

Figure 8.3 The effects on the ground of Hurricane Ivan
Source: Photo by John Rennie Short

Katrina, as it was now officially designated, quickly passed over the southernmost tip of Florida, then moved into the Gulf of Mexico where it strengthened as it soaked up the warm gulf waters. At 11 a.m. on Sunday, August 28, wind speeds exceeded 170 mph. Katrina was now a full-fledged category 5 hurricane, capable of inflicting ferocious damage and enormous harm.

Hurricanes are not unusual events. Geological evidence from lakebed sediments suggest that they have been a regular occurrence for at least 3,000 years and probably much longer. They are an integral part of the environment of the Caribbean, appearing in ancient Mayan hieroglyphics. The ancient Mayans built most of their settlements away from the coast; they knew about hurricanes and organized their settlements accordingly. Cities were built away from the dangers of the strandline. Nature was accommodated in human affairs, a due recognition of the power and intensity of storms and natural calamities. In contrast to the Mayas, today we are building more of our settlements along the coast. Coastal properties and beach locations are prime sites, attracting development and growth as populations and investments congregate close to the shoreline. Rather than conforming to natural systems we take a more arrogant view than the Mayas. Nature is to be subdued, controlled and managed. And in large measure we have been extraordinary successful. We have transformed mangrove stands into beach resorts, swamps into cities and strandlines into suburbia. Coastal development in hurricane zones is just one more manifestation of a deeper and wider reorientation of society and nature. We do not accommodate nature but taunt it by building along earthquake fault lines, placing developments on top of unstable mountains, and constructing cities in the middle of hurricane zones. We ignore our increased knowledge of predictable environmental hazards, accepting short-term benefits and discounting longer-term possible costs. This is part of a deeper and wider sense that the environment is not something that we accommodate, but something we subdue. In part it is a rational position in the rich West. In the past 200 years an environmental revolution has been wrought through science and technology. We have overcome many of the constraints imposed by the environment. Electricity turns night into day, air-conditioning transforms humid, hot air into cool, dry air, heating warms cold air. We have more power to transcend the limits imposed by the world we live in. We live in a human-modified environment. And as more people live in cities and get their food from the grocery store and their weather information from the television, the sense of the environment as a limiting factor gradually disappears. More of people's time and environmental experience, especially in the rich world, is in the socially constructed environment of the home, the office and the shopping mall, where climate is controlled and temperatures are fixed. From this perspective, weather patterns become an external event seen through the car window or on the television, and even hurricanes become an occasional inconvenience, not a determinant of settlement patterns. Where the Mayas listened to the voice of the environment, we ignore it.

The path of Hurricane Katrina was accurately predicted. As early as Friday, 26 August, people knew it would make landfall close to the city of New Orleans in southern Louisiana. Hurricanes are not unusual events in New Orleans. Since 1887 at least 34 hurricanes and 25 tropical storms have passed within 100 miles of the city. Since 1960 17 hurricanes have passed within 100 miles. New Orleans had been spared destruction. By the time Hurricane Isidore reached close to New Orleans in September 2003, wind speed had fallen to 63 mph and it had been downgraded to tropical storm status. Hurricane George in September 1998 reached wind speeds of 110 mph as it passed over Puerto Rico, but by the time it had crossed the Gulf and moved toward New Orleans wind speed had declined to 57 mph. For the people of New Orleans hurricanes were no big deal. And here we encounter a paradox. Major hurricanes are extreme events that will inevitably take place sometime. The longer the period in which they do not occur, the more complacent people become, and the more likely it is that they will occur. Imagine an event that will take place, on average, once every 100 years. You know it will occur within this time but you do not know when. As time passes and the event does not happen the more likely you are to forget about it or ignore it. But the passage of time that creates complacency that it will not occur also makes the event more likely to occur. We believe tomorrow is going to be much like today and yesterday.

Hurricanes cause three main forms of damage: rain, wind and flood. Hurricanes, especially the large, slow-moving ones, can produce downpours of up to 25 inches in 30 hours. The high winds tear buildings and other structures away from their moorings. Everything not securely tied down (and even that which is) becomes subject to the vagaries of wind and even secure structures experience wind damage. The high winds also generate waves. Storm surges associated with the high winds can reach over 20 feet above normal sea level, causing massive flooding. For most coastal cities flooding is a serious problem; for a city below sea level it is potentially catastrophic. Much of New Orleans is below sea level.

Hurricane Katrina made landfall just east of the city at 6:10 a.m. on Monday, August 29. Winds had dipped in strength to 145 mph. It was not the ferocious winds that damaged the city but the storm surge that breached levees in the city. The city was flooded when parts of levees at Seventeenth Street and Industrial Canal collapsed. Almost 80 percent of the city was flooded, in some cases by water over 20 feet deep. An estimated 1,000 people were killed, most of them drowned by the rapidly rising floodwaters. Much of the city was destroyed in the flooding that followed the hurricane.

When is a disaster a natural disaster? At first blush Katrina seemed like a natural disaster. A hurricane is a force of nature. But it was a force of nature whose impacts and effects were mediated through the prism of socioeconomic power structures and arrangements. The flooding of the city was caused by the poorly designed levees that could not withstand a predictable storm surge. It was not Katrina that caused the flooding but shoddy engineering, poor design and inadequate funding of vital public works. Storms surges are neither unknown nor unpredictable in New Orleans. Yet the levees were poorly constructed, with pilings set in unstable soils. They should have been built to withstand 15-foot surges but had settled in many places to only 12–13 feet above sea level.

The storm surge itself was particularly severe because of the loss of wetland. For the previous 20 years wetland was lost in the Gulf Coast at a rate of 24 square miles year. The wetlands have a deadening effect on storms as they absorb much of the energy and water the storm brings. It is estimated that wetlands reduce a storm surge one foot for every three miles of marsh that a surge passes over. The loss of wetlands around New Orleans, something predictable and knowable, added to the potency of the storm surge. The engineered landscape in and around the city exacerbated the storm surge, channeling more water into the city. The Mississippi River Gulf Outlet (MRGO) was built in 1965 to shorten ship travel from the city to the Gulf. Although rarely used it has an enduring ecological effect. It allowed the penetration of salt water into wetlands that destroyed 25,000 acres of marsh, and funneled the storm surge directly into the city. Like the disappearance of the wetlands, the environmental impact of the MRGO was predictable and known.

The mayor of the city issued a voluntary evacuation on Saturday, August 27 and a mandatory evacuation the next day. Those with cars were able to get out, but there was little provision for the most vulnerable; those with no access to private transport were abandoned. While the more affluent could leave, the very poorest, the most disabled, the elderly and infirm were trapped. Between 50,000 and 100,000 were left in the city when the hurricane struck and the levees failed. Some made their way to the Superdome and the Convention Center, which by Wednesday, August 31 housed between 30,000 and 50,000 people. They remained for days, a stunning indictment of social and racial inequality in its starkest and bleakest forms.

The effects of Hurricane Katrina on the city were socially and racially determined. Flooding disproportionately affected the poorest neighborhoods of the city. The more affluent, predominantly white sections of the city, such as the French Quarter and the Garden District, were at a higher elevation and escaped flood damage. The flooded areas were 80 percent non-white. The hardest-hit neighborhoods were non-white and most of the high-poverty tracts were flooded. The racial and income disparities in the city were cruelly reflected in the pattern of flood damage. In that sense the "natural" disaster appears in closer detail as a social disaster. As a Congressional bipartisan report noted, "Katrina was a national failure, an abdication of the most solemn obligation to provide for the common welfare."[24]

By 2012, seven years after Katrina, the damage of the hurricane lives on in abandoned houses, their scattered owners, empty storefronts and ongoing debates about how to rebuild the city. While most of the rest of the country has long moved on to other headlines, the rebuilding of New Orleans is far from over. The good news: almost 150,000 residents have returned. Tourism, an important part of the economy, is close to what it was before the storm. The central tourist area, the French Quarter, escaped flooding. By 2011 the

Figure 8.4 Several years after Hurricane Katrina, vast areas of the Lower Ninth Ward are still not recovered and remain vacant

Source: Google Maps

population was 343,829, compared to 484,674 in 2000. The population of New Orleans is still only two-thirds of what it was before Katrina. Block after block in the Lower Ninth Ward remains empty, overgrown weeds and empty concrete slabs stand in place of the homes that once were there (Figure 8.4). Because of the weakness and slowness of public action to deal with the plight of the poorest, non-profit organizations are also part of redevelopment. The Make It Right NOLA Foundation, established by actor Brad Pitt, is dedicated to rebuilding affordable green homes for working families. By 2012, the Foundation had rebuilt more than 50 eco-friendly homes, allowing more than 200 people to return to this Ward.

Social hazards and vulnerability

The emphasis on floods, earthquakes and hurricanes should not blind us to the fact that the city is also vulnerable to slower chronic hazards such as climate change. This is such an important topic that we devote an entire chapter to the issue. Here, we will restrict our comments to the hazards posed by oil dependency, terrorist threat and infrastructure failure.

Contemporary cities are very much carbon-based, dependent on vast supplies of coal and oil to keep them lit, heated, cooled and their transport moving. However, we are in an era defined as peak oil.

While demand seems infinite the supply is fixed. Estimating oil reserves is difficult. There is always the possibility of discovering reserves as yet unknown. Yet the long-term prospect does not look good. The rate of extraction is now double that of new discoveries. The estimates of when oil supply will peak vary from 2010 to 2015.

What are the urban–environmental consequences of oil supply peaking sometime in the future? In the US, for example, much of its suburbanization is based on cheap and stable oil prices. All those suburban, family homes, out-of-town shopping centers and long-distance commutes arose at a time of steadily cheap gas prices. But as we move into rising and unstable oil prices, this suburbanization, with its heavy reliance on private automobiles, now looks like a landscape created on the basis of cheap oil prices that will never return. The reliance of a built form precariously balanced on one fossil fuel with large and fluctuating costs raises issues of long-term sustainability. The long-term sustainability of low-density, energy-profligate, heavy ecological footprinted suburbs is now a matter of serious consideration.

Jago Dodson and Neil Sipe examine the vulnerability of Australian cities to oil price changes. One significant feature of many Australian cities is that some of the most affluent areas are in the inner city, while many of the moderate-income households are forced to live in outer suburbia. Over 80 percent of trips in Australian cities are made by private car, making citizens very vulnerable to changes in oil prices. Dodson and Sipe create a fascinating VAMPIRE (Vulnerability Assessment for Mortgage, Petrol and Inflation Risks and Expenditure) index that calculates the level of household vulnerability to this mix of costs in Adelaide, Brisbane, Gold Coast, Melbourne, Perth and Sydney (petrol is the term used for what in the US is referred to as oil and gas). They mapped the index and identified four different urban areas. The minimal- and low-vulnerability areas were high-income inner-city neighborhoods. Inner-ring suburbs were moderately vulnerable while the outer suburbs were most vulnerable. The vast outer tracts of low- to moderate-income suburbia were classified as most vulnerable. One of their principal conclusions was the need for public transport to offset suburban oil vulnerability.[25]

Two other social hazards include the terrorist threat and the fragile nature of urban infrastructure. Especially in the wake of the televised spectacularity of 9/11, the threat of a bomb or terrorist attack on a select range of cities is now considered one of the most pressing urban problems. It has led to renewed reordering and surveillance of urban space. The issue of policing urban space is not recent. It emerged with full force in the nineteenth century. However, this is a particularly telling moment in the history of public space. The fear of terrorism provokes substantial changes in security, surveillance, fortress architecture, bunker mentality and spatial policing. The fortress metaphor describes a landscape that is demarcated by physical borders such as gates and walls and by often invisible surveillance devices such as closed-circuit television that watch city streets, parks and gated communities. The events of September 11, 2001 and the subsequent global war on terrorism altered the articulation of security and fortifications of both private and public space in cities around the world. Many accounts present bleak portrayals of future urbanism as the threat of invoking "national security" may trump issues of public access and public space.

Terrorists target cities to attract global media publicity. Since September 11, 2001, however, it is clear that symbolic targets—such as monuments, memorials, landmark buildings and other important public spaces—are increasingly at risk. The responses have been highly intense, visible counter-terrorist measures. Commentators have discussed the costs and benefits of adopting counter-terrorism measures in the face of real or perceived terrorist threats. Until the early 1990s, many cities had no comprehensive security and defensive strategy; attempts to design out terrorism occurred specific target by specific target, and often after an event raised the issue or vulnerability, or a direct threat was made. Fear of terrorism has led to new and dramatic urban counter-responses. Some writers describe this as a new military urbanism.[26]

There is a fragility to urban modernity. As cities rely on ever more complex infrastructure, the possibility of infrastructural failure becomes more possible while the consequences become ever more dramatic. For those of us who live in cities with regular power supplies, the sudden loss of power makes us feel particularly vulnerable as the heating or cooling fails to works, the lights do not turn on and power appliances are frustratingly quiet and inert. It is only then that we realize that as we become more reliant on available power we are acutely vulnerable to power failure. The power grid, for example, is vulnerable to attack and collapse. In 2003 there was a system crash in the power grid across much of the northeastern US. More than 50 million people were in the dark for 36 hours. On closer inspection the collapse was due to the combined defects of deregulation, which led to less investment and thus less shock absorption. This case study also highlights the cascading effects as a local system collapse broaders to a wider regional system failure. The interconnected nature of systems such as transport networks, the internet and banking allows a small failure to cascade through increasingly interconnected systems. The eruption of the Eyjafjallajökull volcano in Iceland in 2010, for example, set off a chain of events that halted or disrupted global mobility flows of people and goods. Airplanes were grounded across much of Europe and the transatlantic for six days. The eruption of a volcano in a small island nation had global ramifications that highlighted the vulnerability of the global aeromobility system that links cities across the world. Cities are interconnected with circuits of flows that can easily be broken or fail. As the infrastructure becomes more complex and more globally interconnected, the possibility of failures cascading through wider systems of interconnection will increase. Cities with brittle or highly connected infrastructure systems are vulnerable to attack, collapse and failure. We should also note that the unreliability and inconsistency of urban infrastructure is a common experience in cities of the developing world, where access to power and other services may be temporary rather than permanent. Again there is unequal access to power and services overlying other divisions of wealth and political power.

BOX 8.5

Fukushima: regulatory capture turns a natural hazard into a disaster

On March 11, 2011 an earthquake on the ocean floor 230 miles northeast of Tokyo, Japan, registered 9.0 on the Richter scale. The resulting tsunami waves lashed over the coastline, eradicating towns and villages. More than 20,000 people were killed and the total damage is estimated at $300 billion.

But this natural hazard also turned into a nuclear disaster. On the day of the earthquake, at the Fukushima Daiichi nuclear power plant, 150 miles north of Tokyo on the Pacific Coast, three of the six power reactors were already closed for routine maintenance. The other three shutdown automatically in response to the first seismic waves that hit at 2:46 p.m. Forty-one minutes later the first tsunami wave hit the area but did not reach over the 33-foot seawall.

Less than ten minutes later a second giant wave 77 feet high and travelling at around 500 mph easily breached the seawall defenses, crashed into the nuclear plant and destroyed the power and the emergency diesel generators that were used to cool the nuclear fuel rods. The rods produce tremendous heat to generate steam to drive turbines that generate electricity. The nuclear reactors were now generating vast amounts of heat with no sources of power to cool them down. That night a decision was made to release contaminated steam in order to avoid a giant explosion. Subsequent explosions also resulted in radiation leakage that spread across 700 square miles and forced the evacuation of 100,000 people.

Japan has long relied on nuclear power plants to provide electricity. Over the years the regulatory authorities had been captured by the nuclear industry, which resulted in optimistic estimates of risk, and so failure to properly regulate. The term regulatory capture is used to define the process that is witnessed around the world as the agencies meant to regulate businesses end up aiding and abetting them or looking the other way in the face of violations. It was an element behind the 2008 fiscal meltdown as regulators failed to monitor, assess and prohibit the riskiness of bank practices; it also played a role in the 2011 Gulf oil spill. In the immediate aftermath of the fallout there was a more critical engagement with the policy of nuclear power reliance in Japan. Germany decided to phase out nuclear power by 2022. Austria, Italy and Switzerland are reconsidering their reliance on nuclear energy.

While the owner of the plant, Tokyo Electric Power Co. (TEPCO) played up the story of the giant unforeseeable tsunami as the culprit, the 2012 official report produced by Japan's Parliament, the National Diet, came to a very different conclusion. The report noted that the accident was the result of collusion between the government, the regulators and TEPCO. The regulators and operators ignored safety risks and disregarded earlier warnings. In 2009 two senior seismologists warned that the plant was very vulnerable to tsunamis.

The report also criticizes the emergency response of TEPCO and the government. Lines of authority were unclear, management chaotic and there was great confusion regarding the timing of evacuation.

The radioactivity released from the Daiichi nuclear power plant will last for many years and Japan's optimistic reliance on nuclear power is now permanently shattered. It was a reliance that ignored Japan's history, giving proof positive that those who ignore history are doomed to repeat it. Evan Osnos reported that along the coast, survivors of the giant waves, "rediscovered gnarled stone tablets, some of them hundreds of years old, which had been left by ancient ancestors at precise points on the shore to indicate the high-water marks of previous tsunamis. The inscriptions implored future generations never to build close to the water again. 'No matter how many years may pass,' read one, 'do not forget this warning.'"

Sources: National Diet of Japan (2012) *Report of Fukushima Nuclear Accident Independent Investigation Committee*.http://naiic.go.jp/en (accessed July 16, 2012). Osnos, E. (2011, October 17) "The fallout." *The New Yorker*, 46–61.

The resilient city

Cities are resilient. Cities have both experienced environmental disasters and transcended them. The term 'resilient city' was coined by urban planners Lawrence Vale and Thomas Campanella. They write of the ability of cities to survive disasters and employ an evolutionary model of recovery consisting of four stages:[27]

1 *Emergency responses*: may last from days to weeks and involves rescue; normal activities cease.
2 *Restoration*: 2–20 weeks; involves the reestablishment of major urban services and return of refugees.
3 *Replacement and reconstruction*: 10–200 weeks; the city is returned to pre-disaster levels.
4 *Development reconstruction*: 100–500 weeks; commemoration and betterment.

Within this general picture, the experience of particular cities may vary with differing equity outcomes. Whose city is destroyed and whose city is commemorated and developed are important questions in assessing the rebuilding of disaster-hit cities. In an interesting study, architect Alireza Fallahi looked at the rebuilding of the city of Bam, Iran after the 2003 earthquake that killed 32,000 people and leveled the old city. Almost 20,000 homes were destroyed. Temporary shelters were often provided inside the plots of land, allowing households to rebuild while also being rehoused. Fallahi makes the point that being homeless does not necessarily mean building less, and describes a process of "experts" learning from local people how to rebuild the city. Rebuilding can range from the top down of Tangshan to the more participatory of Bam.[28]

Vale and Campanella suggest a number of axioms of resilience:

- narratives of resilience are a political necessity;
- disasters reveal the resilience of governments;
- narratives of resilience are always contested;
- local resilience is linked to national resilience;
- resilience is underwritten by outsiders;
- urban rebuilding symbolizes human resilience;
- remembrance derives resilience;
- resilience benefits from the inertia of prior investment;
- resilience exploits the power of place;
- resilience casts opportunism as opportunity;
- resilience, like disaster, is site-specific;
- resilience entails more than rebuilding.

Some cities recover very quickly from disaster. In 1995 an earthquake reduced much of the Japanese port city of Kobe to rubble. Over 64,000 people lost their lives, 300,000 were made homeless and there was more than $1 billion worth of damage. But within a year the port had returned to its position as the sixth largest trading port in the world and within 15 months manufacturing output was back to 98 percent of pre-earthquake levels.[29] Mark Skidmore and Hideki Toy found that after disasters urban economies can upgrade infrastructure and technology, allowing then to quickly reach pre-disaster levels of economic activity, as they now have new and improved assets.[30] The effect is less pronounced in poorer cities in the developing world. Haiti remained in rubble for 15 months after the 2010 earthquake, with an economy still devastated. Yet some developing countries can do well. Consider the case of Indonesia. A registered 9.0 earthquake in the Indian Ocean in 2005 created a tsunami of epic proportions. Although the effects of the tsunami were felt throughout the basin of the Indian Ocean, the islands of Aceh and Naas were hit disproportionately hard. An estimated 170,000 people perished on Aceh alone. Another 500,000 people were displaced when an estimated 110,000 homes were destroyed. School buildings were destroyed (2,000), hospitals were damaged (8), roads were made impassable (3,000 km) and a staggering amount of waste was created (5,765,000 cubic meters). A 2007 report from the United Nations Environment Program concluded that the Indonesian Reconstruction and Rehabilitation Agency did not place enough emphasis on environmental issues and the reconstruction efforts that were taking place were regularly damaging the already delicate ecosystems. Structures were being rebuilt, but with unsustainable materials. Housing was available but there was a lack of attention toward the necessary water systems and sanitation facilities. The report suggested environmental monitoring must be incorporated into the restoration process.[31] Since 2007 specific attempts have been made to incorporate rehabilitation of coastal ecosystems and encourage national monitoring of the environmental improvements.

About $13.5 billion in foreign aid poured into the country from 2005 to 2009, and the government used these funds to employ a unique reconstruction and rehabilitation program centered on a system of community-based development activities. The Indonesian government provided up to $3,000 to every household to rebuild their homes. This assistance was vital in jump-starting the reconstruction effort, and ultimately laid the foundation for a successful recovery. The Asian Development Bank published a report in 2009 outlining the state of progress five years after the tsunami.[32] The results were remarkable. By April 2009, close to 140,000 permanent houses were built, over 950 schools were renovated and/or rebuilt, over 3,000 km of roads were built, over 730 health facilities were rebuilt or created, and close to 100 km of coastland had been restored. The report generally evaluated the five-year rehabilitation project as a success due to the combined efforts of individual Indonesian citizens, the national government and international financial institutions.

The issue of resiliency also raises the question of the role of government. Public policies tend to highlight rescue and repair rather than prevention. The World Bank, for example, suggests that providing disaster response kits in the existing infrastructure, such as schools and public buildings, is one cost-effective prevention strategy. Public policies, unless carefully monitored, can also encourage risky behaviors. In the richer countries government-backed insurance policies for those in flood or disaster zones can perversely mean public underwriting of individual risky behaviors. In the

US such policies have resulted in taxpayer bailouts of affluent households building beachhouses in hurricane zones. Resilience in this case is public underwriting of risky private behaviors. Blythe McLennan and John Handmer note, for example, that the Royal Commission in Australia that examined the Black Saturday bushfires placed responsibility on government rather than on self-reliance of at-risk communities. Government was made responsible rather than the people whose housing choices placed them at risk.[33]

We are also learning that urban resiliency is improved when biophysical systems are employed as risk reducers rather than destroyed as barriers to economic progress. The destruction of the wetlands increased the Katrina storm surge that overwhelmed the levees in New Orleans. In the Netherlands rivers are now allowed to meander across their flood plain rather than being severely channelized. Regular small-scale flooding across the river floodplain is considered a more acceptable risk than the severe flood if the river breaks the banks. The policy can be roughly translated from the Dutch as "better damp feet than a wet head." There is now a growing recognition that the physical world can be used in the protection of the social world.

Conclusion

Cities are susceptible to hazards and are sites of disaster, often of dramatic proportions as well as the slow chronic vulnerabilities of oil dependency and global climate change. They are subject to droughts and floods, heat waves and biting cold, earthquakes, volcanic eruptions as well as terrorist attacks and infrastructure failure. The past is littered with examples of cities that were demolished, from Pompeii to Lisbon to Galveston. Everyone will remember those images from the earthquake in Port au Prince or the tsunami in Southeast Asia as waves full of debris and mud swept people up and away in a deadly torrent. The images were a cruel reminder of the unforeseen forces that can wreak havoc on our cities and our lives. Disasters are social experiences; they affect poor cities more than rich cities and poor residents more than rich residents. Hazards become disasters though the filter of economic inequality and social injustice.

But cities are also resilient. Resilient cities express the power of hope and opportunity in the face of disaster. Cities have risen, phoenix-like from the ashes, a testament to the pulsing life-force that is found in cities. The most human of inventions, cities also express the most human of emotions—hope in the face of adversity.

Guide to further reading

Arnold, M. (eds.) (2006) *Natural Disaster Hotspots: Case Studies*. Washington, DC: The World Bank.

Bankoff, G., Lubken, U. and Sand, J. (2012) *Flammable Cities: Urban Conflagration and the Making of the Modern World*. Madison, WI: University of Wisconsin Press.

Blakely, E. J. (2011) *My Storm*: *Managing the Recovery of New Orleans in the Wake of Katrina*. Philadelphia, PA: University of Pennsylvania Press.

Centre for Research on the Epidemiology of Disasters (CRED): http://www.cred.be.

Graham, S. (ed.) (2010) *Disrupted Cities: When Infrastructure Fails*. New York: Routledge.

Graham, S. (2011) *Cities Under Siege: The New Military Urbanism*. London: Verso.

Hartman, C. and Squires, G. (eds.) (2006) *There is No Such Thing as a Natural Disaster*. New York: Routledge.

Heinrichs, D. and Krellenberg, K. (2012) *Risk Habitat Megacity*. Heidelberg: Springer.

McQuaid, J. and Schleifstein, M. (2006) *Path of Destruction: The Devastation of New Orleans and the Coming of Age of Superstorms*. New York: Little, Brown and Company.

Oxfam (2012) *Haiti: The Slow Road to Reconstruction*. http://www.oxfamamerica. org/press/publications/haiti-the-slow-road-to-reconstruction.

Pelling, M. (2003) *The Vulnerability of Cities: Natural Disasters and Social Resilience*. London: Earthscan.

Pelling, M. and Wisner, B. (eds.) (2009) *Disaster Risk Reduction: Cases from Urban Africa*. London: Earthscan.

Redclift, M. R., Manuel-Navarrete, D. and Pelling, M. (2011) *Climate Change and Human Security: The Challenge to Local Governance under Rapid Coastal Urbanization*. Northampton, MA: Edward Elgar Publishing.

Rozario, K. (2007) *The Culture of Calamity: Disaster and the Making of Modern America*. Chicago, IL: University of Chicago Press.

Smith, K. and Petley, D. N. (2008) *Environmental Hazards: Assessing Risk and Reducing Disaster*. 5th edition. New York: Routledge.

University of Delaware's Disaster Research Center: http://www.udel.edu/DRC.

Vale, L. J. and Campanella, T. J. (eds.) (2005) *The Resilient City: How Modern Cities Recover from Disasters*. New York: Oxford University Press.

Wisner, B. and Pelling, M. (eds.) (2009) *Disaster Risk Reduction: Cases From Urban Africa*. London: Earthscan.

Wisner, B. and Uitto, J. (2009) "Life on the edge: urban social vulnerability and decentralized, citizen-based disaster risk reduction in four large cities of the Pacific Rim," in *Facing Global Environmental Change*, H.-G. Brauch (ed.). Berlin: Springer Verlag, pp. 217–34.

9 Urban political ecology

In recent years an area of study and concern has emerged from the traditionally separate areas of "urban" and "ecology." Previously, most "urban"
studies tended to concentrate on the social and ignore the ecology of the
city, although the Chicago School of the 1920s did employ biological terms
such as invasion and succession in their understanding of the city, and this
strain of social ecology did remain. However, the terms became metaphors,
more analogies than fully theorized concepts. Most of traditional "ecology"
in turn tended to focus on the less urbanized areas. The science of ecology
developed out of a primary concern to understand "natural" processes. The
emphasis was on ecosystems with minimal human connections. The early
models were developed with primary reference to pristine ecosystems.
Between 1995 and 2000, of the 6,157 papers in the nine leading ecological
journals, only 25 (0.2%) dealt with cities. A new urban ecology has emerged
that considers cities as sites of biophysical processes interwoven with
social processes in complex webs of human–environmental relations. We
will use the term urban ecology to refer to the ecology of cities rather
than simply the study of ecology in cities. Specific journals such as
International Journal of Urban Sustainable Development, *Urban Ecosystems*
and *Urban Ecology* publish a wide variety of papers in this new and exciting
field. Researchers in urban ecology are developing both general models and
specific case studies. From the rapidly growing and rich body of work, here
are just three examples. Mary Cadenasso and her colleagues, for example,
theorize the Baltimore metro area as a complex of biophysical, social and
built components. They apply standard ecological approaches such as ecosystem, watershed and patch dynamics in order to answer three general
questions: What is the overall structure of the urban ecosystem? What are
the fluxes of energy, matter, population and capital in the system? What
is the nature of the feedback between ecological information and environmental quality?[1] In a more specific vein, Milan Shrestha and colleagues

show how rapid urbanization affects spatio-temporal patterns of land fragmentation in the Phoenix metropolitan area.[2] Nikos Georgi and K. Zafiriadis look at the impact of trees in parks on urban microclimates. They show, with reference to Thessalonica in Greece, that trees reduce summer temperatures, increase relative humidity and have a general cooling effect.[3]

There is also the development of an urban political ecology that grows out of a long tradition of an accounting of the human impacts on the environment. The classic and foundational work, the 1955 *Man's Role in Changing the Face of the Earth*, deals with the rising dominance of nature by humans and looks at such issues as deforestation, urbanization and the depletion of resources. This work influenced generations of geographers to examine and measure the major impacts of human action on the environment, including destruction of habitats, land-use change, loss of biodiversity, pollution of water and soil and, more recently, global warming and climate change. A number of scholars then sought to link more explicitly these environmental changes to political processes and economic systems in an approach that was termed political ecology. The term was first used in the 1930s but came to greater prominence since the 1980s. Piers Blaikie, for example, looked at soil erosion in developing countries. In his work on Nepal, he showed that rather than the result of overpopulation or mismanagement, soil erosion resulted from marginalization and poverty. Soil erosion was not a function of peasant farmers' mismanagement, but a reflection of their precarious position in commodity markets.[4] Political ecology situates land-use changes in their socio-economic context. Many of the early political ecology studies focused on developing societies and rural land-use changes; more recent work widens the approach to look at the city. Urban political ecology draws upon a wide theoretical discourse that includes post-structuralism, actor–network theory and feminism.[5] Yaffa Truelove, for example, discusses water inequality in Delhi, using a feminist urban political ecology approach.[6] She concludes that women are predominately impacted and spend many hours of their day obtaining water. Urban political ecology is now a vigorous body of work that starts from the premise that the urban environment is not a neutral backdrop but the very field in which power is revealed, contested and enforced.

In the rest of this chapter we will look at some of the main features of urban political ecology by considering circulation and flows, new urban ecological imaginaries, urban footprints, natural capital, biophysical cycles, biotic communities and ecological models of city regions.

Cities as circulation of flows

In a classic paper Abel Wolman introduced the notion of the metabolism of cities. He pictured the city as an entity with circulations of energy, water, material and wastes into and out of a city region of one million.[7]

Flows of energy power a city. We can picture the city as a place where the different energy sources – human muscle power, electricity, nuclear and wind – provide the basic energy for heat, light, power and transport. Different cities use different energy sources across time (the heavier reliance on human power in the ancient cities) and space (locally sensitive sources such as the windmills of Holland). Over the long term in urban history there has been a major shift from human muscle power to a reliance on fossil fuels as a source of energy. Contemporary cities are now an integral part of the carbon civilization; the heavy reliance on fossil fuels and the resultant increase in carbon gases into the atmosphere impact global climate change. We will consider the relationship between climate change and cities in more detail in a later chapter. Figure 9.1 is a photograph of the Ravenswood Power plant in Queens, New York, which opened in 1965 and now produces over 2,480 megawatts of electricity, primarily from natural gas-fueled steam turbines. The plant provides just over 20 percent of New York City's electricity requirements. The plant also emits significant levels of pollutants. While it is considered, according to the regulatory authority of New York State, to be in

Figure 9.1 Ravenswood power plant in Long Island City in Queens, New York

Source: Author Harald Kliems: http://commons.wikimedia.org/wiki/File:Ravenswood_power_plant.jpg

compliance for emissions of nitrogen oxides, sulfur dioxide and carbon monoxide, the result of large investment in emissions-reducing technology, the plant was classified in 2012 as 'severe non-attainment' for the criteria pollutant of ozone.

Urban economies have waxed and waned as energy sources rise and fall. From the coal-field towns of early industrial Britain to the electricity-demanding and oil-thirsty cities of the contemporary world, cities have been shaped by the mix of energy sources they use and, in turn, their growing and incessant demand for power has shaped the search for energy sources.

Energy use varies by the location and wealth of cities. Car-dominated cities in very cold or very hot climates of developed countries, for example, will tend to use a lot of energy. If a city's transport system is dominated by private cars the energy requirement and resultant environmental pollution can be more pronounced. Cities in rich countries use more energy than cities in poor countries.

Focusing on cities as circuits of energy brings into sharp relief the issue of the efficiency of cities. We can picture a more ecologically informed urban study developing around critical energy audits of different types of cities and a more urban-informed ecology developing new models of fuel efficiency in different types of urban environments. Numerous questions are prompted by such an approach: For example, does public transport use less energy than private transport? Are certain building forms and spatial arrangements of the built form more conducive to energy efficiency than others? When walking or bicycling replaces auto transport—albeit only possible for short distances—are there other benefits in addition to reducing energy inputs? Do people who walk more tend on average to be healthier and fitter and thus require less healthcare expenditures?

In an update of the original Wolman paper, Christopher Kennedy and colleagues looked at the changing metabolism of cities across the world, including Brussels, Tokyo, Hong Kong, Sydney, Toronto, Vienna, London and Cape Town.[8] They focused on flows of water, materials, energy and nutrients. The data for cities are not completely comparable since they were collected in slightly different ways at different times; they provide more of a general trend rather than a precise measure. The results suggest that water and wastewater flows increased from the 1970s to 1990s, but may be declining since then due to the reduction in industrial consumption. Material flows are declining but new waste streams are emerging. There is a rising tide of electronic waste (e-waste). E-waste is now a significant element of material flows in the city and between cities. There was little evidence of nutrient recycling while energy consumption increased. The researchers also identified metabolic processes that threaten the sustainability of cities; these included exhaustion of groundwater, accumulation of toxic materials, summer heat islands and irregular accumulation of nutrients.

An urban political ecology does more than just account for the changing level of flows into and out of the city. The redistributional consequences of these flows are a primary consideration. Water flows, to consider just one example, are the subject of numerous studies. Matthew Gandy contextualizes the water supply problems of Mumbai, India as the result of colonialism, rapid urban growth and the dominance of middle-class interests over the needs of the majority poor.[9] Erik Swyngedouw focuses on the city of Guayaquil in Ecuador. He shows that flows of water are deeply bound up with power and influence.[10] Antonio Ioris explores the politicized nature of

BOX 9.1

Long-term studies of urban ecosystems

The US National Science Foundation funds long-term ecological research projects at 24 sites in a Long-Term Ecological Research Network. Only two of them specifically study the ecology of urban systems. The two sites, one in Baltimore, Maryland, the other in Phoenix, Arizona, provide information on urban ecological processes and trends. The program brings together researchers from the biological, physical and social sciences. The Baltimore Ecosystem Study (BES) aims to understand metropolitan Baltimore as an ecological system. The Central Arizona–Phoenix Long-Term Ecological Research focuses on a metropolitan region in an arid-land ecosystem, examines the effects of urban development on the Sonoran Desert and the impact of ecological conditions on urban development.

Visit these websites for updates on research projects: Baltimore Ecosystem Study (http://www.beslter.org); Central Arizona–Phoenix Long-Term Ecological Research (http://caplter.asu.edu).

water scarcity in Lima, Peru.[11] It was not just that there was not enough water, but that the supply problem was caused by discrimination against low-income households. Investment guided by the electoral objectives of the ruling political party and the technocratic nature of water management bureaucracies systematically excluded many local communities. Water scarcity was not a problem of supply, but a political process of exclusion. The privatization of Lima's water supply rather than providing the promised solutions led to more problems of debt financing, unequal access and neglect of environmental quality.[12]

The idea of circulation in the city is widened when using an urban political ecology approach. Jason Cooke and Robert Lewis, for example, look at the construction of Michigan Avenue Bridge in Chicago from 1909 to 1930.[13] They develop the idea of the capitalist economy as a metabolic system for circulating capital. The power of the business elite shaped the new circulation patterns in the city. The construction of the bridge allowed quick and easy access across the Chicago River. The bridge freed up the circulation of traffic in the city as well as the circulation of capital and resulted in increased land values along Michigan Avenue. The bridge was a vehicle for increasing traffic and capital flows in the city.

There are also literal flows in the city as energy, water and traffic flow through infrastructural networks. Figure 9.2, for example, shows the nighttime traffic flows through the city of Seoul. A number of scholars have brought attention to the role of these networked urban infrastructures. A recently edited collection by Stephen Graham highlights their vulnerability.[14] A chapter by Simon Marvin and Will Medd, for example, bring to light the problem of city sewer systems clogging up with discarded fats, oils and grease. They follow the trails of discarded fat from homes, restaurants and fast-food franchises through the sewer system, bringing new meaning to issues of metabolism and flows in the city.[15] In the US three billion gallons of fat, oil and grease are now dumped into the sewer system each and every year.

Jochen Monstadt draws connections between networked infrastructures and urban ecologies.[16] He suggests a fruitful interaction between urban political ecology and technology studies. Networked flows of energy, water, wastewater and transport are both an important part of urban ecologies, as well as sustainers and transformers of urban ecologies.

Imagining the city as flows of food highlights the way cities interact with the humanized agricultural landscapes. Tracing the daily consumption of calories of urban residents from the purchase of food in retail establishments

Figure 9.2 Nighttime traffic in Seoul, South Korea
Source: Photo by John Rennie Short

all the way back to primary producers focuses attention on the web of connections and relations that span across time and space and link changing
agricultural practices with the changing food demands of city dwellers.
These food networks are mediated by complex factors of economy, culture and
society. Following the interconnecting physical and social nexus that links
even the simplest act such as purchasing a morning coffee allows us to see
how urban consumers are linked in complex relations with distributors and
producers. The modeling of these food pathways and their changing configura-
tion, size and character is a novel way to understand people–environment

relations in linkages that connect ecological processes with business cycles and changing patterns of food demand and supply. The supply of food also embodies social inequalities. Researchers have used the environmental term "desert" to refer to areas of the city with limited access to nutritional food. Neil Wrigley's work on "food deserts" in British cities shows that these deserts reflect social exclusions and are an important element in health inequalities.[17]

Following the flows of food waste in a city can also reveal interesting connections. Julian Yates and Jutta Gutberlet, for example, examine the flow of food waste in the Brazilian city of Diadema on the fringe of the São Paulo metropolitan region. Half of the city's population lives below the poverty level, one-quarter live in slums and many are hungry. Close to 60 percent of waste in the region consists of biodegradable food waste. In 2004 the city gave official support to the informal recyclers (*catadores*) associations and to the communal garden scheme, community food banks and people's restaurants. The positive result was a new circulation as biodegradable food waste was diverted from the waste dump to community gardens as compost to produce fruit and vegetables. A new socio-ecological space was created; an alternate urban metabolism was constructed. There were limits to the transformation. The programs did not tackle the production of urban water, or the relations of power that marginalized the *catadores*. However, local people created a new urban ecological imaginary. As the authors note,

> While *catadores* and urban gardeners struggle with formal political negotiations, they are actively reconceiving urban environments through their everyday practices of reclaiming and recirculating the value inherent in food waste, they are contributing to a more even distribution of environmental amenities and resulting benefits.[18]

New urban ecological imaginaries

The *catadores* of Diadema reimagined new flows of food waste and in the process created a new urban political ecology. Reimagining new urban political ecologies is bound up with relations of power and dominant imaginaries, although as the Diadema example reminds us, change can also be initiated from the bottom.

The dominant imaginary has changed. Gene Desfor and Lucian Vesalon describe the production of an industrial form of what they term socio-nature, in this case the Port Industrial district of Toronto in the early twentieth century.[19] The massive infrastructure created a central waterfront area as an industrial landscape. It was a typical product of its time, with a business elite controlling the political regime in a period of rapid industrialization with little thought given to anything but profit and growth. At roughly the same time on the Pacific coast of Canada, another city was renegotiating its relationship to nature in its creation of an industrial socio-nature. In 1855 the Canadian Pacific Railway bought land on the inlet known as False Creek. Railyards and workshops were built and the shallow part of the creek was filled in with rubble and industrial waste to create even more flat spaces for the railway–industrial complex. In 1986, with industry long gone from a heavily polluted site, it was purchased by the government and became the site of the Expo 1986 World Fair. A property company purchased the 66-hectare site and constructed parks and housing. Remediation is ongoing. Some land was given back to the city as public space and the Olympic Village for the 2010 Winter Olympics was built in the southeast part of the city. A brown site became a green shoreline.[20] One hundred years on there are new imaginaries at work in the production of the socio-nature.

Sometimes urban ecological reimaginations are prompted by profound economic change. Deindustrialization, for example, involves the loss of manufacturing jobs and, for cities more reliant on an industrial base, the possibility of economic decline and urban shrinkage. These are not insubstantial problems as the loss of employment opportunities for blue-collar workers in particular can be devastating for household living standards. There are, however, some opportunities from deindustrialization. There is the possibility of the remediation of blighted industrial landscapes into greener areas. In Leipzig, Germany, for example, a former coal-mining area was turned into a theme park. It opened in 2005 and is now a recreational area of 17 interlinked lakes. The old mining area was reimagined and remade as a place of recreation and pleasure rather than pollution and work. While high-paying jobs were lost, these were dangerous jobs that impacted the lungs of coal miners as well as the environmental quality of the surrounding areas.

Two other examples from New York City: first, Gantry State Park in Long Island City. From the mid-nineteenth century to the 1960s railways cars were shipped across the Hudson River and pulled from the boats by giant cranes on gantries. When the railways became obsolete the site lay abandoned and derelict. In 1998 a ten-acre site was turned into a state park, planted with native grasses and plants and opened up as a public walkway with green spaces and play space. Expanded to 12 acres in 2009, the park provides a green space open and accessible to the public. The river is no longer just an impediment to economic transactions; it is now a source for a more livable city as a former industrial site is reimagined as a public park (Figure 9.3).

Figure 9.3 Gantry State Park, New York
Source: Photo by John Rennie Short

The second NYC example is the Highline. In 1934 an elevated rail line opened in Manhattan, New York to deliver goods and material to factories and warehouses along its 13-mile route through Chelsea and the meat-packing district. When the warehouses and factories closed, the rail line was no longer viable. The last delivery was made in 1980. The narrow railway lay empty. In 1999 the Regional Plan Association, commissioned by the railway line owners, CSX Railroad, suggested turning the line into a pedestrian promenade. In 2009 the first section of the linear pedestrian park was opened. Native plant species were planted along its route, growing up beside the old

post industrial landscape

rail tracks. It is wildly popular among locals and visitors. The conversion of an abandoned railroad track into an urban public space for walking is one of the more successful transformations in the shift from industrial to postindustrial (Figure 9.4).[21] In the summer of 2009 construction began on the Beltline in Atlanta, a new park built on 17 acres of vacant industrial land. The $50 million revitalization project creates a postindustrial landscape of parks, lakes and jogging paths on an old industrial site.

There is also reimagining of the socio-natures of the very recent past. Motorways were once thought as fitting symbols of modernity. They conjured

Figure 9.4 The Highline in New York City
Source: Photo by John Rennie Short

up notions of infinite mobility, endless accessibility and the bright promise of a motorized future. More recently they are perceived as blights on the landscape. In the postmodern urban ecology they are being hidden in new urban forms. In Madrid a new park *Madrid Rio* runs along the six-mile line of a former motorway now sunk below ground. On top of the buried highway is the new linear park with green spaces, play places, a greener more accessible place of newly planted trees in a neighborhood's long-denied access to parks and green spaces. In Milwaukee the Park East freeway spur was destroyed to make way for parks as well as new residential areas. Motorways age and decisions have to be made about replacement and renewal. One possibility is for new spaces to be created and opened up by burying the motorways, transforming them into boulevards, part of reimagined urban ecologies that are more green, more equitable and more sustainable than the failed promise of a mid-twentieth century urban modernity. The process is not restricted to the richest countries. There are plans to demolish the Big Worm, a 2.2-mile elevated highway that carries 80,000 vehicles each day through São Paulo.[22]

Much of the urban political ecology literature looks at how power is imposed more than how it is resisted. Yet reimagining urban political ecologies can be progressive and redistributional, not only changing the ecology and creating new socio-natures, but also tweaking relations of power and transforming the very meaning of nature in the city and the nature of cities.

Urban footprints

Cities exert an influence on the environment around them. Recent years have seen the development of various footprints. Three important footprints are now employed: ecological, carbon and water.

The ecological footprint

Ecologists have developed the notion of the ecological footprint to refer to the total area of productive land required to support an ecosystem. The ecological footprint measures how much land and water area a human population requires to produce the resources it consumes and to absorb its wastes.[23] It is measured in global hectares (gha) per capita. The ecological footprint of an urban region includes all the land necessary to support the resource demands and waste products of a city. The global average is around 2.6 gha. The footprint of London was measured at 4.54, slightly lower than the UK average of 4.64. More people in London use public transport than almost any other city in the UK, reducing the relative size of the footprint. The total ecological footprint for London is 34 million gha, 200 times the size of the city. Figure 9.5 shows the variation by boroughs in the city. The highest footprint is in the more affluent boroughs of Richmond, Kensington and Chelsea. The poorer districts, such as Tower Hamlets, Newham and Hackney, have much lower footprints because residents have smaller homes and are less likely to have private automobiles.

Footprints also have spatial ramifications. In Las Vegas, the explosive growth of the city increased the demand for water. The city relies on the Colorado River for much of its water, but the amount is fixed. The city is planning to build a pipeline 250 miles to the north to tap into underground water. But this is the same water source that ranchers and farmers use. Environmental activists argue that this is creating "a sacrifice zone of thousands of square miles."[24]

Measuring the ecological footprint of a city allows us to highlight these spillover effects and further fine-tune concepts such as ecological overshoot and changing carrying capacity and give a better metric to the idea of

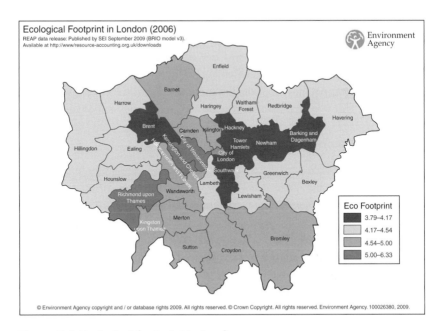

Figure 9.5 Ecological footprint in London

Source: http://www.environment-agency.gov.uk/static/images/Research/Figure-1_sustainability.jpg

long-term sustainability. Developing the idea of a city's ecological footprint will necessitate a more nuanced ecological accountancy than the simple models used for either individual households or for the total global population. There are collective goods and services in cities that can combine with individual patterns of consumption to both maximize and minimize the collective footprint. For example, a city that is more reliant on private autos than public transport has greater energy needs and thus has a larger footprint. In cities, individual and national footprints combine and interact to produce distinctly urban regional effects.

Cities have vast footprints that encompass not only the immediate surrounding area but extend across the nation and the globe. Tracing the footprint of an individual city through time is a fascinating way to connect urban history with ecology.

BOX 9.2

Reducing the carbon footprint: a tale of three cities

Cities around the world are developing policies, devising strategies and implementing ideas to reduce their carbon footprint through reducing greenhouse gas (GHG) emissions. Three examples offer a wide range of policy initiatives.

São Paulo's broad strategy

In 2009 the city of São Paulo set an ambitious target of reducing 2005 levels of GHG emissions by 30 percent by 2012. The plan aims to ensure that all public transport vehicles run with clean and renewable technologies. The city has installed biogas power plants at municipal landfills, improved energy efficiency in public buildings, installed street lights with LED technology, established Bus Rapid Transit corridors and developed a network of bicycle lanes. The establishment of a comprehensive vehicle inspection program resulted in a reduction of over 41,000 tons of carbon monoxide. Since 2009, all new buildings with more than three bathrooms need to use solar power for at least 40 percent of the energy used in water heating.

Philadelphia and energy efficiency

Philadelphia's EnergyWorks is a low-interest revolving loan program that provides finance for improving energy efficiency to meet the goal of reducing city-wide energy consumption by 10 percent. EnergyWorks provides loans of up to $15,000 for energy-efficiency improvements such as installing windows and doors, upgrading heating and cooling systems and installing insulation.

London's low-carbon zones

In 2007 the first London Climate Change Action Plan adopted the ambitious target of reducing 1990 CO_2 emissions by 60 percent by 2025. Ten low-carbon zones (LCZs) were designated. The aim of the LCZs is to help residents and businesses cut their carbon emissions, reduce waste and save energy. "Green Doctors" and "Community Draughtbusters" help residents install energy-efficiency measures. The Brixton LCZ, for example, was launched in March 2010. The area contains around 3,500 properties in a predominantly low- to moderate-income neighborhood.

Source: Bulkeley, H. (2012) *Climate Change and the City*. London and New York: Routledge.

The carbon footprint

A city's carbon footprint is the total set of greenhouse gases that it produces. These include carbon dioxide (CO_2), methane, hydrofluorocarbons, perfluorocompounds and sulfur hexafluoride. The basic unit is kilograms or tons of CO_2. The global average is 1.19 metric tons per person. Benjamin Sovacool and Marilyn Brown measured the carbon footprint of 12 cities: Beijing, Jakarta, London, Los Angeles, Manila, Mexico City, New Delhi, New York, São Paulo,

Seoul, Singapore and Tokyo.[25] They identified four sources: emissions from personal and mass transportation; energy used in buildings and industry; emissions from agriculture; and emissions from waste. From the four possible emissions they found: direct emissions from the metropolitan areas; emissions produced in the metro area but consumed elsewhere; emissions from sources outside the metro area; and emissions that pass through such as freight traffic. They restricted their analysis to the first two sources. Table 9.1 lists the footprints. Four of the cities have footprints smaller than the global average of 1.19, in part a function of income, and the richer cities have a smaller footprint than the national average. What also emerges from these data is the substantial variation between groups of cities with similar income levels. The optimistic public policy message is that the carbon footprint can be reduced with appropriate policies. There were also certain urban forms that reduce the carbon footprint; these included more compact urban growth, more sustainable transportation, more mass transit, and greater use of cleaner, more sustainable energy supplies.

Marilyn Brown and colleagues measured the carbon footprint of the largest 100 metro areas in the US.[26] They measured the residential and transport carbon emissions in each of the metro areas using standardized data. Table 9.2 shows the five metro areas with the largest and smallest carbon footprints. The authors conclude that the largest metro areas have lower carbon

Table 9.1 Carbon footprints

Metro area	Carbon footprint (metric tons per capita)	National carbon footprint (metric tons per capita)
Delhi	0.70	0.27
Manila	1.14	0.48
São Paulo	1.15	2.44
Beijing	1.18	1.00
London	1.19	2.07
Jakarta	1.38	3.28
Seoul	1.59	2.71
Tokyo	1.63	2.59
Mexico City	1.85	1.21
New York	1.94	5.37
Singapore	2.73	2.73
Los Angeles	3.68	5.37

Source: after Sovocool, B. K. and Brown, M. A. (2010) "Twelve metropolitan carbon footprints: a preliminary comparative global assessment." *Energy Policy* 38: 4856–69.

Table 9.2 US metros with the largest and smallest carbon footprints

Metro area	Carbon footprint (metric tons per capita)
Lexington, KY	3.45
Indianapolis, IN	3.36
Cincinnati, OH-KY-IN	3.28
Toledo, OH	3.24
Louisville, KY-IN	3.23
Boise, ID	1.50
New York, NY-NJ	1.49
Portland, OR-WA	1.44
Los Angeles, CA	1.41
Honolulu, HI	1.35

Source: after Brown, M. A., Southworth, F. and Sarzynski, A. (2009) "The geography of metropolitan carbon footprints." *Policy and Society* 27: 285–304.

footprints than non-metro areas, the metro areas with the largest carbon footprint are in the southeast and Midwest, while metros on the west coast and older cities of the northeast are lower, and higher-density metro areas have the smallest footprint. The geographical differences in part refer to densities and the fuel required for heating in the winter and air-conditioning in the summer. The startling conclusion is that average metro per capita footprint is smaller than the average for non-metro US locations. The location, in terms of weather and hence heating and cooling requirements, and the density of cities have important impacts on a city's carbon footprint. Higher-density cities have lower carbon footprints.

The findings also have important implications for policies. Part of a carbon reduction program would be to increase densities, foster more public transport and pursue carbon mitigation programs such as greater energy efficiency to reduce demand for fossil fuels and to promote greater use of non-renewable energy.

The water footprint

The water footprint (WF) is the amount of fresh water used in the consumption of goods and services. It takes almost 500 gallons of water to make a pair of jeans, a figure that includes the water necessary to grow, process and dye the cotton. We use up prodigious amounts of fresh water as we become more affluent and demand more goods. The WF can be measured for a variety of things. A product's WF can be calculated. A two-liter bottle of soda, for example, requires 132 gallons of water. Corporations can also measure their total WF. The drinks company SABMiller estimate that they use 41 gallons of water for every liter of beer they make. Calculating a city's footprint is, like the ecological and carbon footprint, a tricky measure, using imprecise data collected in different ways at various times. The WF of a city is a measure of all the fresh water used to produce all the goods and services consumed in the city. It breaks down into three components: blue, green and gray. The blue WF is the volume of fresh water consumed from surface water and groundwater. The green WF is the volume of rainwater consumed. The gray WF is the volume of fresh water used to dilute the pollutants created by the production of all goods and services for the city.

In the US the average per capita indoor water use is 64 gallons (242 liters) per day. Water conservation measures such as using less water for flushing and making people more aware of water use do have an impact. One study showed that without water conservation measures, total water use in 12 North American cities was 145.5 gallons (551 liters) per day; with water conservation measures, including switching from lawn to native plants, the per capita usage for a single-family home fell to 38 gallons (144 liters) per day.[27] Shengnan Zhao and colleagues calculated the annual per capita WF for Lijiang City, China at 215,300 gallons (815,000 liters). They also calculated a water stress index. Their conclusion was that the city's water efficiency was very low and unsustainable over the longer term.[28]

BOX 9.3

Rethinking New York City's footprint

"Manhattan, not suburbia, is the real friend of the environment. Those alleged nature lovers who live on multi-acre estates surrounded by trees and lawn consume vast amounts of space and energy. If the environmental footprint of the average suburban home is a size 15 hiking boot, the environmental footprint of a New York apartment is a stiletto-heeled Jimmy Choo. Eight million New Yorkers use only 301 square miles, which comes to less than one-fortieth of an acre a person. Even supposedly green Portland, Oregon is using up more than six times as much land a person than New York.

New York's biggest environmental contribution lies in the fact that less than one-third of New Yorkers drive to work. Nationwide, more than seven out of eight commuters drive. More than one-third of all the public transportation commuters in America live in the five boroughs. The absence of cars leads Matthew Kahn, in his fascinating book, 'Green Cities,' to estimate that New York has by a wide margin the least gas usage per capita of all American metropolitan areas. The Department of Energy data confirm that New York State's energy consumption is next to last in the country because of New York City.

… When Manhattan builds up, instead of Las Vegas building out, we are saving gas and protecting land. Every new skyscraper in Manhattan is a strike against global warming. Every new residential high rise means a few less barrels of oil bought from less than friendly nations belonging to the Organization of the Petroleum Exporting Countries.

… The great problem with being reflexively anti-growth is that development in America is close to being a zero-sum game. New homes are going to be built to meet the needs of a growing population. If you stop development in some areas, you are ensuring more development elsewhere. A failure to develop New York means more homes on the exurban edges of America."

Glaeser, E. (2007, January 30) "The greenness of cities." *New York Sun.* http://www.nysun.com/article/47626 (accessed April 6, 2007).

While each of the three footprints has problems in estimation and calibration, they do provide a comparative metric of human impacts. Collectively, they allow benchmarking of demand for renewable resources, the measurement of human contribution to greenhouse gas emissions and the analyses of water usage.[29] More generally, they provide a measure for efficiency of our use of scarce resources, create a yardstick for measuring progress toward greater urban sustainability and open up debates about unequal distribution of resource use. They highlight the important role that cities can play in producing more efficient and equitable resource use.

Natural capital

In recent years ecologists have used the terms ecosystem capital or natural capital.[30] The term is employed to highlight the human use of ecosystems as goods and services that have value. Ecosystem capital is a source of wealth and income generation. Employing the term "capital" in this way highlights the economic use of ecosystems. A 2011 UN study, entitled *Economics of Ecosystems and Biodiversity*, drew attention to the economic benefits of ecosystems.[31] Understanding nature as a set of ecosystem services allows a cost accountancy of economic activities. The report estimated that the cost of the damage to nature in 2008 was between $2 trillion and $4.5 trillion. The wide estimate suggests more of a general indication than a precise measure. Despite the imprecision, the measurement of ecosystem services allows us to estimate, at least in the broadest terms, the costs of using ecosystem services.

Natural capital is not only a fixed asset, such as a rainforest or floodplain; it is also a set of services that provide a stream of benefits. The UN recognizes four types of ecosystem services:

1 *provisioning*: providing timber, wheat, fish, etc;
2 *regulating*: disposing of pollutants, storing carbon;
3 *cultural*: sacred sites, tourism, enjoyment of the countryside;
4 *supporting*: maintaining soils and plants.

Felix Eigenbrod and colleagues measured likely effects of increased urbanization in Britain on ecosystem services, and in particular on three things: agricultural production; carbon storage in soils; and flood mitigation. Increased urbanization reduces agricultural production, and the added construction also disturbs the soil and so removes the stored carbon. Measuring flood mitigation was more problematic. They ran a number of scenarios of different types of urbanization. They found the most effective form of urbanization was high density, as it took up less agricultural land, but allowed enough green spaces to provide benefits to people and mitigate floods.[32]

Viewing nature and socio-nature as capital and a set of ecosystem services allows an estimation of the costs of urbanization and an indication of the more efficient and cost-effective forms of urban living.

The term "capital" involves more than an economic accountancy model of the physical environment. Capital is above all a social relationship. Studies of urban ecologies would be enlivened and propelled by a more critical theoretical view of the physical environment as a social relationship between different groups as mediated through the medium of capital and stream of services. To use the term capital is not simply to provide a metric for the rational use of resources, but also the basis for a critical take on the social use of resources.

Biophysical cycles and social processes

An important theme in traditional ecology is the modeling of biogeochemical cycles. Three important cycles have been recognized: the carbon cycle, phosphorous cycle and nitrogen cycle. These cycles all share a similar characteristic—they are subject to the basic Newtonian principle that matter can neither be created nor destroyed. The cycles are a closed system of fixed stock.

Carbon atoms exist as carbon dioxide in the air and bicarbonate in water. Carbon is taken up by plants through photosynthesis and metabolism. The human intervention in the carbon cycle is most obvious in the burning of fossil fuels. By burning the fossil fuels necessary to power cities we have intervened dramatically in the carbon cycle, adding more CO_2 to the air. Levels of CO_2 are at historic highs and are responsible for the warming of the planet. We can reduce the level of fossil fuel emissions by reducing fossil fuel consumption, using energy more efficiently, developing new forms of energy including nuclear, tide, wind or others.

Cities also play less obvious roles in the carbon cycle. Vegetation in cities acts as important storage sites for carbon. Urban soils, especially in residential areas, sequester large amounts of organic carbon.

The phosphorous cycle is a mineral nutrient cycle. Phosphate in rocks and minerals is taken up by plants then moves through the food cycle. The human-modified phosphorous cycle involves the mining of the mineral and its application as fertilizer and use as detergent. The agriculture that feeds cities around the world is reliant on heavy applications of phosphorus, which leaches into water. Over-fertilized water has resulted in algal blooms that reduce the amount of oxygen in streams, lakes and other water bodies. Cities play an important role in this process, not only because their heavy food demands have prompted greater use of phosphorous, but also the lawns and gardens of towns and cities are in part created by the enhanced application of fertilizers and nutrient supplements that contain phosphorous. The lush green lawns of suburbia are engineered by heavy applications of phosphorous and bound up in what is described as a lawn-chemical economy and its environmental impacts.[33]

Nitrogen is found in the air. Bacteria convert this nitrogen into ammonium in a process known as nitrogen fixing. Legumes are important nitrogen fixers. Plants, which draw on the nitrogen found in mineral form, then pass through the route of herbivores and carnivores to decomposers. Humans also fix nitrogen. The Haber–Bosch process converts nitrogen gas into ammonia. In the combustion of fossil fuels nitrogen is oxidized into nitrogen oxides. The rate of nitrogen fixing has increased dramatically because of the cultivation of legumes such as beans, soybeans and alfalfa, the application of fertilizers rich in nitrogen and the burning of fossil fuels. Excess nitrogen leaches into the soil and is the primary cause of dead zones in rivers and oceans. Nitrogen cycling is altered by urbanization and land-use change. Urban and suburban watersheds have much higher rates of nitrogen loss than forested watersheds.[34]

This very brief view shows how cities play an important role in modifying biogeochemical cycles.[35] Early models of these cycles rarely included human biogeochemical controls such as impervious surface proliferation, engineered aqueous flow paths, landscaping choices and differing levels of human population density, size and growth rates. New models are being developed that try to capture the complex human–environment relationship of biogeochemical cycles in cities. Nancy Grimm and colleagues point to the complex interactive relations between cities and biogeochemical cycles. Urban air pollution, for example, impacts nutrient recycling, and nutrient loads in urban rivers are primary drivers of global biogeochemical cycles. Cities accumulate nitrogen, phosphorous and metals.[36]

There is a complex relationship between biophysical processes and socio-economic behaviors. If we consider individual plots of urban land, for example, it is clear that the knowledge of physical processes influences market choice and policy. To take an obvious example, flood-prone land will be avoided by developers or protected by legislation. Market choices such as the decision to develop certain pieces of land will be determined in part by the policy framework and in part by the opportunities and constraints of the local biophysical processes. In other words there is a complex system of decisions that link ecology and economy, society and policy. Land-use models are being improved by incorporating a more explicit understanding of biophysical processes while ecological models are enhanced by a more explicit consideration of socio-economic processes.

Cities as biotic communities

An important element of traditional ecology is the idea that ecosystems are biotic communities of plants and animals. We can begin to think of urban ecosystems as a distinctive ecological category rather than merely as a disturbance site unfavorably compared to pristine sites. Urban development initially increases local extinction rates and rates of loss of native species. The general tendency is for the replacement of native species by non-native species. However, mature urban environments create a rich range of different ecologies and habitats. Cities contain both open space and built-up sites that vary in vegetation cover and habitat variety. The city is more accurately described as an ecological mosaic than a single category.

The old idea of ecosystems as a set of successions leading to a stable climax is being replaced by a more dynamic perspective that sees change as a constant. Ecological climaxes are always temporary and sometimes short-lived. It is more appropriate to consider ecosystems as ensembles and assemblages of plant and animal species that are constantly adapting to sometimes small-scale change over long periods of time, and at other times to changes that are very large and dramatically quick. The precise form of change depends on a range of environmental factors. Changes in environment can cause shifts in populations of particular species.

Urbanization initially tends to reduce species richness. In suburban areas, however, Michael McKinney's review of the massive literature suggests increasing richness of plant communities, only a moderate decline in invertebrates and a more marked decline for non-avian vertebrates.[37] The impacts extend beyond the city—the decline of migratory birds in the Southern Hemisphere is linked to increased urbanization and associated and land-use change in northern latitudes. However, cities also provide a range of environments for birds, edge species do well and mature city and suburban gardens provide a variety of habitats. Granivorous birds, which eat both seeds and insects, do better than birds that rely just on insects. The city creates new ecological communities as natural predators are removed. The existence of deer in many US suburbs to levels not seen since colonial times is a function of the food-rich environments of gardens and the decline of natural predators. Urban-adapted plants and animals are emerging in this new urban-dominated world.

The urban mosaic provides a variety of ecological niches and offers opportunities and constraints to different species (Figure 9.6). Foxes in Europe and coyotes in the US have shown themselves remarkably adaptable to urban environments. Species that have wide distribution, a large population, a high

Figure 9.6 The mosaic of urban ecosystems: suburban district of Rome, Italy
Source: Photo by John Rennie Short

degree of genetic variation and a large number of offspring will tend to survive and adapt. Cities are bringing about discernible changes in biotic communities that favor some species rather than others. Christina Blewett and John Marzluff, for example, looked at the distribution of cavity-nesting birds in Seattle, Washington. They found that suburban landscapes with highly interspersed land covers had higher densities of black-capped chickadees, nuthatches, flickers and downy woodpeckers, while suburban landscapes with more forest cover had higher densities of creepers, chestnut-backed chickadees, pileated and hairy woodpeckers.[38] In a study of birds in California and Ohio, Robert Blair found species richness and diversity was highest in residential areas compared to preserves. The mosaic of habitats in suburban areas resulted in more species than found in many designated preserves. This study suggests that birds that are urban exploiters successfully reproduce, invade locally and have multiple broods while urban avoiders do not reproduce.[39]

We can identify specific urban biotic communities. Plants in cities, for example, are exposed to more pollutants, warmer temperatures, higher levels of CO_2 and nitrogen deposition than plants in comparable rural areas. Certain tree species can tolerate and survive heavy air pollution. The London plane tree, for example, thrives in the polluted air of London. Similarly, in the northeast US maple and crab apple trees are more tolerant of air pollution than many other species. Jillian Gregg and colleagues grew the same cottonwood clone in urban and rural sites. They found that the urban biomass was double that of rural sites. The principal reason was that higher levels of nitrous oxides in the urban sites suppressed the ozone levels that reduce plant growth.[40]

Cities are engineered landscapes. Urban plant communities, for example, reflect socio-economic and cultural factors, as well as the traditional ecological factors. Cultural preferences are interconnected with the physical factors. In addition to elevation and land-use history, factors such as family income and housing explain plant diversity in Phoenix. Along with biotic factors, human variables such as household preferences and family wealth now structure plant community formation in residential areas.[41]

Urban biotic communities can also be engineered. It is now well recognized that planting trees and shrubs around buildings can reduce summer heat. A study of the Chinese city of Harbin compared the dust removal capacities of 28 tree species. The results showed that *Pica koraiensis* and *Juniperus rigida* were the ideal conifers for dust removal, and *Populus alba x berolinensis*, *Lonicera maackii* and *Prunus maackii* were the ideal deciduous tree species. These tree species had deep channels of dense hair on their leaf surface that were more effective in dust removal. The authors proposed that dust removal capacity could be an important basis in the choice of tree planting in the city.[42]

Reimagining the city as a biotic community will allow us to explore the idea that the city is an important ecological category where plants and animals live and where new ecological niches are being opened and closed as urbanization increases and changes. Urbanization has transformed ecosystems around the world and has encouraged the development of new forms of more urban-tolerant plant, animal and bird life.

We live in a world of urbanized nature, a socio-nature whose production is a nexus of socio-political and ecological processes. Parks, for example, represent a form of socio-nature that engineers landscape to suit various and changing models of nature. From the tight geometries of the Enlightenment gardens to the sweeping curve of the Romantic retreat to the urban parks of the nineteenth century, designed to provide a green lung, to the activity-themed play spaces of today, parks represent not an uncovering of nature but the production and reproduction of ideologies of preferred socio-nature. In their detailed study of vegetable gardens in the Barcelona urban region of Catalonia in Spain, Elena Domene and David Sausi write of class-produced natures.[43] They highlight the vegetable plots of retired workers for whom the plots were places of socialization, ways to support their families and, for many migrants, a link with their rural past. These garden squatters occupy the interstitial areas of the city with vegetable plots that on the one hand embody a form of urban sustainability but on the other hand represent a less politically acceptable form of urban greening than public parks and private gardens in the city. Golf courses are another form of socio-nature that in many cities also contain complex sets of messages about wealth, power and social exclusivity. Even wildlife corridors are less objective facts of nature than socially produced and politically contested socio-natures. In a case study of Birmingham, UK, James Evans shows how designated corridors are part social, part natural hybrids that are also claims to political legitimacy. The ecological assessment was a highly charged political process.[44] Nature was not so much uncovered and revealed as manufactured. Nature is not so much natural as political. Nature is best understood as a socio-natural hybrid.

Ecologies of a model city region

Using the models of ecology is not new in urban studies. In the first third of the twentieth century the Chicago School theorists Robert Park and E. W. Burgess developed models of city structure and social change that used biological terms to describe the city and the processes of social differentiation over space and time. While later critics have accused them of a crude social Darwinism, their biological referencing was more rhetorical and metaphorical than causal. Drawing on their work we can suggest the following simple urban political ecological model of business district, inner older suburbs, newer suburbs and the urban–rural zone of transition.[45]

Capital flows through these zones transforming the urban ecologies. In the case of the US, for example, there is a profound revalorization of the metropolis. From the immediate post-World War II to around 1980 there was massive public and private investment in the suburban fringes. Capital investment was fixed into the suburban landscapes in the form of houses, roads, factories, stores and infrastructure. From 1950 to around the mid-1970s, the primary dynamic of the US metropolis was a suburban shift. But since the 1970s there has been a revalorization into a new metropolitan form marked by complex patterns of growth and decline, expansion and contraction. This dynamic consists of at least four investment/disinvestment waves: reinvestment in selected central cities; disinvestment from certain inner-ring suburban neighborhoods; new rounds of housing investment in affluent neighborhoods; and the decline of former boomburbs as the housing market, especially in the newest suburban areas, has collapsed.

BOX 9.4

Urban gradients

Ecologists often use urban gradients to investigate sample sites. Sampling sites along the gradient allow researchers to estimate the effects of urbanization. Here is a gradient proposed by Marzluff *et al.* It is a very coarse-grained scale and finer-grained divisions are necessary for more locally based studies.

Term	Percentage built up	Building density	Population density
Wildland	0–2	0	<1/ha
Rural/exurban	5–20	<2.5/ha	1–10/ha
Suburban	30–50	2.5–10/ha	>10/ha
Urban	>50	>10/ha	>10/ha

Sources: Marzluff, J. M., Bowman, R. and Donnely, R. (eds.) (2001) *Avian Ecology in an Urbanizing World.* Norwell: Kluwer Academic; see also Boone, C. G., Cook, E., Hall, S. J., Nation, M. L., Grimm, N. B., Raish, C. B., Finch, D. M. and York, A. M. (2012) "A comparative gradient approach as a tool for understanding and managing urban ecosystems." *Urban Ecosystems* 1-13, doi: 10.1007/s11252-012-0240-9.

If we conceptualize the city as consisting of a central business district, an inner ring of inner-city residence tending toward lower density and a suburban frontier where non-urban land is being transformed into urban land uses then we have a simple structure for suggesting important urban ecological zonal processes. Figures 9.7, 9.8 and 9.9 together provide a photographic transect from downtown, through inner suburbs to the suburban fringe. These ecologies need to be connected to the waves of investment and disinvestment as well as the changing political geographies as space, place and nature are continually created and contested as both lived experience and political process.

Figure 9.7 Urban transect 1: downtown Seoul, South Korea
Source: Photo by John Rennie Short

The most significant feature of the central business district is as a site for massive investment and possible disinvestment: a dense collection of buildings and the agglomeration of artificial surfaces and urban land uses. Here, because of the density of impermeable surfaces, flooding is at its most pronounced and the urban heat island is the hottest. Flooding is only contained by elaborate artificial drainage systems that can deal with the abrupt peaking of water discharges. The urban heat island is the distinct warming of urban areas compared to rural areas. In the average US metro region, for example, urban areas are 2–10°C warmer than surrounding rural areas. The heat island is reduced by open space and water bodies and heightened by the greater density of urban buildings. Temperatures increase toward the dense city center not only because trees and vegetation have been cleared, but also because tall buildings and narrow streets restrict air flow and lead to the heating of trapped air. There are in fact two heat islands: a canopy layer at the ground level and a higher boundary-layer heat island. A recent mapping of the canopy surface temperature in the city of Baltimore showed that there was an 8°C difference between local forested areas and the urban areas, with a major heat island clearly identifiable in the city center. In the city center there was an island of cooler temperatures caused by the canyon shading of high-rise buildings. In these shadowed city canyons temperatures were 5–10°C cooler than the rest of the city.[46]

The urban heat island can be lessened by covering the roof of buildings with vegetation to reduce heating while also improving air quality. The city of Chicago has an active policy of encouraging the vegetation of commercial roofs through grants, tax breaks and other financial inducements to developers and builders.

The city is in the process of continual change. The Burgess mode assumed only one zone of transition, an area of housing deterioration. We can identify a variety of zones of transition, including the abandonment of older housing, the deindustrialization of industrial factory zones and areas of new building and gentrification. In each of these zones new urban ecologies are being created. Take the case of the former factory areas. The decline of manufacturing, especially in the older inner-city areas, is creating the brownfields of an industrial legacy with issues of soil contamination and pollution. The trajectory from factories to brownfields to greenfield sites is not only a social process but also an ecological transformation.

Figure 9.8 Urban transect 2: inner city, Queens, New York
Source: Photo by John Rennie Short

Within the residential areas there are a variety of ecological processes. Expanding cities leave in their wake fragments of natural vegetation and also create new ecologies. One of the most obvious is the creation of engineered green spaces of gardens. This can involve the application of fertilizers, creating increased levels of nitrogen and phosphorous in the soil and water run-off. In many cities there is a move for more "natural" gardens, especially in dry and arid climates where the lush green lawn is only possible by constant applications of water and fertilizer. But on the more positive side, city gardens can also protect threatened species. Many city gardeners consciously grow plants that attract threatened bird and insect life. The role of domestic gardens in ecological transformation and the creation of new ecological niches is often overlooked and underappreciated. Suburban areas are complex systems of built form and green space that can contain refuges for a whole range of plant and animal life.[47]

At the edge of the city the dominant ecological change is the creation of the built form across formerly non-urban land uses. A 200-year story of changing land use in the Baltimore Washington corridor is depicted at http://biology. usgs.gov/luhna/hinzman.html. The pace of change picks up dramatically in the latter half of the twentieth century. At http://pubs.usgs.gov/circ/2004/ circ1252/2.html sprawl in the Boston area from 1973 to 1992 is clearly visible as red areas of developed land spread around the coastline and extend inland, from 515 square miles to 764 square miles. More recent changes are recorded by satellite-based land-surface mapping techniques. From 1986 to 2000 in the mid-Atlantic region of the US, the amount of developed land increased from around 600,000 acres to over one million acres. Satellite imagery reveals that almost 20 percent of the total land surface is now developed land and at current rates this will increase to 1.8 million acres or 36 percent of the region's land surface by 2030. Sprawl provides some benefits. It gives middle-income households access to a wide range of safe, affordable homes in tight housing markets. A wide scatter of employment, retail and commercial development spreads economic opportunities throughout the metropolis. But there are costs to sprawl. The first is the reliance on oil-based private transport. Sprawl is a form of development that is too low density to support public transport. The heavy, and in some case total, reliance on private auto transport imposes a heavy environmental price in terms of air pollution, and the increasing dedication of space for roads and parking. The reliance of a built form precariously balanced on one fossil fuel with large and fluctuating costs raises issue of sustainability and affordability.[48] The long-term sustaining benefits of green space are not yet fully understood. The transformation of mixed land use into a suburban landscape can result in the loss of species habitat;

even when green spaces are kept they are often as isolated islands that are unable to function as true niche ecologies.

Urban development consists of the replacement of permeable land with impermeable surfaces. More paved surfaces means less area for water to drain into soils, thus bringing more water into fewer drainage systems, increasing erosive power and sediment load. The replacement of permeable land with impermeable tarmac creates problems from more flooding to increased run-off that collects toxins in the groundwater. A study of

Figure 9.9 Urban transect 3: periurban, Anne Arundel County, Maryland
Source: Photo by John Rennie Short

non-tidal streams in Maryland, US found that when 10–15 percent of areas are paved then increased sediment and chemical pollutants reduces water quality; at 15–20 percent there is markedly reduced oxygen levels in streams; and at 25 percent many organisms die. Many studies decisively document the local impairment of streams with an increase of urban land use.[49]

Even this very brief description of exemplar ecological processes in the different zones of the city reveals the possibility of new models of cities, models that incorporate both biophysical and socio-economic processes. Integrating the ecology into urban models and the urban into ecological models will enrich both areas of traditional study.

Guide to further reading

Alberti, M. (2009) *Advances in Urban Ecology*. New York: Springer.

Douglas, I., Goode, D., Houck, M. and Wang, R. (2011) *The Routledge Handbook of Urban Ecology*, New York: Routledge.

Galli, A., Wiedmann, T., Ercin, E., Knoblauch, D., Ewing, B. and Giljum, S. (2012) "Integrating ecological, carbon and water footprint into a 'Footprint Family' of indicators: definition and role in tracking human pressure on the planet." *Ecological Indicators* 16: 100–12.

Goldman, M. J., Nadasdy, J. P. and Turner, M. D. (eds.) (2011) *Knowing Nature: Conversations at the Intersection of Political Ecology and Science Studies*. Chicago, IL: University of Chicago Press.

Grimm, N. B., Faeth, S. H., Golubiewski, N. E., Redman, C. L., Wu, J., Bai, X. and Briggs, J. M. (2008) "Global change and the ecology of cities." *Science* 319: 756–60.

Kareiva, P. M., Tallis, H., Ricketts, T. H., Daily, G. C. and Polasky, S. (2011) *Natural Capital*. New York: Oxford University Press.

McDonnell, M. J., Hahs, A. K. and Breuste, J. H. (2009) *Ecology of Cities and Towns: A Comparative Approach*. Cambridge: Cambridge University Press.

Niemela, J. (2011) *Urban Ecology: Patterns, Processes, and Applications*. New York: Oxford University Press.

Peet, R., Robbins, P. and Watts, R. (eds.) (2010) *Global Political Ecology*. New York: Routledge.

Richter, M. R. and Weiland, U. (eds.) (2012) *Applied Urban Ecology: A Global Framework*. Oxford: Wiley-Blackwell.

Robbins, P. (2012). *Political Ecology: A Critical Introduction*. 2nd edition. Chichester: Wiley-Blackwell.

Shulenberger, E., Endlicher, W., Alberti, M., Bradley, G., Ryan, C., ZumBrunnen, C., Simon, U. and Marzluff, J. (2008) *Urban Ecology: An International Perspective in the Interaction Between Humans and Nature*. New York: Springer.

Stefanovic, I. and Scharper, S. B. (eds.) (2012) *The Natural City: Re-envisioning the Built Environment*. Toronto: University of Toronto Press.

Wackernagel, M. (ed.) (2009) "Methodological advancements in footprint analysis." *Ecological Economics* 68: 1903–2178.

Wright, R. T. and Boorse, D. (2010) *Environmental Science: Towards a Sustainable Future*. 11th edition. Boston, MA: Addison Wesley.

On urban footprints: the EPA has posted interactive "footprints" for several US cities. You can view them at: http://www.epa.gov/watertrain/smartgrowth/02animation/chspk.htm

Part IV
Urban environmental issues

10 **The environmental revolution: a brief context**

In Chapters 2 and 3 we showed that cities have long confronted environmental issues such as pollution, disease and degradation. They have also responded to these problems with policies, pollution proclamations and other legislative efforts. However, the 1970s saw a profound transformation as concern for the environment became legitimized and institutionalized. Environmental protection and management is no longer on the margins, or occasionally dealt with by municipalities: it is now integrated into mainstream public consciousness and public policy. The subject matter of Chapters 10–16 is best understood within the context of the modern environmental movement. This chapter provides a brief overview of the movement as it applies largely to the US, although we draw on examples from other countries around the world.

The major shift in regulatory context is a result of a new environmental awareness. There were farseeing advocates and isolated pieces of legislation prior to this shift, but after 1970 environmental protection and pollution control became mainstream, and connected to national and international issues. Importantly, the catalyst for the environmental movement came as a result of environmental deterioration in cities.

A movement galvanizes

In 1948, in Donora, Pennsylvania, close to 40 people died from an episode of air pollution inversion. Dubbed the "smog tragedy," it focused attention on the growing problem of air pollution in urban areas. A similar smog disaster occurred in New York City in 1966 when 80 people died. In the 1950s, numerous beaches were closed due to water contamination, and by the early 1960s the Animas River in New Mexico reported radioactive content in the water, while the Passaic River in New Jersey was so polluted that thousands of dead fish washed up on the shores. Cape Cod beaches were fouled by oil spills, as was the York River in Virginia.

The year 1969 appeared to be the "year of disasters." Lake Erie was declared "dead"—so polluted it was devoid of fish and aquatic life. In Washington, DC, the Potomac River was clogged with blue-green algal blooms that were both a nuisance and a public health threat. In the nation's capital, the rivers were little more than open sewers. Off the shores of Santa Barbara in California a large oil spill contaminated miles of shoreline, killing sea otters, birds and other marine animals while television crews filmed frantic volunteers trying to wash the oil off the dying animals and birds. On June 22, railroad sparks set fire to the Cuyahoga River in Cleveland. The river, saturated with oil, kerosene, debris and other flammable chemicals was engulfed in a five-story-high blaze of flames. The fire burned for five days. The Cuyahoga became the symbol of urban water pollution and the need for the federal government to become involved in clean-up and regulation. Jonathan H. Adler, Associate Professor of Law at Case Western Reserve University remarks on the legacy of the Cuyahoga:

> The Cuyahoga fire was a powerful symbol of a planet in disrepair and an ever-deepening environmental crisis, and it remains so to this day. That a river could become so polluted to ignite proved the need for federal environmental regulation. Following on the heels of several best-selling books warning of ecological apocalypse and other high-profile events such as the Santa Barbara oil spill, the 1969 Cuyahoga fire spurred efforts to enact sweeping federal environmental legislation. The burning river mobilized the nation and became a rallying point for the passage of the Clean Water act.[1]

The "Mistake by the Lake," as the city of Cleveland was called, and the series of fires became a poster child for the environmental movement. The memorable 1969 fire sparked legislation for the Clean Water Act of 1972.

These and other episodes around the US served as a rallying point for the emergence of the modern environmental movement, and the passage of the 1969 National Environmental Policy Act. This watershed legislation established the Environmental Protection Agency (EPA) and charged the agency to study pollution and recommend new policies. The creation of the EPA was just one of many new progressive laws that decade that included the Civil Rights Act, Medicare, Federal Fair Housing, OSHA, Title 9 and Medicaid.

The years 1969–73 saw the onset of dramatic legislative and regulatory changes in the US, a profound moment when rapidly rising public concern articulated the need for alternative attitudes, policy and practice. The government responded quickly to this intense political and social pressure, passing dozens of environmental policies, and creating new institutions such as the EPA, which merged environmental responsibilities previously scattered among dozens of offices and programs. New departments, agencies and boards for protecting the environment proliferated at the federal, state and city level. A flurry of congressional activity produced a flood of legislation. Within a few years of establishing the EPA (1969), Congress passed the Clean Air Act (1970) and the Clean Water Act (1972) and Endangered Species Act (1973). These three pieces of legislation form the backbone of the modern environmental regulation. Table 10.1 shows a history of key environmental legislation. The 1970s marked an important shift in the role of government as a consensus developed around the proposition that environmental problems are public problems that cannot be solved only through private action and an unregulated market.

Environmental policy and regulation is also influenced by numerous environmental organizations that emerged and are now involved in formal politics.

Table 10.1 Timeline of environmental revolution in the US

Date	Event
1969	National Environmental Protection Act (NEPA), creates EPA
1970	Millions participate in first Earth Day
1970	Clean Air Act
1972	Clean Water Act
1973	Endangered Species Act
1974	Clean Water Drinking Act
1976	Resource Conservation and Recovery Act (RCRA)
1976	Toxic Substances Control Act (TSCA)
1977	Clean Water Act Amended
1980	Comprehensive Environmental Response, Compensation and Liability Act (CERCLA). Created the Superfund program
1986	Safe Water Drinking Act Amendments
1986	Superfund Amendments and Reauthorization Act (SARA)
1987	Water Quality Act Amended
1989	Montreal Protocol on ozone-depleting chemicals enters into force
1990	Clean Air Act Amended; set new automobile emissions standards, low-sulfur gas
1990	Oil Pollution Act of 1990
1994	Executive Order 12898 defined Environmental Justice
1996	Safe Water Drinking Act Amended
1997	Kyoto Protocol Signed (but not ratified)

BOX 10.1

Public right to know

One of the most important evolutions in environmental regulation is freedom of information acts, or community right-to-know laws. In open and transparent democracies, these laws allow the public to access information about the types of pollutants and who has permits to pollute in their communities. "Right-to-know" laws take two forms: community right-to-know (environmental) and workplace right-to-know (health and safety). Each grants certain rights.

The catalyst for mandating right-to-know laws in the US actually occurred in the developing world. In 1984 in Bhopal, India, leakage from the Union Carbide pesticide manufacturing plant created a lethal cloud of methyl isocyanate gas (MIC) which eventually killed approximately 4,000 people and injured more than 120,000. The only other place in the world that Union Carbide manufactured MIC was at its plant in the Kanawha Valley of West Virginia. On August 11, 1985, just a few months after the completion of a safety improvement program at this plant, 500 gallons of highly toxic MIC leaked from the plant. Although no one was killed, 134 people living around the plant were treated at local hospitals. These incidents underscored demands by industrial workers and communities in several states for information on hazardous materials. Public interest and environmental organizations around the country accelerated demands for information on toxic chemicals being released "beyond the fence line"—outside of the facility. These events helped the passing of the Emergency Planning and Community Right-to-Know Act (EPCRA) in 1986. The act mandated the US EPA to require that factories and other business develop plans to prevent accidental releases of highly toxic chemicals.

One of the EPCRA's primary purposes is to inform citizens of toxic chemical releases in their areas. EPCRA Section 313 requires the EPA and all states to collect data annually on releases and transfers of certain toxic chemicals from industrial facilities, and make the data available to the public.

The law was passed to increase the public's knowledge and access to information on chemicals at individual facilities, their uses and releases into the environment. States and communities, working with facilities, can use the information to improve chemical safety and protect public health and the environment. Below is a list of some of the information that the EPA provides the public:

- Emergency Planning and Community Right-to-Know Act (EPCRA);
- information on toxic substances and releases;
- community environmental issues;
- Food Quality Protection Act (FQPA) of 1996;
- air pollution;
- water quality;
- lead program;
- hazardous waste.

The most widely used is the Toxics Release Inventory (TRI). The TRI is a database containing data on disposal or other releases of over 650 toxic chemicals from thousands of US facilities and information about how these facilities manage those chemicals through recycling, energy recovery and treatment. Since the creation of right-to-know laws, there are more ways of accessing the information that corporations with excess pollutants used to withhold. The annual publication of the Toxic 100 is a list that includes the 100 "worst" industrial air polluters in the US, ranked by the quantity of pollution they produce and the toxicity of the pollutants.

Many other countries have committed to the public right-to-know. In 2005 the UK passed the Environmental Information Regulations (EIRs) along with the Freedom of Information Act. These two laws clarify and extend previous rights to environmental information. According to the UK's Environmental Agency:

> We have always recognised the vital role that access to information plays in helping us achieve our goals. Such access is essential to the credibility of our regulatory

functions. As we rely on your power and influence to help us achieve sustained environmental improvements, we will ensure that you have up-to-date environmental information available. The public has a right of access to environmental information held by public authorities and some other organisations.

Canada and Australia also have environmental right-to-know laws. In the EU, the Aarhus International Convention of 1998 established three crucial rights of the public: to access environmental information held by public authorities on request, to participate in environmental decision making and to access justice through the courts. In 2000 the EU created a European Pollutant Emission Register, the first European-wide register of industrial emissions into air and water, and since 2006 has extended it to include more emitting facilities, require more substances reported, wider coverage and public participation.

The public right-to-know laws provide the public and the community of NGOs an important role in environmental protection. Groups and individuals can signal infractions, collect information, run projects and educate citizens about their rights and obligations. Right-to-know laws are critical tools in environmental regulation and management.

Source: UK Environment Agency (2012) "Your right to know." http://www.environment-agency.gov.uk/aboutus/35684.aspx (accessed June 2012); US EPA (2012) "Learn about your right to know." http://www.epa.gov/epahome/r2k.htm (accessed June 2012).

Organizations such as the Sierra Club and Natural Resources Defense Council are quite effective at lobbying at the national, regional, state and local levels.

The Environmental Revolution has also changed cultural values. Since the emergence of the modern environmental movement in the 1970s, a majority of the US population sees environmental issues as a serious problem.

Even when environmental protection is contrasted with economic well-being, there is still strong support for environmental regulations. This could be because the regular reporting of environmental hazards and pollution continues to convince the US public that there is potential and actual danger to human life and well-being.

Economic structural changes, such as deindustrialization, have left many US cities with abandoned warehouses, factories and toxic waste sites. Many business elites, citizens groups and civic leaders agree that the solution to these problems is central to the city's economic recovery. Even in healthy, economically strong cities questions of environmental quality are part of efforts to attract more business investment. Issues of urban economic growth and decline are now bound up with issues of environmental quality.

A global environmental revolution

The modern environmental movement was not limited to the US. In the 1970s the environmental movement occurred simultaneously in many countries around the world. Canada, for example, created the Department of Environment in 1971 with a similar set of responsibilities as the US EPA. Like the US, Canada also passed key environmental legislation that includes the Canadian Environmental Assessment Act, the Pest Control Products Act, the Canada Shipping Act, the Arctic Waters Pollution Prevention Act, the Fisheries Act and the Transportation of Dangerous Goods Act. Other countries also established similar laws so that today the UK has its Environment Agency, Sweden its Ministry of Environment and Hong Kong its Environmental Protection Department.

Countries in the developing world have also adopted environmental laws and have created national-level or federal environmental agencies. In 1972 Mexico created the Secretariat of Environment and Natural Resources (Secretaría del Medio Ambiente y Recursos Naturales, SEMARNAT). In 1973 Brazil created the Ministry of the Environment (at that time called the Special Secretariat for the Environment) as the agency in charge of coordinating, supervising and controlling the Brazilian Environmental Policy Act. Pakistan has its Environmental Protection Agency, Egypt has an Environmental Affairs Agency. China passed its national law in 1984, creating what is now called the Ministry of Environmental Protection. Today many, if not most, countries have some national agency that is responsible for national environmental protection and regulation. Similarly, most countries have laws on pollution and environmental protection.

Unlike the US, in many other countries, formal green politics have played a more important role in policy-making. In political systems with proportional representation, green parties have achieved some success in Australia, New Zealand and Germany.

Pollution does not respect political boundaries, making international law an important aspect of environmental law. Numerous legally binding international agreements encompass a wide variety of issue areas, from terrestrial, marine and atmospheric pollution through to wildlife and biodiversity protection. In addition to national-level regulations, countries agree to discuss and create international agreements and treaties on a variety of environmental topics.

BOX 10.2

International environmental timeline

1968   Experts from around the world meet for the first time at the UN Biosphere Conference in Paris to discuss global environmental problems, including pollution, resource loss and wetlands destruction.

1972   The Club of Rome, a group of economists, scientists and business leaders from 25 countries, publishes *The Limits to Growth*, which predicts that the Earth's limits will be reached in 100 years at current rates of population growth, resource depletion and pollution generation.

1972   Participants from 114 countries go to Stockholm for the UN Conference on the Human Environment. Only one is an environment minister, as most countries do not yet have environmental agencies. The delegates adopt 109 recommendations for government action and push for the creation of the UN Environment Programme.

1973   The Convention on International Trade in Endangered Species of Wild Fauna and Flora (CITES), which eventually restricts trade in roughly 5,000 animal species and 25,000 plant species threatened with extinction, is adopted. While the treaty has a broad mandate, inadequate enforcement allows a billion-dollar black market in wildlife trade to flourish.

1979 The Convention on Long-Range Transboundary Air Pollution, which helps combat acid rain and regulate pollution traveling across national borders, is adopted. Later protocols regulate emissions of nitrogen oxides, sulfur, heavy metals, persistent organic pollutants and several other pollutants.

1982 The UN Convention on the Law of the Sea sets a comprehensive framework for ocean use and outlines provisions on ocean conservation, pollution prevention and protecting and restoring species populations.

1983 The US Environmental Protection Agency and the US National Academy of Sciences release reports concluding that the build-up of carbon dioxide and other "greenhouse gases" in the Earth's atmosphere will likely lead to global warming.

1987 The Montreal Protocol on Substances that Deplete the Ozone Layer is adopted to support the phasing out of production of a number of ozone-depleting chemicals.

1987 The World Commission on Environment and Development publishes *Our Common Future* (The Brundtland Report), which concludes that preserving the environment, addressing global inequities and fighting poverty could fuel—not hinder—economic growth by promoting sustainable development.

1989 The Basel Convention, which controls movement of hazardous wastes across international borders, is adopted to prevent "toxic traders" from shipping hazardous waste from developed to developing countries.

1992 Most countries and 117 heads of state participate in the groundbreaking UN Conference on Environment and Development (Earth Summit) in Rio de Janeiro. Participants adopt *Agenda 21*, a voluminous blueprint for sustainable development that calls for improving the quality of life on Earth.

1997 The Kyoto Protocol strengthens the 1992 Climate Change Convention by mandating that industrial countries cut their carbon dioxide emissions by 6–8 percent from 1990 levels by 2008–12. But the protocol's controversial emissions-trading scheme, as well as debates over the role of developing countries, cloud its future.

2002 104 world leaders and thousands of delegates meet at the World Summit on Sustainable Development in Johannesburg and agree on a limited plan to reduce poverty and protect the environment.

2012 Rio +20 Summit marks the twentieth anniversary of the 1992 United Nations Conference on Environment and Development (UNCED) in Rio de Janeiro. The Conference focuses on two themes: (1) how to build a green economy to achieve sustainable development and lift people out of poverty, including support for developing countries that will allow them to find a green path for development; and (2) how to improve international coordination for sustainable development. The seven priority themes include decent jobs, energy, sustainable cities, food security and sustainable agriculture, water, oceans and disaster readiness.

Source: WorldWatch: http://www.worldwatch.org/brain/features/timeline/timeline.htm (accessed July 2012).

A range of scales—global, national, local—are now subject to environmental regulation and concerns. Many countries have laws that are stricter than those at the international level, while others must work to achieve international standards.

BOX 10.3

Millennium Development Goals

In 2000, building on a decade of major United Nations conferences and summits, world leaders came together at the United Nations Headquarters in New York to adopt the United Nations Millennium Declaration, committing their nations to a new global partnership to reduce extreme poverty and to set out a series of time-bound targets—with a deadline of 2015—that have become known as the Millennium Development Goals (MDGs).

The eight MDGs range from halving extreme poverty to halting the spread of HIV/AIDS and providing universal primary education, all by the target date of 2015. They have galvanized unprecedented efforts to meet the needs of the world's poorest.

The eight goals are:

1 eradicate extreme poverty and hunger;
2 achieve universal primary education;
3 promote gender equality and empower women;
4 reduce child mortality;
5 improve maternal health;
6 combat HIV/AIDS, malaria and other diseases;
7 ensure environmental sustainability;
8 develop a global partnership for development.

Several of the MDGs have a direct urban connection. The goal to reduce child mortality, for example, includes vaccination programs, but also has a strong focus on improving sources of drinking water, a topic we discuss in more detail in

Chapter 11. Similarly, the goal to achieve sustainability has targets that focus on improving urban environmental quality, particularly for slum dwellers. The targets for this goal are:

- integrate the principles of sustainable development into country policies and programs and reverse the loss of environmental resources;
- reduce biodiversity loss, achieving, by 2010, a significant reduction in the rate of loss;
- halve, by 2015, the proportion of the population without sustainable access to safe drinking water and basic sanitation;
- achieve, by 2020, a significant improvement in the lives of at least 100 million slum dwellers.

A 2012 report notes that three important targets on poverty, slums and water have been met three years ahead of 2015. The proportion of people using improved water sources rose from 76 percent in 1990 to 89 percent in 2010, translating to more than two billion people currently with access to improved sources such as piped supplies or protected wells. The report also notes the share of urban residents in the developing world living in slums declined from 39 percent in 2000 to 33 percent in 2012. More than 200 million have gained access to either improved water sources, improved sanitation facilities or durable or less crowded housing. This achievement exceeds the target of significantly improving the lives of at least 100 million slum dwellers, also ahead of a 2020 deadline. However, the report acknowledges that improvements in slum conditions are failing to keep pace with the growing ranks of the urban poor. Even though the share of the urban population living in slums has declined over the last ten years, the absolute number of slum dwellers in the developing world is growing and will continue to increase in the near future. The number of urban residents living in slum conditions in the developing world is now estimated at some 828 million, compared to 657 million in 1990 and 767 million in 2000.

Enormous progress has been made toward achieving the MDGs. Global poverty continues to decline, more children than ever are attending primary school, child deaths have dropped dramatically, access to safe drinking water has been greatly expanded and targeted investments in fighting malaria, AIDS and tuberculosis have saved millions. Ambitious goals set in 2000 are being realized, proving that the international community can help cities addresses their most pressing environmental and social challenges.

Source: United Nations (2012) "Millennium Development Goals." http://www.un.org/millenniumgoals (accessed July 2012); United Nations (2012) *Millennium Development Goals Report 2012*. New York: United Nations Press. http://www.un.org/millenniumgoals/pdf/MDG%20Report%20 2012.pdf (accessed July 2012).

At the urban scale, cities are subject to state and/or federal regulations on a variety of environmental issues. In the US, for example, cities abide by all federal regulations and are accountable to the EPA. However, cities are also in a system of networks that extend beyond the federal level. For example, many US cities belong to the International Council for Local Environmental Initiatives (ICLEI). ICLEI is an international association of local governments as well as national and regional local government organizations who have made a commitment to sustainable development. In 2012 ICLEI members included over 1,220 local government members from 70 different countries. US cities have adopted many best practices from cities around the world. For example, Los Angeles' South Coast Air Quality Management District (AQMD) imposed stringent air quality controls and is a leading example of the importance of the metropolitan-based approach to emissions reduction being pursued in the United States today, albeit nested in a broader state and federal government policy.[2]

Greening cities

Cities—and their residents—in both the developed and developing world confront environmental issues on a daily basis—and have created de facto environmental policies where there are no national regulations. For example, long before there was the Clean Water Act of 1972, US cities had regulations on water pollution, although many were not as far-reaching as the eventual federal legislation would be. Environmental regulation is not a temporary inconvenience, but a permanent part of many societies. For these reasons, the environmental movement in cities has become so successful and pervasive that it is legitimate to speak of the "greening" of cities around the world. This greening has occurred in a number of ways.

- *Regulatory reform.* This involves the clean-up of pollution and its prevention. National or federal regulation on air pollution, water pollution, soil contamination and toxic hazards have direct impacts on cities. The clean-up of cities is now seen as good economics, good ecology and good public relations. Issues of urban economic growth and decline are now intimately connected with issues of environmental quality.
- *Growth of environmental organizations.* Since the 1970s, hundreds and thousands of environmental organizations have emerged at local, national and international levels. At the global level, the United Nations Environment Programme and the International Panel on Climate Change are just two examples. Some environmental organizations have an international agenda—such as the WorldWatch Institute or the Nature Conservancy. Others may focus on national or regional issues such as the Chesapeake Bay Foundation or the Adirondack Mountain Club. At the urban scale, the influence of environmental organizations can include national or regional organizations (which often have local chapters in key cities) or organizations that focus primarily on local issues—such as Casey Trees, a nonprofit organization whose mission is to restore, enhance and protect the tree canopy in Washington, DC. Similarly, dc greenworks is dedicated to leadership in the emerging field of green roofs, rain gardens and other techniques of bioretention, and works with low-income communities in DC to install green infrastructures. City policies are often influenced by the mission of these various organizations. Environmental organizations can also form coalitions which can be politically powerful in informing municipal policies and priorities.
- *Sustainable cities.* Moving toward sustainability is truly reforming the city. Here, urban environmentalism is part and parcel of other social movements covering a range of social issues, including social justice, gender and racial equality. Protests against incinerators, waste disposal sites and polluting factories have incorporated issues of social justice and environmental quality. In some cases advancing sustainability can be accomplished through efforts to "green" the city through the creation of urban

farms and gardens from derelict sites, tree planting along meridians and in parks, green roofs and other forms of green infrastructure. In other cases, advancing sustainability requires more than just reform, it requires a social and perhaps economic transformation.

Resistance

As strong and pervasive as the environmental movement has been in the past 40 years, there has been resistance. Some argue that the last 20 years have seen a backlash against environmental regulation in the US. They cite the first term of Ronald Reagan (who tried to eliminate the EPA), Vice President Dan Quayle's Council on Competiveness and the 1994 Contract with America, which reduced the budget and personnel of the EPA. This backlash was also notable during the first ten years of the twenty-first century, with a retreat from ratifying the 1997 Kyoto Protocol and the emergence of climate change deniers. This is a disconcerting trend: in 2010 the EPA water budget was only slightly higher than the 1990 budget, and a survey of young Americans taken in 2012 shows they do not favor environmental protection as strongly as their parents did 30 years ago.

Much of the success—and resistance to—the environmental movement is from the constant repetition of dire predictions. Images of environmental apocalypse have haunted us for decades, action and attention often prompted by forecasts of environmental degradation and social collapse. Some question these predictions, and this has become easier, ironically, as environmental clean-up and regulation has meant that some of the worst predictions seem to be receding. Gregg Easterbrook, in his book *A Moment on Earth*, suggests that many pollution problems have been cleared up, environmental catastrophes are overplayed in the media and technological developments, rather than leading to more problems, are being used to solve them. He argues for more environmental optimism in political discourse. Some of the backlash criticizes current environmental regulation for impeding job growth, costing money and adding to the debt. This critique can be seen in a broader context which sees government as "too big" and "the problem." Those who criticize environmental regulation see it as another example of a liberal conspiracy and the insatiable demand for more government involvement. They argue that current regulation is too costly and too inefficient and we should leave it to the market to improve environmental quality. Despite these polarized ideologies, and a backlash against environmental regulation, public opinion in the US is still more positive about environmental laws and legislation than almost any other area of government.

After 40 years of institutionalized environmental regulation, many cities have better water quality and air quality, despite decades of rising populations. More stringent regulations, better technology and more enforcement have resulted in improvements in the quality of the environment. On the other hand, climate change is now emerging as a serious challenge that many cities are reckoning with. While many countries have adopted environmental regulation and strive to achieve higher standards, some countries have yet to pass or enforce environmental regulation. Forty years after the start of the environmental revolution cities still confront a range of challenges with regard to water, solid waste, air and climate change. Some pollutants have been reduced; others have increased; still new pollutants have been created. These issues remain critical as cities undertake sustainability plans and revise their future.

Guide to further reading

Benton, L and Short, J. R. (1999) *Environmental Discourse and Practice*. Oxford: Blackwell.

Benton, L. and Short, J. R. (eds.) (2000) *Environmental Discourse and Practice: A Reader*. Oxford: Blackwell.

Fiege, M. (2012) *The Republic of Nature: An Environmental History of the United States*. Seattle, WA: University of Washington Press.

Kline, B. (2011) *First Along the River: A Brief History of the U.S. Environmental Movement*. 4th edition. New York: Roman & Littlefield.

McNeill, J. R. and Kennedy, P. (2001) *Something New Under the Sun: An Environmental History of the Twentieth-Century World*. New York: W. W. Norton.

Shabecoff, P. (2003) *A Fierce Green Fire: The American Environmental Movement*. San Francisco, CA: Island Press.

Simmons, I. G. (2008) *A Global Environmental History*. Chicago, IL: University of Chicago Press.

Steffen, A., McKibben, B. and Jones, V. (2011) *Worldchanging:A User's Guide for the 21st Century*. Revised and updated edition. New York: Abrams.

Uekoetter, F. (ed.) (2010) *Turning Points of Environmental History*. Pittsburgh, PA: University of Pittsburgh Press.

11 Water

A fresh and dependable supply of water is critical to sustaining life and supporting healthy communities, economies and environments. In every city there are two broad water issues (Figure 11.1): water supply and water quality, both of which rely on water infrastructure. In the case of water supply, water infrastructure includes the systems of delivery (aqueducts, pipes). Water quality is an issue associated with pollution and the removal of contaminated or dirty water. Here, water infrastructure includes sewerage systems and treatment facilities. In this chapter we examine both supply and quality issues. We will first consider these water issues in developed cities, with a particular focus on US cities. Then we will examine these issues in developing cities.

Water in developed cities

Water supply

Drinking water comes from surface water or groundwater. It goes into a water treatment facility where it is purified to certain standards. In cities, an underground network of pipes delivers drinking water to homes and businesses served by a public water system.

In the US there are approximately 155,000 public water systems that range in size from small (serving populations in the hundreds) to extra large (serving populations greater than 100,000). Each day water utilities in the US supply nearly 34 billion gallons of water. Water suppliers use a variety of treatment processes to remove contaminants from drinking water. These processes include coagulation, filtration and disinfection. Some water systems also use ion exchange and absorption. Water utilities select the treatment combination most appropriate to treat the contaminants found in the source water of that particular system. Water utilities must test their water

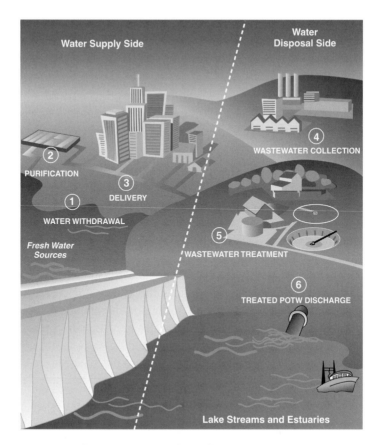

Figure 11.1 Visualizing water supply and water treatment systems

Source: EPA Office of Water (2004) "Primer for municipal wastewater treatment systems."
http://water.epa.gov/aboutow/owm/upload/2005_08_19_primer.pdf

frequently for specified contaminants and report these results to the states. If a water system does not meet the minimum standards, it is the water supplier's responsibility to notify its customers. Many water suppliers are now also required to prepare annual reports for their customers, and the use of the web has allowed the public to stay better informed.

Both the federal government and states have responsibilities for providing safe water. The Safe Drinking Water Act was passed by Congress in 1974 to protect public health by regulating the nation's public drinking water supply. The EPA has also set standards for some 90 chemicals, microbes and other physical contaminants in drinking water. The law was amended in 1986 and again in 1996; generally, amendments to an original law involve increasing standards, not lowering them. The amendments require cities and states to protect drinking water *and* its sources: rivers, lakes, reservoirs, springs and groundwater wells.

While most populations in the developed world take safe drinking water for granted, there remain a number of threats to drinking water. Possible contaminants include lead, arsenic and chromium. Improper chemicals disposal, animal and human wastes, wastes injected underground and naturally occurring substances also have the potential to contaminate drinking water. Drinking water that is not properly treated or disinfected, or that travels through an improperly maintained distribution system, may also pose a health risk. In the post-9/11 world, drinking water utilities find themselves facing new responsibilities due to concerns over water system security and threats of infrastructure terrorism.

The average American uses about 90 gallons of water each day in the home, and each American household uses approximately 107,000 gallons of water each year (compare this to the average European who uses 53 gallons each day; and the average citizen of sub-Saharan Africa who uses 3–5 gallons).[1] Today it costs about $2 to provide 1,000 gallons of clean water, which equates to approximately $300 per household per year. While this is a small cost compared with fueling our automobiles, the cost of making water safe continues to rise. In the US and Canada and many other developed countries much of the existing drinking water infrastructure (the underground networks of pipes, treatment plants and other facilities) was built during the sanitation revolution of the late nineteenth century. This infrastructure is now more than 100 years old, and in many cases cannot handle the volume of demand due to population growth, or the pipes may be leaking or deteriorating. A recent EPA *Drinking Water Infrastructure Needs Survey* estimated that US cities will need to invest $150.9 billion over a 20-year period to ensure the continued source development, storage, treatment and distribution of safe drinking water.[2] In most developed cities, clean and safe drinking water is a given, but there are still challenges ahead from protecting the supply and the need to upgrade or expand infrastructure.

Issues about water supply are often highly political. Take the case of New York City. The city's water supply system today is the result of a long history. Ninety percent of the city's water comes from the Catskill Mountains located 100 miles north of the city. The watershed is composed of 19 reservoirs and three controlled lakes that have a combined capacity of 580 billion gallons. Each day, the water system delivers one billion gallons to New York City residents. The city's water supply is among the nation's highest-quality drinking water. But getting this high-quality water supply is not without its

BOX 11.1

Returning to the tap?

There was a time when brands like Evian and Perrier conjured up images of purity and luxury. Bottled water is now the new normal. It is also big business. In 2010 Coca-cola sold about 293 million cases of its top brand, Dasani, while rival Pepsi sold 291 million of Aquafina. The biggest player is Nestlé, which sells such brands as Poland Spring, Zephyrhills and Pure Life.

But there is now a backlash. Critics of bottled water point to economic, regulatory and environmental issues associated with drinking bottled water. Economically, for example, it costs only $10 for 1,000 gallons of tap water, while consumers spend $1,000 for 1,000 gallons of bottled water. A recent NRDC report noted that their study found that an estimated 25 percent or more of bottled water was really just *tap water in a bottle*. And while there is a $23 billion dollar gap in investments for water infrastructure, its costs cities $42 million each year to dispose of the tons of discarded PET plastic water bottles. In addition, since some 25–30 percent of "bottled water" actually comes from municipal water sources, bottlers are profiting from underfunded public water systems.

In terms of regulation, the EPA requires cities to disclose to consumers the drinking water conditions, while bottled suppliers are not required to report water quality results to consumers. Although bottled water can cost up to 10,000 times more per gallon than tap water, the reality is that tap water is actually held to more stringent quality standards than bottled water, and some brands of bottled water are just tap water in disguise. For example, in the US, it is not the EPA but the US Food and Drug Administration (FDA) that regulates bottled water used for drinking. While most consumers assume that bottled water is at least as safe as tap water, there are still potential risks. Although required to meet the same safety standards as public water supplies, bottled water does not undergo the same testing and reporting as water from a treatment facility. Water that is bottled and sold in the same state may not be subject to any federal standards at all. Some point out that municipal water systems were delivering excellent water long before plastic became all the rage. And yet bottled water manufacturers encourage the perception that their products are purer and safer than tap water.

The environmental costs of bottled water are many. Critics contend that bottled water is wasteful, contributes to ballooning landfills and is being marketed as a necessity by an industry making billions on what consumers used to happily get from a kitchen tap or public fountain. Americans buy an estimated 30 billion plastic water bottles every year. Nearly 90 percent of those water bottles are not recycled and wind up in landfills, where it takes thousands of years for the plastic to decompose. Approximately 1.5 million barrels of oil—enough to run 100,000 cars for a whole year—are used to make plastic water bottles, while transporting these bottles burns even more oil. The growth in bottled water production has increased water extraction in areas near bottling plants, leading to water shortages that affect nearby consumers and farmers. In addition to the millions of gallons of water used in the plastic-making process, two gallons of water are wasted in the purification process for every gallon that goes into the bottles.

A 2009 film, *Tapped*, examined the role of the bottled water industry and its effects on our health, climate change, pollution and our reliance on oil. A growing coalition of cities and environmental and public health organizations now advocates for "a return to the tap."

New York-based TapIt, a non-profit group launched in 2008, works to promote the use of tap water over bottled water. They encourage restaurants to provide free refills of tap water to patrons that have their own reusable bottles. They have also worked with hundreds of US colleges to install water fountains known as hydration stations so students can refill water bottles rather than buy new ones (these fountains have taller faucets to allow tall bottles to be refilled). Some campuses are even banning the sale of bottled water. Hydration stations are also popping up in airports, parks, office buildings and restaurants. In 2011 40 local Salt Lake City food service businesses signed up to be official TapIt Partners, pledging to offer free tap water to the public. In that same year the District of Columbia and TapIt recruited more than 60 eateries in the District to offer free water refills to those who bring their own reusable bottles. TapIt represents an example of a grassroots organization that leverages the power of the internet, social media and mobile telephones to drive social change. TapIt offers an iPhone app, mobile website and TapIt stickers on the windows of participating restaurants.

Sources: "Tapped" film: http://www.tappedthemovie.com; for more information on TapIt, go to http://www.tapitwater.com; EPA Office of Water (2009) "Water on tap: what you need to know." http://water.epa.gov/drink/guide/upload/book_waterontap_full.pdf (accessed May 2012).

controversies. The water supply system was organized in the early 1900s and numerous towns and communities were evacuated or dismantled to make room in the valleys for the new reservoirs. There has been a long history of conflict between the city and the watershed residents. More recent clashes have occurred over the way water operations can negatively impact the recreation and tourism that some watershed communities rely on for their economic livelihood.

Similarly, Atlanta faces ever-increasing water problems due to its distance from major waterways. The metro area had a population of 2.9 million in 1990; it was 5.2 million in 2012. An increase in population impacts its daily draw on the water reserve; in 1990 the city consumed 320 million gallons, by 2010 it was 510 million gallons. With two million more residents projected by 2030, water use is expected to rise to more than 700 million gallons a day.[3] In 2007, a major drought brought Atlanta's water supply to the brink of failure. Officials admitted that there was only three months left of stored fresh water to supply Atlanta. Many cities in the west such as Los Angeles, Las Vegas and Denver have ushered in water conservation measures—including offering incentives, installing high-efficiency toilets and low-flow shower heads, and increasing monthly water bills for big water users—but Atlanta has been slower to enact such policies. Instead of reducing usage, Atlanta has focused its efforts on increasing supply. Since 1956, Atlanta's primary source of water has been Lake Lanier. But the city has met resistance from the state, which has attempted to limit the amount of water released from the lake. The river system also serves the states of Alabama and Florida. These states fear that Atlanta's increasing use of water upstream will harm ecosystems in their states. In 2009 a federal judge declared it illegal for Atlanta to use Lake Lanier for its drinking supply. In June 2011 the 11th Circuit Court reversed the ruling, stating the city could withdraw water from the lake. In early 2012 both Florida and Alabama petitioned the Supreme Court to review the decision and resolve the issue of water usage from the Lake Lanier system.[4] The case is still under review, but if Atlanta is denied use of the lake its water supply could be reduced by roughly 40 percent. While we tend to think of "water wars" in the arid west, this is an example of how increased demand for water, even in areas that tend to have high rainfall, can be politically charged issues.

Water quality

Many cities in Europe, Australia, Canada and the US built sewer systems to collect wastewater, as well as wastewater treatment facilities in the nineteenth and early twentieth centuries. However, population growth has meant that by the twenty-first century, the volume of sewage and stormwater exceeded the processing ability of most treatment plants. This is particularly noticeable during heavy rains.

Combined sewage overflow (CSO) refers to the temporary direct discharge of untreated water (Figure 11.2). CSOs occur most frequently when a city has a combined sewer system (CSS) that collects wastewater, sanitary wastewater and stormwater run-off in various branches of pipes, which then flow into a single treatment facility. CSSs serve about 772 communities containing 40 million people in the US. Most communities with CSSs (and therefore with CSOs) are located in the northeast and Great Lakes regions, and the Pacific northwest.

During dry weather, CSSs transport wastewater directly to the sewage treatment plant. However, rainwater or urban storm run-off is not directed separately, but co-mingled with household wastes and industrial wastes. When it rains, few facilities can handle the sudden increase in the volume of water, and as a result, the excess volume of sewage, clean water and stormwater is discharged untreated into rivers, lakes, tributaries and oceans (Figure 11.3).

As the joke goes: CSO doesn't stand for "crap spewing out" but it might as well. CSOs contain raw sewage from homes, businesses and industries, as well as stormwater run-off and all the debris and chemicals that wash off the street or are poured in storm drains. This toxic brew is unappealing and quite dangerous. CSOs contain untreated human waste, oxygen-demanding

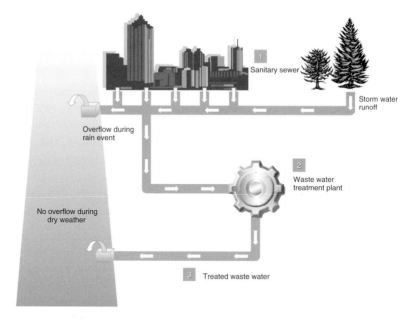

Figure 11.2 Combined sewer systems. In cities with combined sewer systems, stormwater run-off from rain combines with sewage in the same pipe system and is discharged, untreated, directly into rivers, creeks or estuaries. The discharge consists of many pollutants, untreated sewage and debris

Source: Lisa Benton-Short

Figure 11.3 On the Cincinnati waterfront, residents are reminded that the Ohio River is subject to combined sewer overflows

Source: Photo by Lisa Benton-Short

substances, ammonia, pesticides (such as malathion, sprayed on the city to fight West Nile virus), nutrients, petroleum products (from sources such as gas stations, auto repair shops and garages) and other potential toxins and pathogenic microorganisms associated with human disease and fecal pollution. Forty types of disease-causing pathogens have been found in raw sewage that discharges into CSOs.

In 2010, approximately 850 billion gallons of raw sewage were discharged from CSOs into the nation's waters. In addition, some 51 million pounds of toxic chemicals were released from municipal sewage treatment plants. In just New York Harbor, more than 27 billion gallons of raw sewage and polluted stormwater is discharged out of 460 CSOs each year.[5] Although water quality in New York Harbor and throughout the Hudson River Estuary has improved significantly over the last few decades, many parts of the waterfront and its beaches are still unsafe for recreation after it rains. As little as one-twentieth of an inch of rain can overload the system. The main culprit is outmoded sewer systems, which combine sewage from buildings with dirty stormwater from streets. Unless investment in reducing CSO events increases significantly, the EPA predicts that by 2025 sewage pollution will exceed 1968 levels—the highest in US history.

CSOs are among the major sources responsible for beach closings and shellfish restrictions and the contamination of drinking water. Residents of these cities are often warned to avoid contact with river water or beach water for several days after periods of heavy rainfall.

In 2012 the National Resources Defense Council (NRDC) issued a report noting US beachwater continued to suffer from serious contamination and pollutants by human and animal waste. As a result, America's beaches issued the third-highest number of closings or advisories in the report's history in 2011, with the second-highest number occurring one year before in 2010.[6]

Many of the worst beaches are found in Los Angeles, New York, New Jersey, Illinois and Louisiana. Southern California beaches have struggled with water quality, mostly a result of sewage pollution. The NRDC study found that fecal contamination at Los Angeles and Orange County beaches causes between 627,800 and 1,479,200 excess gastrointestinal illnesses each year. In Louisiana, the BP oil disaster, which began with the April 20, 2010 explosion on the Deepwater Horizon rig and ended when the well was capped on July 15, 2010, continues to affect beaches along the Gulf of Mexico in Louisiana, Mississippi, Alabama and Florida. Oil spill inspection and clean-up efforts continued throughout 2011 and into 2012, even at beaches whose oil spill closures, advisories and notices were lifted.

In Europe the situation is better. A 2012 report from the European Environment Agency noted that 91 percent of coastal waters, rivers and lakes met the minimum water quality standards set by the bathing water directives, improving significantly since the 1990s. Although the UK has some 22,000 combined sewer outfalls, the country has done a better job than the US at investing in infrastructure upgrades. By the end of 2008 more than 6,000 overflows posing the highest risk had been improved, rebuilt or removed altogether. In 2000 only 24 beaches had achieved the highest quality standards; by 2008 this had increased to 82. Water and sewage companies have pledged to invest more than £1 billion between 2010 and 2015 in continued improvements. However, some European cities such as Venice, continue to confront water pollution issues (see Figure 11.4).

Figure 11.4 Venice, a city of islands and canals, has never maintained a main sewage system. For this reason, a large portion of the wastes generated in the historic center of Venice have always been discharged directly into its channels. Today, water quality is quite poor

Source: Photo by Lisa Benton-Short

In 2011, a UK citizens' group, Surfers Against Sewage, launched a crowd-sourcing map "Sewage Alert Service" to inform beach users in real time when raw sewage is discharged into the sea via CSOs.[7] Providing beach users with these warnings constitutes a dramatic improvement in the provision of public information on sewage spills at some of the UK's best-loved beaches, and it also shows the power of new technology to inform and mobilize the public.

The other major type of public sewer system is the sanitary sewer systems (SSSs), built in cities since the start of the twentieth century, and almost exclusively installed since the 1970s. SSSs have separate pipes that collect sewage and stormwater separately. However, when it rains the excess stormwater can overload the system and so stormwater is discharged untreated (see Figure 11.5) At least this prevents the direct discharge of sewage, but recent studies have found that during the initial rainfall, the concentration of pollutants in urban stormwater rivals and in some cases exceeds sewage

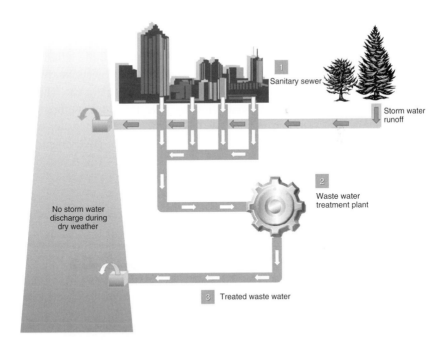

Figure 11.5 Separate sanitary systems. For cities that have separate sanitary and stormwater systems, stormwater run-off during rains are collected separately from sewage. The stormwater run-off is diverted from the wastewater treatment plant and discharged, untreated, into rivers, creeks or estuaries

Source: Lisa Benton-Short

plants and large factories as a source of damaging pollutants, beach closures and shellfish decline.[8]

Waste treatment infrastructure

Once sewage has reached a treatment plant, physical processes can separate biological solids from wastewater. Screens or filters catch raw sewage and debris, channeling them to sludge ponds where the organic materials are broken down by bacteria. Often sludge that has dried is shipped to landfills.

By the early twentieth century, some cities began to install secondary treatments in their facilities. Secondary treatment involves aeration and finer filtration. It is a more controlled way of producing bacterial growth to break down biological solids. Secondary treatment eliminates more than 95 percent of disease-carrying bacteria from the water. Table 11.1 outlines some of the current challenges to wastewater treatment.

Municipalities in the US are undertaking projects to mitigate CSO and have been since the 1990s. Prior to 1990, the quantity of untreated combined sewage discharged annually to lakes, rivers and streams in southeast Michigan was estimated at more than 30 billion gallons per year. In 2005, after spending $1 billion of a planned $2.4 billion on stormwater projects, untreated discharges were reduced by more than 20 billion gallons per year. This investment has yielded a 67 percent reduction in CSOs. It includes numerous sewer separation, CSO storage and treatment facilities and wastewater treatment plant improvements constructed by local and regional governments. Many other cities are undertaking similar projects to address CSOs, but these are

Table 11.1 Challenges in wastewater treatment

Many of the wastewater treatment and collection facilities are now old and worn, and require further improvement, repair or replacement to maintain their useful life.
The character and quantity of contaminants presenting problems today are far more complex than those that presented challenges in the past.
Population growth is taxing many existing wastewater treatment systems and creating a need for new plants
Increasing urbanization provides additional sources of pollution not controlled by wastewater treatment
One-third of new development is served by decentralized systems (e.g., septic systems) as population migrates further from metropolitan areas (particularly in the exurbs)

Source: EPA Office of Water (2004) "Primer for municipal wastewater treatment systems." http://water.epa.gov/aboutow/owm/upload/2005_08_19_primer.pdf (accessed May 12, 2012).

billion-dollar projects and cities look to states and the federal government for help in funding these large-scale infrastructure projects.

While many cities in the US struggle with finances to upgrade their wastewater treatment plants, the majority of wastewater plants in northern and central Europe now apply tertiary treatment. Elsewhere in the EU, particularly in the southeast, the proportion of primary and secondary treatment is even higher.

Water quality legislation: the US Clean Water Act of 1972

By the 1960s many states in the US had permit programs in place, but they set different standards. Although state efforts were underway, progress was not fast enough. In addition, because water flows through streams, rivers, lakes and estuaries without regard to political boundaries, broad federal protection was important. There were a series of legislative precedents for water quality, including the 1948 Federal Water Pollution Control Act, the 1956 Water Pollution Control Act and the 1965 Water Quality Act of 1965. These were among the first comprehensive statements of federal interest in clean water, but mounting frustration over the slow pace of pollution clean-up efforts along with increased public interest in environmental protection set the stage for the 1972 law. Most environmental experts point to the 1972 Federal Water Pollution Control Act Amendments (commonly called the 1972 Clean Water Act) as setting the framework for the last 40 years of water pollution policy. The Clean Water Act (CWA) is the cornerstone of surface water quality protection in the US. (It is important to remember that the Act does not deal directly with groundwater nor with water quantity/distribution issues—these are under state control and regulation.) Today, many developed countries and some developing countries have similar regulations and ambitions.

The CWA was remarkable for several reasons. It is the principal law governing pollution of the nation's surface waters. For the first time, a single federal agency was placed in charge of water pollution, taking some control from the states and establishing basic federal standards to which all states had to comply. The CWA also provided technical tools and financial assistance to address the many causes of water pollution. Another revolutionary aspect to the legislation was that it required reporting of discharge information and making it publicly available. This gave citizens a strong role to play in protecting water resources and the tools to help them do so. In today's information economy, discharge release and toxic release inventory are available to citizens using the internet, and citizens have access to even more information such as permit documents, pollution management plans and inspection reports.

The provision for public information also allows citizens to bring action (or sue) to enforce the Clean Water Act. This has proved instrumental in making progress in cleaning up water pollution.

The CWA was highly ambitious: its major goals were to eliminate the discharge of water pollutants, restore water to "fishable and swimmable" levels and to completely eliminate all toxic pollutants. Today these three aspirational goals of the CWA of 1972 have not been achieved, yet they continue to provide the framework for how the federal government (and municipalities) deal with water pollution.

Two major components of the CWA impact cities directly. Title II and Title VI authorize federal financial assistance for municipal sewage treatment plant construction and set out regulatory requirements that apply to industrial and municipal dischargers. Both of these are critical to understanding the relationship between cities and water pollution reform. With federal assistance, many cities were able to upgrade sewage treatment plants, adding secondary or tertiary treatment, or increasing the volume the treatment plants could process. Title VI, which applied mainly to industrial and municipal discharge, resulted in monitoring and management programs for point source pollution. The result is a decrease in industrial point pollution.

For its time the CWA was genuinely revolutionary. It made it illegal for any point source—that is a specific source of pollution—to discharge any pollutant into the waters unless specifically authorized by permit.

Assessing the CWA 40 years later

In 2012 the CWA celebrated its fortieth anniversary. There have been successes and failures. We now have a more complete picture of the volume and types of municipal pollution, but the results are mixed. Point-source pollution from industrial sources has been reduced. In many rivers and lakes, oxygen levels have recovered due to the filtering out of organic wastes. Some pollutants have declined, but others are on the increase. In 1972 two-thirds of the lakes were too polluted for swimming or fishing, by 2012 two-thirds of the lakes, rivers and waterways were safe for swimming and fishing. In 1970, more than 70 percent of industrial discharge was not treated at all; today 99 percent is. Lake Erie, declared "dead" in 1969, has recovered. And in the Cuyahoga River, once a stark symbol of the plight of America's rivers, the blue herons have returned and the city now boasts new marinas lining a river walk and upscale sidewalk cafes (see Box 11.2).

BOX 11.2

Cuyahoga 40 years after the fire

In June 1969, the Cuyahoga River in the city of Cleveland caught fire. The river, filled with kerosene and other flammable material, was probably ignited by a passing train that provided the spark. Although it burned for only 30 minutes, the incident and the famous photograph of the river on fire became a pivotal part of an emerging environmental movement. *Time Magazine* described the Cuyahoga as the river that "oozes rather than flows" and in which a person "does not drown but decays." The memorable 1969 fire sparked legislation for the Great Lakes Water Quality Agreement and the Clean Water Act of the 1970s. The federal government directed large amounts of money, dealt with industrial polluters and enforced regulations: the river was on track to get cleaner.

The 100-mile river is composed of an upper half and a lower half (near Cleveland). Today, the lower half of the river is no longer the incendiary sewer of the dark days of 1969. Industrial discharge has been controlled significantly. However, the river still receives discharges of stormwater, CSOs and incompletely disinfected wastewater from urban areas upstream of the park. Non-point sources are the result of urbanization, more specifically suburbanization, in metro Cleveland. People are moving into suburbs that were once former open spaces and farmland. And with suburban settlement comes an increase in impervious surfaces such as rooftops, driveways, parking lots, sidewalks and lawns.

Although much healthier than 40 years ago, some sections of the river are far from healthy. Most of the river remains unacceptable for recreational use due to the high concentrations of *Escherichia coli* (*E. coli*), a fecal-indicator bacterium. Contaminant and bacteria levels can still be high, especially after periods of rain. Here is a summary of some of the river's lingering problems:

- trash
- toxics
- bacteria
- fish tumors
- lack of aquatic diversity
- beach closings.

For these reasons, the EPA recently classified portions of the Cuyahoga River watershed as one of the 43 Great Lakes Areas of Concern.

On a more optimistic note, in 2009, a water quality analyst was the first person in more than half a century to find a living freshwater mussel in that long-polluted portion of the river. Somehow the freshwater mussel managed to survive the last century through the worst of the pollution on the Cuyahoga and the mussel population is on the increase.

Source: Adler, J. H. (2004) "Smoking out the Cuyahoga fire fable: smoke and mirrors surrounding Cleveland." *National Review*, June 22. http://old.nationalreview.com/adler/ adler200406220845.asp (accessed July 2012); Scott, M. (2009, April 12) "Scientists monitor Cuyahoga River to adhere to Clean Water Act," *Cleveland.com*. http://www. cleveland.com/science/index.ssf/2009/04/scientists_ monitor_cuyahoga_ri.html (accessed July 2012)

However, much remains to be done. Almost half of the nation's waters are still impaired—too polluted to serve as sources of drinking water or to support good fish and wildlife. Wetlands continue to be lost to pollution and development. Today, non-point-source pollution—run-off from farms and from cities—is the leading cause of water pollution today, but is inadequately addressed by the CWA. Table 11.2 compares the Clean Water Act priorities in 1972 and 2012. The CWA thus addressed the most visible and obvious problems in 1972, and those seen as having a greater public health risk; after 40 years, however, it is clear that while some improvements have been made,

Table 11.2 Comparing Clean Water Act Priorities, 1972–2012

Addressed in 1972 Clean Water Act	Not Addressed in 1972 Clean Water Act, but seen as a critical twenty-first century issue
Point source pollution	Non-point source pollution
Emphasis on Navigable waters	Problems with smaller, non-navigable water
Surface water	Ground water
Degradation of quality: pollution	Drinking water: supply/quantity

more complex issues that the legislation had ignored, such as non-point pollution, are now the most pressing issues of the twenty-first century.

In 1970 only 85 million Americans had a wastewater treatment plant. By 2008, 226.4 million people received centralized collection and wastewater treatment (more than 74 percent of the US population).[9] And yet, more than 15 large cities continue to discharge some 850 billion gallons of untreated sewage and run-off each year during heavy rains and floods. In some cities, beach closures have increased in places where the combined sewers are older and inadequate for the population. And while the total releases of toxic pollutants have decreased (particularly with regard to mercury and DDT), more than 47 states still post fish consumption advisories for contamination due to mercury, PCB, dioxins and DDT.

The success of the CWA at controlling point pollution contrasts with its failure to address non-point sources of pollution. Because it falls outside the CWAs permit requirements, and because it was not addressed in a significant way in the original legislation, non-point sources are now the dominant cause of water pollution. According to the EPA, by 2010, 43 percent of lakes and 37 percent of estuaries were impaired due to non-point pollution and municipal discharge. Addressing non-point pollution remains underfunded. Since 1990, the EPA has spent some $90 billion on water but only 44 percent of that on non-point. In addition, while the EPA has set aside $8 billion for water projects over the next several years, only $200 million is slated for non-point pollution. Andrew Karvonen has shown that often supposedly "apolitical" engineers wield considerable power with regards to infrastructure, while engaged citizens are often not politically savvy enough to effect alternative solutions. He notes that while urban run-off is frequently discussed in terms of technical expertise and environmental management, it actually encompasses many non-technical issues such as land use, quality of life, aesthetics and community identity.[10] Today, debates about what to do with urban non-point pollution are part and parcel of larger debates about land-use planning, urban development and sustainability.

BOX 11.3

Point and non-point pollutants

There are two main sources of water pollution, point and non-point. Point sources are those where there is a clear discharge mechanism such as effluent pipes or outfalls. According to the US EPA, point sources are defined as any "discernible confined and discrete conveyance including but not limited to any pipe, ditch, channel, tunnel, conduit, well, discrete fissure, container, rolling stock, or concentrated animal feeding operation, or vessel or other floating craft from which pollutants are or may be discharged." These stationary devices can be measured for the amount of pollution discharged. The main point sources of pollution in cities are industrial and municipal facilities. Point sources of water pollution are relatively easy to monitor and control. The National Pollution Discharge Elimination System is a permit program of the Clean Water Act. As a result, point source pollution in the US has declined significantly since the implementation of the Clean Water legislation.

Non-point sources are any sources from which pollution is discharged which is not identified as a point source. There are two main types of non-point sources of water pollution in cities. The first is urban run-off, a term that refers to the various pollutants that accumulate in soil and on roadways that are washed into the sewerage systems during floods or rains. As rainwater makes its journey over roads, parking lots and other urban structures, it picks up a variety of pollutants. The EPA notes that urban run-off is now the largest source of pollution to estuaries and beaches. The second type of non-point source include wastes and sewage from residential areas. These wastes are collected in the larger sewer system, and it is impossible to tell their origin. These pollutants are harder to regulate and reduce. In addition, non-point sources are often intermittent and diffuse, making it hard to quantify individual contributions. Section 319 of the Clean Water Act 1997 Amendments requires states and territories to develop programs to deal with non-point source pollution. The EPA has set aside $370 million in funds to implement non-point source pollution controls.

Point sources (primarily industrial wastes)

Oil and grease
Heavy metals
Organic chemicals
Acids and alkalides
Salts
Solvents
Organic matter
Suspended solids
Heat/thermal pollution.

Non-point sources

Salt
Oil
Gasoline
Antifreeze
Floatables and debris (plastics, cans, bottles)
Pesticides and fertilizers
Organic matter (including human and animal wastes)
Microbial pathogens
Solids and sediment
Sewage
Heavy metals (chromium, copper, lead and zinc)

Sources: US EPA (2012) "What is nonpoint source pollution?" http://water.epa.gov/polwaste/nps/whatis.cfm (accessed July 2012). US EPA (2007) "Polluted runoff (nonpoint source pollution: managing urban runoff," Document EPA841-F-96-004G. www.epa.gov/owow/nps/facts/point7.htm (accessed March 2007). EPA (2006) "Polluted runoff: nonpoint source pollution: the nation's largest water quality problem." www.epa.gov/owow/nps/facts (accessed March 2007).

Over the last 40 years, the US has spent some $335 billion on drinking water improvements and $300 billion on sanitation infrastructure, including some 13,000 wastewater treatment plants. During this time the EPA estimates that some 22 billion tons of pollution have been eliminated from water bodies. The reduction in discharge is important, yet we have a long way to go to reach the goal of "zero" discharge.

Another alarming trend is the condition of water infrastructure. A 2011 report by the American Society of Civil Engineers rates the US water infrastructure as a D minus. This is the lowest grade of all infrastructure—even transportation was given a higher grade. The report notes that by 2020 the inefficient, problematic water infrastructure will cost the US some $416 billion in productivity and a loss of 7,000 jobs if we do not significantly increase investment in infrastructure.[11]

The mixed results of the CWA reflect several factors. First, new technologies impact our ability to measure pollution in smaller and smaller amounts. Scientists are able to document levels of pollution in parts per million or billion, levels that we were unable to detect previously. In addition, our knowledge and understanding of the impact of pollution on human health and environmental health is improving, and this adds new "pollutants" to the list and can require a readjustment of allowable exposure standards. There are now approximately 80,000 chemicals registered with the EPA: few have been comprehensively tested, many resist breakdown and some accumulate in fat tissues. Another factor is that the EPA has tended to focus on "end of the pipe" solutions—that is, regulating discharges at the source of the polluter rather than focusing on ways to encourage "front end" controls or preventive measures. The EPA has been less successful at encouraging the reduction of waste creation in the first place. While millions of dollars have been available to assist in the development and adoption of technological solutions, less is available for programs that encourage reduction or reusing of pollutants in the first place. This is disappointing as even the EPA admits prevention is cheaper than clean-up: for every $1 dollar spent on protecting clean water sources, a city can save about $27. But there is little money available for protecting clean waters.

Since 2000, the EPA has established a set of federal policies and incentives for reducing urban storm run-off and to help smaller cities (of under 100,000 population) develop adequate wastewater treatment facilities. In addition, the 1987 Clean Water Act amendments directed states to develop and implement non-point pollution management programs. Under section 319 of the Act, $400 million in grants was made available for states to assess the extent of non-point source water quality impairments and to develop and implement plans for managing non-point sources.

Responding to the new realities of water issues, in 2012 the EPA announced it would in future focus on:

- urban water;
- green infrastructure to reduce pollution;
- integrated planning for urban stormwater management;
- funding for water projects;
- a renewed interest on the so-called iconic water bodies: Gulf of Mexico, Chesapeake Bay, Puget Sound, Everglades, Great Lakes;
- mountain-top mining;
- fracking (hydrofracturing for natural gas);
- adapting to climate change.

Let us consider two of these critical twenty-first-century issues: green infrastructure and integrated planning.

Green infrastructure and integrated planning for urban stormwater run-off

The CWA has shifted from a program-by-program, source-by-source, pollutant-by-pollutant approach to more holistic watershed-based strategies. Under the watershed approach equal emphasis is placed on protecting healthy waters and restoring impaired ones. This has inspired a rethinking of how to deal with non-point pollution in a more integrated and systematic way. Cities like Philadelphia are realizing this potential.

Philadelphia's two rivers, the Schuylkill and the Delaware provide drinking water for the city and also receive the city's stormwater run-off and CSO events. There are some 160 sewage outfalls around the city, and like many northeastern cities with older infrastructure, Philadelphia experiences numerous storms and rains that contribute to CSO events that violate its obligations under the Clean Water Act requirements.

In 2011 Philadelphia's Water Department's "Green Cities, Clean Waters" plan was approved. The plan "will stimulate tourism, recreation, and riverfront development, along with achieving economic benefits and creating jobs. Cleaner rivers create increased civic pride in the riverfront area, higher property values, and greater potential for valuable riverfront projects."[12] The plan requires Philadelphia to reduce its annual CSO volume by almost 8 billion gallons.[13] It aims to reduce the excess run-off and CSO issues not by reworking its current sewage infrastructure, but by reshaping the way the city interfaces with nature.[14] The city has pledged to invest $1.2 billion over the next 25 years.

The plan has been praised by environmental organizations and others because it presents an innovative "land–water–infrastructure" approach to achieve its watershed management and CSO control goals. One approach is to focus on green infrastructure and the physical reconstruction of aquatic habitats. In 2006 Philadelphia passed stormwater management regulations for new development and redevelopment that requires the capture of the first one inch of rainfall, reducing pollutant loads through infiltration and/or detaining and controlling run-off rates at levels that minimize flow into the sewer system. Developers and designers are encouraged to use land-based elements such as redirecting run-off from impervious surfaces to green areas, bioretention (vegetation that is water-loving) and swales, subsurface storage, green roofs and the tree canopy. These green approaches reduce demand on the sewer infrastructure, but they have a secondary benefit of protecting open space. Seventy percent of the budget for the 25-year "Green City, Clean Waters" plan is to be used on green projects. Funds will encourage the development of green infrastructure, while also meeting the goal of reducing impervious surfaces by one-third.[15]

In addition to these green land-based projects, Philadelphia is also undertaking water-based improvements. The city plans to reconstruct and stabilize stream beds and banks, create aquatic habitat, remove plunge pools, improve fish passages and reconnect floodplains. Restoring these waterways will remove them from the state's list of "impaired waters" and increase the level of compliance with the Clean Water Act.

Finally, the city is also using traditional means to reduce CSOs. These engineering and construction projects include traditional storage, conveyance, and treatment measures within the combined sewer collection and treatment system. Conventional measures involve the installation of inflatable dams, underground sewage storage tanks, and storm relief sewers. But the emphasis of the plan is on green infrastructure, not gray infrastructure. In contrast, for example, the first impulse of Washington, DC, was to address CSO problems primarily through engineering and high-cost capital investments (Box 11.4). Philadelphia plans to keep rainwater out of the sewer system to begin with, as opposed to implementing more expensive and high-tech sewage treatment mechanisms.

Following the example of Philadelphia, New York City also took a balanced approach. In 2012 the city announced it would invest $2.4 billion in green

BOX 11.4

Gray or green solutions?

Washington, DC is like many cities in the developed world—it confronts the challenge of non-point source pollution in its waterways. It is subject to the requirements set out in the Clean Water Act, and because Washington, DC is part of the Chesapeake Bay watershed, it is also subject to the plan for improving water quality for the Bay.

In theory, all sewers in the District flow into the Blue Plains Treatment Plant, located at the southern tip of the city. Luckily, some two-thirds of the District has an SSS, but this means about one-third of its sewers are CSSs, built over 1870–1910. A rain event of even one half-inch causes the CSS to overflow into the Anacostia and Potomac Rivers, which then flow into Chesapeake Bay. In DC, there are some 53 CSO outfalls, many of which are in the older neighborhoods. CSOs can adversely affect water quality by introducing bacteria and trash to the water. The high level of organic debris in the water can also lead to low dissolved oxygen in the water, which can stress or kill fish. In the mid-1990s, DC averaged about 3.2 billion gallons per year of CSO overflow. Improvement projects which have included upgrades to various pipes and existing tunnels helped to reduce CSOs to about two billion gallons. But this is still in violation of the Clean Water Act and the EPA has told DC to solve this problem by 2015.

Solving the problem of stormwater run-off has led to a debate about which type of solutions to implement: gray or green infrastructure?

Gray solutions are man-made, more technological answers to non-point source pollution. For example, the gray solution being proposed for DC consists of three large concrete tunnels that will serve as storage for stormwater run-off. These tunnels can store flow from many sewer connections. Because a tunnel can share capacity among several outfalls, it can reduce the total volume of storage that must be provided for a specific number of outfalls. These tunnels store combined sewage but do not treat it. When the storm is over, the flows are pumped out of the tunnel and sent to a wastewater treatment plant. The estimated costs is $2.5 billion for the three tunnels. Two tunnels will be 23 feet in diameter and will hold 157 million gallons of stormwater run-off; the third will be smaller. But is investing $2.5 billion to build concrete storage tanks a good solution?

Another possibility is a green solution. A green solution manages stormwater before it enters the city's CSS. Such solutions can include a variety of approaches that use resources to hold, retain and slowly release the water into the sewerage system. These can include bioretention or rain gardens, which store and filter water. Rain gardens planted with water-loving vegetation can be small—in backyards, parks or along greenways, for example. Other small-scale green infrastructure includes rain barrels, which store water that is running out of gutters on homes and other buildings. Bioswales are another example. A bioswale uses an existing low area to retain water. Finally, green roofs can also retain and store water. The main idea behind green infrastructure is not to eliminate the stormwater run-off, but to slow it down, retain it and release it slowly over a longer period of time, thus allowing the existing sewerage infrastructure to handle the volume of water once the rain event is over.

The challenge in using green infrastructure, however, is that a city will need many different types across many different locations. This may be why the simpler solution of building massive concrete tunnels is appealing. Yet the benefits of a green solution to stormwater management connects to other types of initiatives, such as new requirements for green buildings. Investing in green infrastructure instead of large tunnels and detention tanks is a smarter approach to stormwater management; it can save the city money, and at the same time improve green space and the aesthetic quality of the city.

Figure 11.6 Rain garden in Washington, DC helps capture stormwater run-off. Sunken tree boxes store water; boxes and planters are filled with water-loving plants, and the bricks allow water to run between and underneath the paving, to be absorbed by sand and stored in underground cisterns

Source: Photo by Lisa Benton-Short

infrastructure and $1.4 billion in gray infrastructure to deal with CSO releases. New York City plans to add green roofs, expand the greenbelt in low-lying areas, plant trees, install green traffic meridians, and add more pervious sidewalks, all of which will absorb water. New York City is betting that if it installs green infrastructure, it won't have to spend billions of dollars to construct man-made storage tanks.

Water in developing cities

Water supply

Just as cities in the developed world face issues of water supply, so do those in the developing world. But the scope and scale of problems are often more pronounced in the developing world. One major issue is water scarcity. In many developing cities, particularly megacities or primate cities, the building of water supply systems has failed to keep pace with continued rapid population growth. Many cities were not only unprepared to manage such high population growth rates, but also did not have the necessary financial, technical or managerial capacities to do so. These cities fell behind in constructing and properly managing water supply and wastewater treatment systems.[16]

Continued urban population growth means tremendous stress on already inadequate resources, while also exacerbating the social and economic challenges to infrastructure investment. The right to safe water and adequate sanitation remains a promise unfulfilled. In 2000 the United Nations came out with the Millennium Development Goals (MDGs), one of which was to increase the number of people who have access to clean water. The new goal is to reduce by half the number of people lacking access to clean drinking water by 2015 (however, it should be noted that for whatever reasons, the MDGs did not include any sanitation targets.) Progress has been made, and the goal is nearly in reach. In 2000, some 170 million urban residents lacked access to safe drinking water. In 2010 the World Health Organization estimated that 780 million urban residents live without improved sanitation worldwide and 141 million urban residents live without improved sources of drinking water.[17] The number of those without access to safe drinking water is declining, but it is still high.

Unsafe drinking water continues to be responsible for more than 80 percent of diseases and 30 percent of deaths in the developing world.[18] And at any given time, close to half the people in the developing world (three billion plus) are suffering from one or more of the main water-linked disease such as diarrhea, cholera, enteric fevers, guinea worm and trachoma. Some experts argue that drinking water and proper wastewater treatment during the past 15 years has been progressively deteriorating, not improving.[19]

Every day in developing cities across the world, residents across the city depend on a variety of informal, and often illegal, techniques and practices to access water and sanitation. Although most cities have water supply infrastructure, it is often unreliable and inconsistent. Water runs only intermittently, or under irregular or insufficient pressure, it may be contaminated, and the system can suddenly breakdown, causing water to cease flowing for days or week at a time. This particularly impacts the city's poor and slum dwellers, many of whom have no legal access to piped water supply. They are particularly vulnerable to water access. As a result, the poor must spend considerable time each day to access water and sanitation facilities. This can include missing work to procure water, walking miles in search of sanitation or clean water, and buying water from illegal and informal sources. In addition, some will tap illegally into the water supply. Such water practices are predominately carried out by girls and women, since the responsibility of household management usually falls to them. Sometimes, for example, girls will be kept out of school in order to help with procuring tanker water delivery. Because many cities have criminalized practices such as tapping into existing pipes poor women are in an even more vulnerable position. Thus water access in developing cities is more than just about water: the time spent trying to get water curtails other opportunities, such as generating income or getting an education. It can also be dangerous.

Issues of urban water scarcity and water quality have attracted growing attention. In discussions on urban water, there are some who believe that privatization is the solution for providing clean water for everyone. But others are concerned with what they see as an alarming lack of debate on this issue, noting that water companies seem to dominate international water doctrine and wield powerful influence. They also note that while it may be tempting to see water scarcity as a physical phenomenon such as low rates of rainfall or groundwater depletion, research has revealed it is also the result of poverty and inequality. The work of urban political ecologists has dissected how water is connected to social power in the city. Unequal water distribution is often the outcome of discrimination, poor policies and inequality.

Urban political ecology has been important in offering a counter-narrative to those who believe privatization of water is the solution. Urban political ecology has revealed the meanings and consequences of water and the ways in which it is tied with politics, power and identity, not just natural or scientific factors. Eric Swyngedouw's work on the city of Guayaquil, Ecuador, shows that the meanings and consequences of water practices shape power, rights and citizenship in the city. He notes there are socio-environmental processes that produce water inequality in the city which are often linked to city-wide structures of water governance. In other words, water politics. Matthew Gandy's work on water in Mumbai reveals that the city has been dominated by middle-class interests to "modernize" by building elevated highways and new information technologies, while the urban poor and their need for clean, safe water has been ignored.[20] Similar problems exist in Lima, Peru, where Antoinio Ioris found that discriminatory practices against low-income residents have exacerbated water scarcity.[21] He also notes that improvements in water services had tended to concentrate in higher-income areas. Peru embraced water privatization, but, on average poor families are spending 48 percent of their income on food and drinking water. The research of feminist political ecologist Yaffa Truelove has shown that in Delhi, India, water is closely linked with gender, class and health. She notes that water inequality is also a result of everyday water practices that are produced by gender and class and that women and girls are disproportionally impacted by the lack of water access.

BOX 11.5

Phnom Penh: a success story

A good example of what is possible is the case of the city of Phnom Penh in Cambodia. Located on the banks of the Mekong River, the city of Phnom Penh in Cambodia is home to two million people. In 1993 it had a dismal water supply situation. Some 83 percent of water was lost due to leakages and unauthorized connections. In 1993 only 25 percent of the population had access to running water. Water-related diseases accounted for 30 percent of all hospitalizations and diarrheal disease was pervasive. In addition to the years of neglect as Cambodia recovered from the destruction of the Khmer Rouge and Vietnamese occupation, corruption and inefficiency were ubiquitous in government agencies. The Phnom Penh Water Supply Authority (PPWSA) was corrupt, incompetent and bankrupt. An estimated 80 percent of the staff worked less than two hours per day.

With new leadership, by 2000, water losses were reduced to about 35 percent, and by 2009 to about 6 percent. Everyone pays for the water consumed, and the PPWSA's profit each year has increased significantly. In an interview,

Ek Sonn Chan, who took over as Director of the PPWSA, stated "It's not the problem of scarcity of water resources; it's not the lack of financing, but because of the lack of good governance." In 20 years, water supply in the city was transformed and the PPWSA is completely turned around. Its annual water production increased by 437 percent, the distribution network by 557 percent, pressure of the system by 1,260 percent and its customer base by 662 percent. In 2011, 92 percent of Phnom Penh's households had running water. In addition to large-scale upgrades and expansion of service, most households pay for their water use, and are metered. Unaccounted water (from leaking pipes or illegal tapping) is now below 7 percent.

The water authority has successfully eliminated corruption and made its governance system as transparent as in London or New York. It has less management and technical expertise than many other developing countries like Brazil, Egypt or India, and no private sector to which specific activities can be outsourced. Yet it has accomplished its goals for providing clean water to all its inhabitants, rich or poor, and at affordable prices. Phnom Penh is an example of how good governance can create better outcomes in the cities of the developing world. It also underscores that water scarcity is not just a physical phenomenon, but is strongly related to socio-political forces that create problems of water management, distribution and unequal access. If clean water is possible in Phnom Penh, it is possible throughout the developing world.

Sources: Biswas, A. (2010) "Water for a thirsty urban world." *The Brown Journal of World Affairs* 17 (1): 146–62; Biswas, A. and Tortajada, C. (2010) "Water supply of Phnom Penh: an example of good governance." *International Journal of Water Resources Development* 26 (2): 157–72; Gifford, R. (2011, June 2) "Phnom Penh's feat: getting clean top water flowing." *National Public Radio.* www.npr.org (accessed February 2012).

In addition to water inequality, water scarcity is also a product of demand that exceeds supply. Take the case of Mexico City.

Mexico City: sinking

While many are familiar with Mexico City's infamous air pollution problem, an equally pressing issue centers on its water supply. Approximately 72 percent of Mexico City's water supply comes from its aquifers.[22] Mexico City is built on top of vast underground aquifers. Four of the 14 aquifers in the Valley of Mexico Basin are overexploited. One study reported that the per capita rechargeable water available for the Valley of Mexico in 2010 is 163 m³, but in 2030 it is predicted that rechargeable water per capita will be 148 m³.[23] The city consumes at least 60,000 liters of water per second, 80 percent of which is groundwater.[24]

Aquifers can be an excellent source of clean water since many pollutants are filtered out as the water passes through the soil and rocks. However, aquifers are slow to replenish. Today, water is being extracted twice as fast from the aquifer as it is being replaced. As a result, land subsidence has become a serious problem for Mexico City. It is estimated that over the past century, Mexico City has sunk by as much as 10 meters in some areas.[25] Since the 1970s, the rate of sinking in the central part of the city is about 6 cm each year, although this varies geographically—some areas of the city are sinking as much as 40 cm each year. Some of these are home to the urban poor.

Land subsidence weakens foundations to buildings and other infrastructure, including damaging the water and sewerage pipes. Pipes are ruptured or develop small leaks. It is estimated that 30 percent of water is lost in the pipes from leaks before reaching users—enough to provide water to more than four million people.[26] Leaking sewerage pipes also contaminate the groundwater with heavy metals and microorganisms.

Land subsidence has also caused flooding in the valley, particularly after heavy rains. Originally, Texcoco Lake was nine feet (three meters) lower than central Mexico City; today the lake is over 6.5 feet (two meters) higher than the city.[27] Dikes had to be built to confine the stormwater flow and pumps are needed to lift the drainage water under the city to the level of the drainage canals. In addition, because of subsidence, gravity no longer takes sewage and run-off to the Grand Canal; the city had to install pumps in order to remove sewage to Texcoco Lake. The cost of pumping water into, within and out of the Basin of Mexico amounts to almost $900,000 per day.[28]

Because of irregular water supply, many households use water storage tanks, located on rooftops. These are often left uncovered and are not always cleaned regularly, which enables bacteria to flourish.

Mexico City also has a significant part of the population with no access to sewage facilities. Some 30 percent of Mexico City residents do not have toilets, and as much as 93 percent of wastewater is discharged untreated.[29] About 75 percent of the city's residents have access to the city's current wastewater system of unlined sewer canals, sewers, rivers, reservoirs, lagoons, pumping stations and deep drainage systems. During the rainy season, when domestic wastewater and industrial wastewater are mixed with stormwater run-off, more than 1.5 million tons pass through the city's sewerage system untreated.[30] A plan to build four treatment plants has floundered for lack of financing and no serious proposals for the construction of new treatment plants have been made since the 1990s. As a result, the city lacks the capacity to reuse significant volumes of water for industrial and agricultural production within the valley, a practice that would also alleviate the demand for fresh water.[31]

Mexico City, like many developing cities, is challenged by problems related to water management that may cause social conflicts. The wealthier population of Mexico City consumes up to 40 times more water than that used by the poor. Compounding the issue is that many of the poorest areas of the city not only receive water of unacceptable quality, they also suffer the inconvenience of rotating schedules for water deliveries, whether through the water network or from tanker trucks.[32] The poor pay private vendors, on average, 6–25 percent of their daily salaries. General distrust of tap water quality has led to much of the population purchasing drinking water; Mexico was ranked the third largest consumer of bottled water in 2009.

Water quality

While there is considerable progress in providing water to the urban areas of the developing world, the quality of water provided remains a serious, critical and somewhat neglected issue. Attention on wastewater management has proved to be a poor cousin to water supply. The discharge of untreated or partially treated wastewater has significantly contaminated water sources in and around urban centers of the developing world, ranging from Delhi, Lagos and Cairo to Mexico City. Water expert Asit Biswas argues that the real water crisis the world is likely to face in the coming decades will not come from absolute physical scarcity, but rather from widespread water contamination.[33]

Many cities in the developing world have much more serious non-point sources of water pollution than cities in the developed world because large sections of their population are not served by sewers, drains or solid-waste collection. In some cases, environmental regulations on point source pollution, such as those from industrial manufacturing, are not always enforced—so developing cities confront significant point pollution too. The Ganges River is a good example of the ways in which water quality issues can differ significantly in the developing world and how the simplistic application of Western technology is not always successful.

The Ganges: a sacred river

The Ganges winds 1,500 miles across northern India, from the Himalaya Mountains to the Indian Ocean, and through 29 cities with populations over 100,000. Known as Ganga Ma (Mother Ganges), the river is revered as a goddess whose purity cleanses the sins of the faithful and helps the dead on their path toward heaven. Hindus believe that if the ashes of their dead are deposited in the river, they will be ensured a smooth transition to the next life or freed from the cycle of death and rebirth. It is said that a single drop of Ganges water can cleanse a lifetime of sins. In cities along the Ganges, daily dips are an important ritual among the faithful. Many cities are pilgrimage sites.

Despite its spiritual importance, the physical purity of the river has deteriorated dramatically. While industrial pollutants account for some of the river's pollution, the majority of the Ganges' pollution is organic waste: sewage, trash, food, human wastes and animal remains. Today nearly half a billion people live in the basin of the Ganges and more than 100 cities dump their raw sewage directly into the river. Some pollutants are measured at 340,000 times permissible levels. Not surprisingly, waterborne illnesses are common killers, accounting for the deaths of some two million Indian children each year.

In the city of Varanasi, one of India's oldest and holiest cities, some 40,000 cremations are performed each year; those unable to afford the cost of traditional funerals often dump the body into the river. In addition, the carcasses of thousands of dead cattle, which are considered sacred to Hindus, are tossed into the river each year. The city also pumps some 200 million gallons of sewage waste daily into the river.[34] And yet thousands of Hindus continue to come to Varanasi and the Ganges.

The Ganges presents an interesting case study. Because the river is holy, it attracts tens of thousands of pilgrims each day for ritual bathing, exposing large numbers of people to untreated, contaminated water. As one man commented:

> There is a struggle and turmoil inside my heart. I wanted to take a holy dip. I need it to live. The day does not begin for me without the holy dip. But, at the same time, I know what is B.O.D and I know what is fecal coliform.[35]

It is estimated that 40 percent of the people who take a dip in the river regularly have skin or stomach ailments.

In 1985 the Indian government initiated the Ganga Action Plan, to clean up the river in selected areas by installing sewage treatment plants. A 2012 World Bank report found that only one-third of the sewage generated by the towns and cities on the mainstem of the river is treated; and untreated or poorly treated industrial wastewater is responsible for 20 percent of all wastewater inflows into the river. In 2011, India announced the $1.5 billion National Ganga River Basin Project, with $1 billion in financing from the World Bank.[36] The project will support the National Ganga River Basin Authority (NGRBA), which was created in 1999. The NGRBA will fund investments (like sewage treatment plants, sewer networks, etc.) that are critical for reducing pollution in the Ganga but, importantly, the project will help build the capacity of city-level service providers responsible for running these assets and also modernize their systems for doing so.

Some remain skeptical of the new plan, and the adoption of expensive, multi-million-dollar Western-style treatment plants. The Indian government spent close to $500 million over 1985–2000 on treatment plants, with mixed results. Many of the treatment plants malfunctioned, were designed improperly or could not handle the bacterial load. In addition, corruption and ineffective monitoring contributed to the problems. There is growing criticism over the adoption of Western-style technology to solve developing world issues. Western-style technology tends to be very expensive, relies on highly trained engineers and workers to maintain the technology, and requires a stable and consistent supply of electricity. In addition, Western-style waste treatment plants were engineered for use in countries where there are no monsoon rains, and where the population does not drink directly from the water source. Few considered the radically different ways that people use rivers in India.

As an alternative to the high-technology treatment plants, Veer Bhadra Mishra, a Hindu priest and civil engineer, collaborated with University of California Berkeley engineer William Oswald to develop a non-mechanized, low-tech sewage treatment plan. This plan is more compatible with the climate of India and replaces the high-tech solution with a wastewater oxidation pond system that would store sewage in a series of ponds and use bacteria and algae to break down waste and purify the water. The ponds allow waste to decompose naturally in water. Bacteria grow on the sewage and decompose it; the algae feed on the nutrients released by the bacteria and produce oxygen for the water. This alternative treatment does not require electricity, but relies on sunshine to speed up the decomposition. The pond system is much less expensive than mechanical treatment plants.

The debate between Western-style technology and lower-cost alternatives in the case of the Ganges highlights general themes applicable to many of the solutions to pollution in developing cities. While the 1980s and 1990s saw large-scale investments by governments in high-tech solutions, the lack of results has caused many to seek alternatives that are locally sensitive and economically affordable. More recently, experts in both the developing and developed world contend that solutions should respond to local demands and should be as simple, sturdy and inexpensive as possible. Low-cost, low technology solutions such as the pond system or pour-flush latrines or even improved pit latrines have been successful. An important element in pollution control is the role of public participation. The involvement of the local community and households is a crucial component to success.

Conclusions

Some cities have seen dramatic decreases in certain types of pollutants, while other pollutants have increased. Some cities have successfully implemented pollution laws and regulations, while many cities (and countries) lack the resources to adequately enforce such measures. The twin pressures of population growth and increasing consumption of resources have, in some cases, offset new laws and regulations designed to decrease or prevent pollution. Finally, an important trend is that cities are beginning to plan for climate adaptation, which involves dealing with water supply, water quality and flooding. We explore the connection between climate change and its impact on urban water in more detail in Chapter 13.

Lastly, water issues in developing cities are not all problematic. Some cities are developing and protecting vital water resources. Figure 11.7 shows the river gardens of Beijing. The river gardens attract tourists and at the same time provide an important ecosystem within the rapidly growing city. Today most countries and municipalities have enacted pollution reform measures. The aspiration for cleaner water is there.

Figure 11.7 Beijing Bamboo Park. Many cities in the developing world have restored important water features, such as this water garden in downtown Beijing

Source: Photo by Michele A. Judd

Guide to further reading

Bakker, K. (2010) *Privatizing Water: Governance Failure and the World's Urban Water Crisis*. Ithaca, NY: Cornell University Press.

Biswas, A. (2010) "Water for a thirsty urban world." *The Brown Journal of World Affairs* 17 (1): 146–62.

Biswas, A., Tortajad, C. and Izquierdo-Avino, R. (2009) *Water Management in 2020 and Beyond*. Berlin: Springer-Verlag.

Cisneros, B. J. and Rose, J. B. (2009) *Urban Water Security: Managing Risks*. Leiden: Taylor & Francis.

Craig, R. K. (2009) *The Clean Water Act and the Constitution: Legal Structure and the Public's Right to a Clean and Healthy Environment*. 2nd edition. Washington, DC: Environmental Law Institute.

Engel, K., Jokiel, D., Kraljevic, A., Geiger, M. and Smith, K. (2011) *Big Cities, Big Water, Big Challenges: Water in an Urbanizing World*. Berlin: World Wildlife Fund.

Gumprecht, B. (2001) *The Los Angeles River: Its Life, Death and Possible Rebirth*. Baltimore, MD: Johns Hopkins University Press.

Jones, J. A. A. (2010) *Water Sustainability: A Global Perspective*. London: Hodder Education.

Karvonen, A. (2011) *Politics of Urban Runoff: Nature, Technology, and the Sustainable City*. Cambridge, MA: MIT Press.

Lewin, T. (2003) *Sacred River: The Ganges of India*. Boston, MA: Houghton Mifflin/Clarion Books.

Melosi, M. V. (2011) *Precious Commodity: Providing Water for America's Cities*. Pittsburgh, PA: University of Pittsburgh Press.

Pearce, F. (2007) *When the Rivers Run Dry: Water—the Defining Crisis of the Twenty-first Century*. Boston, MA: Beacon Press.

Solomon, S. (2010) *Water*. New York: Harper Perennial.

Swyngedouw, E. (2004) *Social Power and the Urbanization of Water: Flows of Power*. Oxford: Oxford University Press.

Uitto, J. and Biswas, A. (eds.) (2000) *Water for Urban Areas: Challenges and Perspectives*. Tokyo and New York: United Nations University Press.

World Health Organization and UNCIEF (2012) "Progress on sanitation and drinking-water." http://www.wssinfo.org/fileadmin/user_upload/resources/JMP-report-2012-en.pdf.

For a good account of environmental policymaking, see:

Desfor, G. and Roger, K. (2004) *Nature and the City: Making Environmental Policy in Toronto and Los Angeles*. Tucson, AZ: Arizona University Press.

On the web, see:

Watch an interesting short video on combined sewer overflows at: http://www.riverkeeper.org/campaigns/stop-polluters/sewage-contamination/cso.

12 Air

The harmful effects of air pollution are a major issue in cities in both the developed and developing world. The chapter is divided into two parts. In the first part, we draw upon the experience of the US as an exemplar of an industrial country; in the second part we examine a range of developing cities.

Air pollution in the US

Air pollutants are composed of either visible particulates (ash, smoke or dust) or invisible gases and vapors (fumes, mists and odors). The US Environmental Protection Agency (EPA) has identified more than 189 air pollutants; however, the regulatory measures are primarily directed at only a few. These dozen air pollutants account for a majority of the volume of air pollutants. Most of the primary air pollutants are generated when fossil fuels are used. Throughout the twentieth century developed and developing societies have increasingly relied on three primary fossil fuels—coal, natural gas and petroleum—for energy and transportation. Because all fossil fuels are carbon-based, they release carbon monoxide and carbon dioxide when combusted. Coal also contains sulfur oxides and nitrogen oxides that are released during combustion. New sources of pollution, combined with increased use of fossil fuels for a myriad of needs, means that air pollution for many cities has continued to increase despite regulatory efforts.

The six criteria pollutants

Nearly all of the so-called "most common" or most prevalent pollutants are generated in urban areas. In the US, the EPA has classified six "criteria pollutants" as those which pose the most ubiquitous threat to health and the environment (Box 12.1). These pollutants are called criteria pollutants because the EPA uses them as the basis for setting permissible levels.

BOX 12.1

The common air pollutants

Ozone (ground-level ozone is the principal component of smog)

Source: chemical reaction of pollutants, VOCs and NOx

Health effects: breathing problems, reduced lung function, asthma, irritated eyes, stuffy nose, reduced resistance to colds and other infections, may speed up aging of lung tissue

Environmental effects: ozone can damage plants and trees; smog can cause reduced visibility

Property damage: damages rubber, fabrics, etc.

*VOCs** (volatile organic compounds); smog-formers

Source: VOCs are released from burning fuel (gasoline, oil, wood, coal, natural gas, etc.), solvents, paints, glues and other products used at work or at home. Cars are an important source of VOCs. VOCs include chemicals such as benzene, toluene, methylene chloride and methyl chloroform

Health effects: In addition to ozone (smog) effects, many VOCs can cause serious health problems such as cancer and other effects

Environmental effects: In addition to ozone (smog) effects, some VOCs such as formaldehyde and ethylene may harm plants

* All VOCs contain carbon (C), the basic chemical element found in living beings. Carbon-containing chemicals are called organic. Volatile chemicals escape into the air easily. Many VOCs, such as the chemicals listed in the table, are also hazardous air pollutants, which can cause very serious illnesses. The EPA does not list VOCs as criteria air pollutants, but they are included in this list of pollutants because efforts to control smog target VOCs for reduction.

Nitrogen dioxide (one of the NOx); smog-forming chemical

Source: burning of gasoline, natural gas, coal, oil, etc. Cars are an important source of NO_2.

Health effects: lung damage, illnesses of breathing passages and lungs (respiratory system)

Environmental effects: nitrogen dioxide is an ingredient of acid rain (acid aerosols), which can damage trees and lakes. Acid aerosols can reduce visibility

Property damage: acid aerosols can eat away stone used on buildings, statues, monuments, etc.

Carbon monoxide (CO)

Source: burning of gasoline, natural gas, coal, oil, etc.

Health effects: reduces ability of blood to bring oxygen to body cells and tissues; cells and tissues need oxygen to work. Carbon monoxide may be particularly hazardous to people who have heart or circulatory (blood vessel) problems and people who have damaged lungs or breathing passages

Particulate matter (PM-10); (dust, smoke, soot)

Source: burning of wood, diesel and other fuels; industrial plants; agriculture (plowing, burning off fields); unpaved roads

Health Effects: nose and throat irritation, lung damage, bronchitis, early death

Environmental effects: particulates are the main source of haze, which reduces visibility

Property damage: ashes, soots, smokes and dusts can dirty and discolor structures and other property, including clothes and furniture

Sulfur dioxide

Source: burning of coal and oil, especially high-sulfur coal from the eastern United States; industrial processes (paper, metals)

Health effects: breathing problems, may cause permanent damage to lungs

Environmental effects: SO_2 is an ingredient in acid rain (acid aerosols), which can damage trees and lakes. Acid aerosols can also reduce visibility

Property damage: acid aerosols can eat away stone used in buildings, statues, monuments, etc.

Lead

Source: leaded gasoline (being phased out), paint (houses, cars), smelters (metal refineries); manufacture of lead storage batteries

Health effects: brain and other nervous system damage; children are at particular risk; some lead-containing chemicals cause cancer in animals; lead causes digestive and other health problems

Environmental effects: lead can harm wildlife

Source: US Environmental Protection Agency: http://www.epa.gov/air/oaqps/peg_caa/pegcaa11.html (accessed August 2006).

The EPA has developed one set of limits (called primary standard) to protect health; another set of limits called secondary standard are intended to prevent environmental and property damage. Almost all air quality monitors are located in urban areas, so air quality trends are more likely to track changes in urban emissions rather than changes in total national emissions.

Criteria pollutants are the most regulated air pollutants and policy has focused on limiting (but not outlawing) their use and discharge into the environment. One of the most difficult aspects to limiting the production of criteria pollutants is that they are emitted from a variety of sources—the generation of energy, the use and combustion of fossil fuels in various forms of transportation and agricultural activities. For many cities, however, much of the air pollution comes from motor vehicle exhaust, a problem that has increased due to more vehicles being on the road and more miles being driven each year. For example, hydrocarbons—the precursor to smog or surface-level ozone—come primarily from cars, motorcycles, trucks and gasoline-powered equipment (such as lawnmowers and blowers). Diesel-powered vehicles and engines contribute more than half the mobile source of particulate emissions. Almost one-third of nitrogen oxides come from motor vehicles, and 95 percent of the carbon monoxide in US cities comes from mobile sources. The significant connection between vehicles and urban air pollution is well established.

In the US, a city that does not meet the primary standard for a given pollutant is called a non-attainment area. Cities in non-attainment must create an action plan for each pollutant in non-attainment. Action plans include: targets for reduction of pollutants; ways to increase use of public transportation and decrease the use of single-occupancy vehicles; securing voluntary reductions; and developing a variety of outreach and educational tools. Although the EPA has been regulating criteria pollutants since the 1970 Clean Air Act was passed, many urban areas are still classified as non-attainment for at least one criteria pollutant. The American Lung Association's 2012 Report *State of the Air* examines three key pollutants—smog, short-term particulate matter (stays in the air for under 24 hours) and long-term particulate matter. The report shows that air quality in many cities has improved, but that over 127 million people—41 percent of the nation—still suffer pollution levels that are too often dangerous to breathe.[1] Approximately 5.7 million Americans live in cities that received "F"s for all three pollutants.

Photochemical smog

Smog occurs when VOCs react with nitrogen oxides and oxygen in the presence of heat and sunlight.[2] Pollutants undergo reactions that form ground-level ozone, or smog. Hence smog is not emitted directly into the air, but is created through a series of chemical reactions. It is produced by the combination of pollutants from many sources, including smokestacks, cars, paints and solvents. The EPA estimates that almost 60 percent of smog is produced by transportation sources—cars, trucks and trains. Smog is found primarily in urban areas and is often worse in the summer months when heat and sunshine are more plentiful. Table 12.1 lists the worst US cities for smog in 2012. Although Californian cities lead the list, nearly all cities have improved air quality since 2000.

Table 12.1 Ten worst ozone-polluted cities, 2011

2012 Rank	Metropolitan area
1	Los Angeles–Long Beach–Riverside, CA
2	Bakersfield–Delano, CA
3	Visalia–Porterville, CA
4	Fresno–Madera, CA
5	Sacramento–Arden–Arcade–Yuba City, CA
6	Hanford–Corcoran, CA
7	San Diego–Carlsbad–San Marcos, CA
8	Houston–Baytown–Huntsville, TX
9	Merced, CA
10	Charlotte–Gastonia–Salisbury, NC–SC

Source: American Lung Association, Annual Report, 2011. http://www.stateoftheair.org/2011/city-rankings/most-polluted-cities.html (accessed July 2012).

Ozone smog represents the single most challenging pollution problem for most US cities. More than 81 million Americans live in urban areas that exceed air quality concentrations for ozone, and progress on ozone reduction is the slowest for all criteria pollutants. But progress has been made. In 2012, 22 of the 25 cities with the most ozone pollution improved their air quality over 2011. Still, nearly four in ten people in the US live in areas with unhealthy levels of ozone pollution.[3]

Smog is highly corrosive to rubber, metals and lung tissue. Short-term exposure can cause eye irritation, wheezing, coughing, headaches, chest pain and shortness of breath. Long-term exposure scars the lungs, making them less elastic and efficient, often worsening asthma and increasing respiratory tract infections. Because ozone penetrates deep into the respiratory system, many urban residents are at risk, including the weak and elderly, but also those who engage in strenuous activity. For those who were born and have lived in smoggy cities such as Los Angeles, Houston and Washington, DC, long-term exposure may be breaking down the body's immune system, increasing the chances of suffering respiratory illness and harming the lungs in later life. In many US cities, doctors are reporting increasing frequency of asthma in children. A recent Southern California Children's Health study looked at the long-term effects of particulate pollution on teenagers. Tracking 1,759 children between 10 and 18, researchers found that those who grew up in more polluted areas faced increased risk of having underdeveloped lungs, which may never recover to their full capacity. The average drop in lung function was 20 percent below what was expected for the child's age, similar to the impact of growing up in a home with parents who smoked.[4] Ozone is not just a US problem. Changes in air pollution from the reunification of Germany proved a real-life laboratory. East and West Germany had different levels and sources of particles. Outdoor particle levels were much higher in East Germany, where they came from factories and homes. West Germany had higher concentrations of traffic-generated particles. After reunification, emissions from the factories and homes dropped, but traffic increased. A German study explored the impact on the lungs of six-year-olds from both East and West Germany. Total lung capacity improved with the lower particle levels. However, for those children living near busy roads, the increased pollution from the increased traffic and ozone exposure kept them from benefiting from the overall cleaner air.[5]

BOX 12.2

Geography and disparities in air pollution

Sometimes it is all about location. Motor vehicles emit large quantities of carbon dioxide (CO_2), carbon monoxide (CO), hydrocarbons (HC), nitrogen oxides (NOx), particulate matter (PM) and substances known as mobile source air toxics (MSATs), such as benzene, formaldehyde, acetaldehyde, 1,3-butadiene and lead (where leaded gasoline is still in use). Recent studies have found that people who live near heavy traffic or a major road may be at higher risk. Growing evidence shows that the vehicle emissions coming directly from those highways may be higher than in the community as a whole, increasing the risk of harm to people who live or work near busy roads.

The number of people living near a busy road may include 30–45 percent of the population in North America. Another study found an increase in risk of heart attacks from being in traffic, whether driving or taking public transportation. In one study, urban women living in Boston experienced decreased lung function associated with traffic-related pollution. But this is not just limited to the US. A Danish study found that long-term exposure to traffic air pollution may increase the risk of developing chronic obstructive pulmonary disease (COPD). They found that those most at risk were people who already had asthma or diabetes. Studies have found increased risk of premature death from living near a major highway or an urban road.

In January 2010, the Health Effects Institute published a major review of the evidence by a panel of expert scientists. The panel looked at over 700 studies from around the world, examining the health effects. They concluded that traffic pollution causes asthma attacks in children, and may cause a wide range of other effects, including: the onset of childhood asthma, impaired lung function, premature death and death from cardiovascular diseases, and cardiovascular morbidity. The area most affected, they concluded, was roughly 0.2–0.3 miles (300–500 meters) from the highway.

Sources: American Lung Association (2012) "Highways may be especially dangerous for breathing," in *State of the Air, 2012.* http://www.stateoftheair.org/2012/health-risks/health-risks-near-highways.html (accessed July 2012). Health Effects Institute (2010) "Traffic-related air pollution: a critical review of the literature on emissions, exposure, and health effects—a special report of the HEI Panel on the Health Effects of Traffic-Related Air Pollution." http://pubs.healtheffects.org/view.php?id=334 (accessed July 2012).

Photochemical smog is one pollutant that is often exacerbated due to geography. Cities located in basins and valleys—such as Los Angeles, Mexico City and Denver—are particularly susceptible to the production of smog. Denver, the Mile High City, suffers from smog and other air pollutants that are made worse by its elevation in the Rocky Mountains. Because of Denver's high altitude, the city experiences frequent temperature inversions when warm air is trapped under cold air and cannot rise to disperse the pollutants to a wider area. As a result, smog may hover in place for days at a time, generating a "smog soup" that envelops the city. Denver has worked to turn around its air pollution problem, but despite these efforts, as of 2011 Denver remained out of compliance with EPA standards. Another issue of "smog geography" occurs in Montreal and Toronto. These cities are upwind of major industrial cities in the Midwest, and some of the air pollutants that generate smog in Toronto and Montreal originate across the border. In Canada, Windsor, Toronto, Montreal and Vancouver are cities where acceptable ozone levels are exceeded an average of ten or more days in the summer.

Cities with high concentrations of fossil-fuel-burning power plants, metal smelters, cement and fertilizer factories and with high densities of cars and trucks are also likely to experience smog production.

Since the 1970s the US and Canadian governments have taken steps to control pollution emissions from automobiles and factory stacks. Catalytic converters capture much of the automobile exhaust, vapor traps on gas pumps help prevent the evaporation of carbon dioxide into the air. Recent efforts to develop hybrid and zero-emission vehicles (such as electric cars) are other ways both private and public interests are using technology to alleviate air pollution. However, new sources of pollution, combined with increased use of fossil fuels for myriad needs, has meant that air pollution for many US and Canadian cities has continued to increase despite regulatory efforts.

The EPA developed an "Air Quality Index" to educate people about how to respond to high levels of smog (Table 12.2). When ozone concentrations are high, smog alerts are issued to warn people with asthma and those with chronic respiratory disease to stay indoors and for healthy people not to exercise. For many urban residents, smog forecasts are part of their daily

Table 12.2 The US EPA's Air Quality Index

Air Quality Index (AQI)	Numerical value	Meaning	Color
Good	0–50	Air quality is considered satisfactory, and air pollution poses little or no risk.	Green
Moderate	51–100	Air quality is acceptable; however, for some pollutants there may be a moderate health concern for a very small number of people who are unusually sensitive to air pollution.	Yellow
Unhealthy for sensitive groups	101–150	Members of sensitive groups may experience health effects. The general public is not likely to be affected.	Orange
Unhealthy	151–200	Everyone may begin to experience health effects; members of sensitive groups may experience more serious health effects.	Red
Very unhealthy	201–300	Health alert; everyone may experience more serious health effects.	Purple
Hazardous	>300	Health warnings of emergency conditions. The entire population is more likely to be affected.	Maroon

Source: US EPA.

planning. On the worst smog concentration days in Washington, DC officials encourage the use of public transportation by making the metro and metro buses free to all users. In most cities, major newspapers feature a daily "Air Quality Index," noting which air pollutants are likely to be above federal standards. An important change is the development of air quality apps for smart phones, which allow users to view their local air quality for a range of pollutants (Box 12.3).

BOX 12.3

There's an app for that

New technologies are changing the way people access information about environmental quality. In the US, for example, the American Lung Association offers State of the Air, a free smartphone app for iPhone and Android with current air quality information. According to their website, it is considered "a life-saving resource for people living with lung disease like asthma and chronic obstructive pulmonary disease (COPD), people with heart disease or diabetes, as well as older adults and children." With State of the Air, users can enter their zip code or use the geo-locator function on their smartphones to get current and next-day air quality conditions. The app provides levels of both ozone and particle pollution, and sends out alerts if air quality is unhealthy. If the day's air pollution is particularly severe, the app will provide specific recommendations—advising that outdoor activities should be rescheduled or that people who work outdoors should limit extended or heavy exertion.

The EPA also offers a free app. Their AIRNow mobile application provides real-time air quality information that can be used to protect your health when planning your day. This app allows a user to get location-specific reports on current air quality and air quality forecasts for both ozone and fine particle pollution (PM2.5). Air quality maps from the AIRNow website provide visual depictions of current and forecast air quality nationwide, and a page on air quality-related health effects explains what actions people can take to protect their health at different AQI levels, such as "code orange."

In the UK, the London Air Android app 3.0 displays the latest air pollution levels recorded at over 100 monitoring locations in the London Air Quality Network. The London app was designed by the Environmental Research Group at King's College London. There are also apps for Beijing and one for the city of Chengdu. The Chengdu Air Quality widget shows the real-time Air Quality Index (AQI), measured by the US Embassy in Chengdu.

Also under discussion is the idea of putting air quality sensors in cell phones, with the idea that these would provide a crowd-sourced way of gathering numerous location-specific, real-time air quality measurements. If developed, this would be an innovative and interesting way for scientists to gather data at a microscale.

To download Air apps, see websites for: American Lung Association (http://www.lung.org/healthy-air/outdoor/state-of-the-air-app.html); EPA air app (http://m.epa.gov/apps/airnow.html).

Toxic and hazardous pollutants

Air pollutants also include those that are either toxic or hazardous. Table 12.3 lists several toxic or hazardous air pollutants. Hazardous and toxic air

Table 12.3 Toxic air pollutants

Mercury
Lead
Polychlorinated biphenyls (PCB)
Dioxins
Benzene
Pesticides such as DDT
Cadmium compounds
Chloroform
Formaldehyde
Methyl chloride
Arsenic

Source: US EPA.

pollutants can cause cancer or can kill swiftly. The 1990 Clean Air Act amendment requires factories and other businesses to develop plans to prevent accidental releases of highly toxic chemicals.

Currently the EPA has identified 188 toxic or hazardous chemicals, many of them not well-researched and poorly understood in terms of their impact on human health and the environment. In urban areas, toxic air pollutants are of special concern because of the concentration of people close to sources of emissions. In response to growing concerns about the long-term exposure to toxic or hazardous chemicals, the EPA has singled out 33 of the 188 toxic air pollutants as the greatest threat to public health in cities. A 2010 EPA report noted that the risks of cancer from toxic air pollutants, such as formaldehyde and benzene, are higher in urban areas. These toxic air pollutants contributed nearly 60 percent of the average individual cancer risk. The report noted that urban areas and major transportation corridors have a higher risk than the national average.[6]

The EPA's strategy is to target these pollutants in cities by using a variety of national and local controls. Policy approaches to toxic or hazardous pollutants differ from criteria pollutants in two important ways. First, while states were charged to develop and implement their own plans to meet criteria pollutants, there is a federal plan to deal with toxics. Second, EPA policy aims to eventually outlaw or completely eliminate the discharge of toxic pollutants. Since 2003 the EPA, working with state and local partners, has nationally monitored air toxic pollutants through the National Air Toxics Trends Station (NATTS) program. From 1990 to 2005, emissions of air toxic pollutants declined by approximately 42 percent. The principal objective of the NATTS network is to provide long-term monitoring data across representative areas of the country for NATA priority pollutants (e.g., benzene, formaldehyde, 1,3-butadiene, hexavalent chromium and polycyclic aromatic hydrocarbons [PAHs] such as naphthalene) in order to establish overall trends. A 2010 report showed a decrease in these concentrations in some sites. There is still much to be done to achieve the goal of "zero" discharge of toxins.

Indoor air pollution

Another type of pollution that is often overlooked, but is a threat to public health nevertheless, is indoor air pollution. Radon is an example of an indoor air pollutant. It is a radioactive gas that results from a natural breakdown of uranium in soil, rocks and water. Although radon gas can be found in any type of building, it is more likely that a person will be exposed to it in their homes.[7]

The radon gas enters into a home primarily through the foundation and basement. It can come in through cracks in basement floors, drains, sump pumps and construction joints, among other places.[8] Radon cannot be seen or smelled, so without testing people may not realize they are exposed to unsafe radon levels in the home. This is of particular concern because the EPA estimates that radon is currently linked to 20,000 lung cancer deaths per year, with the risk increasing substantially if the person smokes.[9] Naturally occurring radon in indoor air has been identified as the second leading cause of lung cancer after tobacco smoking. In 1988 Congress added Indoor Radon Abatement to the Toxic Substances Control Act,[10] thus drawing attention to the danger it poses as an indoor air pollutant and toxic substance. The EPA estimates that 1 in 15 homes have elevated radon levels.[11]

Radon is a problem all over the US and Canada, but it varies in prevalence and intensity. In Canada, for example, a 2010 report found that radon levels in Ottawa were three times the national average; cities in the provinces of Manitoba, Nova Scotia and Saskatchewan have the highest proportion of homes estimated to be above the guideline, but there is great variation even within the cities and provinces.[12] Radon measurements can vary due to geology, aerial radioactivity, soil permeability and foundation type.[13] The magnitude of the problem, its geographic variability and the fact that it can vary in measurements in the short term and long term makes it difficult to manage at the national and state level. States differ in their programs; some provide free radon testing kits or regulate providers of radon testing and mitigation.[14]

The obligation to test for radon lies with the homeowner and it is particularly recommended when buying, building or selling a house. The EPA has focused most on providing information and data.[15] Testing for radon can be done using charcoal canisters, alpha track detectors, ion detectors and more advanced electrical devices.[16]

Fixing radon problems and testing for radon occurs most frequently when developers are building new homes or renovating older ones. During construction it is easy and relatively inexpensive to incorporate radon-resistant features. These include putting down gravel, plastic sheeting and vent pipes. Although costs vary, it will typically be less expensive to do it during construction as opposed to fixing problems later.[17]

Another indoor pollutant is carbon monoxide. Carbon monoxide is a product of incomplete combustion of organic matter due to insufficient oxygen supply to enable complete oxidation to carbon dioxide (CO_2). Symptoms of mild to acute poisoning include lightheadedness, confusion, headaches, vertigo and flu-like effects; larger exposures can lead to significant toxicity of the central nervous system and heart, and even death. Carbon monoxide is a toxic gas, but being colorless, odorless, tasteless and initially non-irritating, it is very difficult for people to detect. In many industrialized countries carbon monoxide is the cause of more than 50 percent of fatal poisonings. Each year, more than 400 Americans die from unintentional CO poisoning, more than 20,000 visit the emergency room and more than 4,000 are hospitalized due to CO poisoning.[18]

Sources of indoor carbon monoxide include cigarette smoke, house fires, faulty furnaces, heaters, wood-burning stoves, internal combustion, vehicle exhaust (in an enclosed garage), generators and propane-fueled equipment such as portable stoves. Carbon monoxide can be generated when such equipment is used inside buildings or semi-enclosed spaces. Poisoning is typically more common during the winter months due to increased domestic use of gas furnaces, gas or kerosene space heaters and kitchen stoves, which if faulty and/or used without adequate ventilation may produce excessive carbon monoxide. Carbon monoxide poisoning also occurs after natural disasters—such as hurricanes and ice storms—that interrupt power service. Homeowners use generators or portable stoves for heat, power and cooking, but if these devices are located too close to the indoors they can cause a build-up of CO inside the home.

Prevention is relatively easy and inexpensive. For about $25 a homeowner can purchase a carbon monoxide detector which is usually installed around heaters and other equipment. If a relatively high level of carbon monoxide is detected, the device sounds an alarm, giving people the chance to evacuate and ventilate the building.

Many of us are surrounded daily, in offices and in our homes, by products that outgas chemicals. Polyester shirts and water bottles contain toxic dyes and catalysts; computers, video games and hair dyes all offgas teratogenic and/or carcinogenic compounds. Carpets, wallpaper adhesives, paints, insulation and other items in our homes release chemicals, impacting indoor air quality. Allergies, asthma and "sick building syndrome" are on the rise. Yet legislation establishing mandatory standards for indoor air quality is practically non-existent, and there is no easy way for the average homeowner to test for the hundreds of possible chemicals in their home.

Air quality trends

Similar to water pollution, the most significant air pollution problems today stem not from point sources, but from non-point sources. While much of the earlier legislation focused on point sources (such as factories and smoke-stacks), regulation today confronts increased non-point pollutants. Common non-point sources of air pollution include automobiles, trucks, buses, air-planes, lawn and garden equipment and even charcoal-burning backyard BBQs. The most common non-point sources are mobile. Although twenty-first-century automobiles produce 60–80 percent less pollution than cars did in the 1960s, more people are driving more cars more miles. In 1970 Americans traveled one trillion miles in motor vehicles; by 2000 they drove more than three trillion miles. But there has been good news; vehicle miles driven seems to have flattened somewhat, peaking in 2008. While it used to be that vehicle miles grew some 2–3 percent per year, it has declined to about 1 percent over the last several years. In 2012 it was estimated the Americans drove approximately 3.05 trillion miles. But we walk, bike and use public transport more often. Despite these positive developments, vehicle emis-sions remain a significant part of the air pollution picture.

The first US federal law dealing with national air quality standards was the 1970 Clean Air Act (CAA). It focused on point-source air pollutants such as energy plants, factories and other stationary emitters. Since the 1970 CAA there have been several amendments, including the 1990 and 1997 Clean Air Act Amendments. The 1990 Amendments have a direct impact on cities. For the first time, the EPA established five categories for cities in non-attainment (those that did not meet federal regulations for criteria pollutants). The cat-egories were marginal, moderate, serious, severe and extreme. Those cities originally classified as severe or extreme were mandated to become attain-ment areas by 2005. In 2010, no city had violations of SO_2, NO_2 and CO. Downward trends in annual NO_2, CO and SO_2 are the result of various national emissions control programs. For example, an abrupt decline in NOx emissions was brought about by the EPA's NOx SIP Call program which began in 2003 and was fully implemented in 2004.

Non-attainment areas for ozone have improved: between 2001 and 2010, ozone non-attainment areas showed a 9 percent improvement in ozone con-centration levels. Nearly all US cities experienced fewer unhealthy days in 2010 compared to 2002.[19] Today, ozone and particle pollution are the primary contributors to unhealthy Air Quality Index days. Los Angeles, labeled "extreme" in 1990 remains in non-attainment, although its notoriously smoggy air is the cleanest it has been in 50 years. Still, in one out of every three days, millions of Los Angelenos breathe in dirty air. In addition, the 1990 Clean Air Act amendment established stricter tailpipe standards, finally addressing the growing problem of non-point pollution. Figure 12.1 shows a "smog station" where cars are required to be tested for tailpipe emissions

Figure 12.1 A smog station in Los Angeles. In order to maintain a car's legal registration, owners must periodically test the tailpipe emissions of their cars. Los Angeles is a city that imposes more stringent emissions standards that US federal standards

Source: Photo by John Rennie Short

every few years. This downward trend in air quality concentrations is expected to continue and will have profound health benefits for residents in US cities.

Pollution is a social construction. As new technologies allow researchers to measure and assess different pollutants at lower levels, and as medical and environmental science understands the impact of pollutants on public health and ecosystem integrity, standards of acceptability shift. For example, recent research has shown health impacts of exposure to particulate matter at lower levels than detected before. This may cause regulations to tighten. As a social construction, pollution and regulation are also subject to the political process. In the past, many of the EPA standards have been challenged in the federal courts. In 1998, several businesses and state groups challenged the 1997 CAA standards, claiming the EPA misinterpreted the Act and gave itself unlimited discretion to set air standards. In 2001, the US Supreme Court unanimously upheld the constitutionality of the EPA's long-standing interpretation that it must set these standards based solely on public health considerations and not on the consideration of costs.

In the US, policy approach to urban pollution has been two-fold. First, to impose regulatory measures that restrict and measure the quantities of pollutants released. Second, to establish economic measures that include pollution "taxes" or other financial disincentives. Yet after more than four decades since the Clean Air Act, air pollution reform in US cities is as mixed as water pollution. Some pollutants—primarily those coming from point sources—have decreased. Total emissions of toxic air pollutants have decreased by approximately 42 percent between 1990 and 2005. The EPA says control programs for facilities such as chemical plants, dry cleaners, coke ovens and incinerators are primarily responsible for these reductions. The six criteria pollutants have also decreased. Air pollution was lower in 2010 than in 1990 for: ozone by 17 percent; particulate matter by 38 percent; lead by 83 percent; nitrogen oxides by 45 percent; carbon monoxide by 73 percent; and sulfur dioxide by 75 percent.[20] Overall, according to the EPA, between 1980 and 2010 total emissions of the six criteria pollutants have decreased 67 percent.[21]

However, parallel to water pollution, much of the improvement in air pollutants has been offset by increased populations that demand more energy as well as increased automobile use. Despite decades of regulation and good intention, cities in the developed world remain a long way from eliminating the threat of air pollution to both public health and environmental quality.

Air pollution in developing cities

Developing cities have many of the same air pollutants as developed cities. High concentrations of ozone, carbon monoxide, carbon dioxide, sulfur oxides and nitrogen oxides, as well as particulate matter significantly impact human health in developing cities. Table 12.4 lists the ten smokiest cities for particulate matter, and Table 12.5 lists several cities where particulate matter is often above standards. Many of the most polluted cities in the world report high levels of all six criteria pollutants. It is not uncommon for cities such as Bangkok, Beijing, Kolkata, Delhi and Tehran to experience 30–100 days or more of poor air quality in a year. Jakarta exceeds health standards about 170 days per year; Mexico City exceeds acceptable levels of air pollution more than 330 days per year. A recent study shows that each of the 20 largest megacities has at least one major air pollutant that exceeds WHO health protection guidelines. Beijing, Cairo, Jakarta, Los Angeles, Mexico City, Moscow and São Paulo are facing a variety of air pollution problems requiring comprehensive solutions. The study shows that the ambient air

Table 12.4 Ten smokiest cities, particulate matter 2011

Rank	City	Annual mean PM10 ($\mu g/m^3$)
1	Ahwaz, Iran	372
2	Ulan Bator, Mongolia	279
3	Sanadaj, Iran	254
4	Ludhiana, India	251
5	Quetta, Pakistan	251
6	Kermanshah, Iran	229
7	Peshawar, Pakistan	219
8	Gaberone, Botswana	216
9	Yasouj, Iran	215
10	Kanpor, India	209

Source: World Health Organization (2011), Database: Outdoor Air pollution in cities. http://www.who.int/phe/health_topics/outdoorair/databases/en/index.html (accessed May 12, 2012).

Table 12.5 Annual mean suspended particles (PM10 $\mu g/m^3$) in selected cities

City	PM10 $\mu g/m^3$*	Data date
Amsterdam	**24**	2008
Athens	**41**	2008
Beijing	**121**	2009
Berlin	**26**	2008
Brussels	**28**	2008
Mumbai	**132**	2008
Cairo	**138**	2008
Kuala Lumpur	**49**	2008
London	**29**	2008
Los Angeles	**25**	2009
Mexico City	**52**	2009
Milan	**44**	2008
Montreal	19	2008
Moscow	**33**	2009
New York	**21**	2009
Singapore	**32**	2009
Sydney	12	2009
Tokyo	**23**	2009

Source: http://www.who.int/mediacentre/factsheets/fs313/en/index.html; http://www.who.int/phe/health_topics/outdoorair/databases/en (accessed May 12, 2012).
Note* Numbers in bold exceed WHO guideline level, which is 20.

quality in a majority of megacities is getting worse as population, traffic, industrialization and energy use increase. According to the World Health Organization, 1.3 million people in cities die prematurely each year from outdoor air pollution.[22] Another 340,000 people in cities die each year from diseases caused by indoor pollution.[23] A 2012 OECD report predicted that by 2050, there could be 3.6 million premature deaths each year from exposure to particulate matter, most of them in China and India. The report argues that urban air pollution could become the top environmental cause of mortality worldwide by 2050, ahead of dirty water and lack of sanitation.[24]

The magnitude of toxic emissions is also a serious and growing problem in the developing world. In contrast to cities in the Global North, cities in the developing world face rapid increases in population, high-density slums, dramatic increases in cars and trucks, less fuel efficiency and often limited governmental control. Although the highest number of vehicle owners (per 1,000 in a population) are found in the rich countries such as the US, Australia, Italy, New Zealand and Canada, the fastest rate of increase in vehicle ownership is in the developing world. Between 1980 and 1998 ownership in several developing countries increased five-fold—in South Korea vehicles per 1,000 people increased 1,514 percent; in Thailand 692 percent; Nigeria 550 percent; China 300 percent; and Pakistan 300 percent.[25]

There are several reasons why developing cities are confronting among the worst air pollution situations in the world. One reason is the lack of regulations or enforcement that would lower air pollutants. In some instances, national or municipal regulations are less stringent than in the US. Few developing countries require catalytic converters on automobiles, as the US does, for example. Or, as we have discussed previously, a lack of resources makes enforcement of laws difficult. Unlike many point source polluters in the US or Europe, many developing countries have little data on industrial polluters. A second reason is that high rates of urbanization have led to the growth of slums. As slums increase, the lack of infrastructure such as electricity and vented stoves means more people turn to other forms of energy, such as wood, biomass and coal. Another reason is that some cities have older equipment and vehicles which burn inefficiently and produce higher levels of emissions. For example, many cities have imported older vehicles without functioning catalytic converters. In Egypt it is estimated that 25 percent of the vehicle fleet is over 30 years of age. These reasons help explain the magnitude of air pollution and its more serious impact on public health in developing cities. We provide examples below to highlight the ways in which air pollution in developing cities presents a pressing health challenge.

Ulaanbaatar: rapid urban growth and air pollution

Ulaanbaatar, Mongolia, is now ranked as one of the worst cities in the world for air pollution. Although pollutants such as SO_2 are higher than international standards, particulate matter (PM) is the largest and relatively most severe air pollution problem in Ulaanbaatar and ranks the worst in the world. A recent World Bank Report noted that Ulaanbaatar's particulate matter was 14 times the recommended levels. In order to meet Mongolian air quality standards Ulaanbaatar would have to reduce air pollution by 80 percent, a significant undertaking, and definitely a long-term project. Urban growth and the emergence of informal settlements is a major contributor to increasing air pollution. Between 1989 and 2006, the city's population doubled; today the city's population is just over one million. Currently 60 percent of the city's residents live in informal settlements called *ger* districts. The traditional dwelling, the *ger*, is partly responsible for poor air quality. The *ger* is a felted tent structure that many have erected because there is not enough formal housing. The *ger* has inefficient coal-fueled stoves with very short stacks; as a result, air emissions do not travel high enough to leave the neighborhoods and thus settle on the ground. The *ger* household heating systems use raw lignite coal or wood for heating and cooking. The type of coal used is particularly soft and smoky. The city also confronts emissions from older power plants and boilers, as well as numerous old cars with inefficient combustion. Unpaved roads also add dry dust to the air.

Ulaanbaatar has long and cold winters. Heating is necessary for about 7–8 months of the year. As a result, the pollution is visibly worse during the winter, as stoves and boilers used for heating and cooking produce toxic black smoke plumes that hover like a blanket over the city. The chokingly thick pollution is a result of a combination of factors: the poor combustion of coal in what are essentially wood stoves, the congested road traffic, the dry ground conditions and industry. One of the worst sources of the pollution is dust. The dust originates from the *ger* heating appliances, the desert and dry ground conditions. With few trees and hardly any parkland in the city, the regularity and severity of windstorms in the city is increasing, creating dangerous levels of airborne dust. Strong winds, particularly in spring, also allow dust from the Gobi desert and other arid regions of Mongolia to reach the city.[26] The air pollution also affects the visibility to such an extent that airplanes on certain occasions are prevented from landing at the city airport.

Indian cities: older vehicles

In India most cities exceed the air quality standards set by the World Health Organization. In Delhi, 2,000 metric tons of air pollutants are emitted into the atmosphere every day. Vehicular sources account for 65 percent of air pollutants; industrial emissions account for 29 percent and domestic emissions 7 percent. Particulate matter is also a problem, and the smaller diameter particulate matter lodges deeply into the lungs. Indian and Chinese cities are among the smokiest cities in the world. In cities such as Delhi the rapid rate of population growth combined with increased resource consumption compounds air pollution problems. Between 1970 and 1990 the number of vehicles in India increased from two million to 21 million. The majority of the vehicles are found in urban centers, with about one-third of the 21 million vehicles concentrated in 23 metropolitan cities. The Centre for Science and Environment, an environmental advocacy group based in New Delhi, estimates that the city is adding 300,000 cars per year. Nearly three-quarters are three-wheeled vehicles. For India as a whole the vehicle population will more than double between 2005 and 2015, from 50 million to 125 million in ten years.[27]

Unlike the US, which mandates tighter emissions standards on both automobiles and two- and four-stroke engines (such as ATVs, lawnmowers, motorcycles), India does not. A high percentage of two-wheeled vehicles in India, while generally more fuel-efficient than passenger cars, have no

BOX 12.4

Autorickshaws in Delhi

In Delhi, there are an estimated two million vehicles, three-quarters of which are two- or three-wheeled vehicles (called tempoes) (Figure 12.2). Many of them are old and poorly maintained. As a result, the large number of two-wheelers contribute greatly to carbon monoxide, sulfur dioxide and nitrogen oxide emissions. In general motor vehicles in developing countries tend to be less fuel efficient and more polluting than those in the North because of a lack of access to new technologies, a greater proportion of older vehicles, poorly surfaced or badly maintained roads, weaker environmental legislation or weak enforcement of the regulations and the dominance of low-quality fuels such as diesel, with a high sulfur content.[28]

An estimated 80 percent of Delhi's taxis and autorickshaws (powered by four- or two-stroke engines) are older than 15 years; many are 30-plus years old. More than 40,000 autorickshaws are on the roads of Delhi and many of these are two-stroke engines that contribute significantly to emissions of carbon monoxide and suspended particulate matter. Autorickshaws were identified as the major culprits in air pollution. In 1998, in response to pressure from environmentalists to check the growing pollution in almost every major city, the Indian government imposed stiff new emission norms and announced it would phase out all commercial vehicles older than 15 years, blaming the worsening air quality on the sharp rise in vehicles. Newer autorickshaws run on compressed natural gas and can be distinguished by the yellow/green paint livery (older models were painted black and white). While there are thousands of autorickshaws on the road, suspended particulate matter has decreased.

Figure 12.2 Traditional and autorickshaws in Hyderabad, India
Source: Photo by Michele A. Judd

pollution controls. Unfortunately a recent report noted that while some air pollutants had improved in other Indian cities, Delhi's air quality had worsened.

Air pollution is also damaging India's most famous monument, the Taj Mahal at Agra. Surrounding cities contain more than 2,000 polluting industries, ranging from brick kilns to an oil refinery as well as traffic that uses high-sulfur diesel fuel. Acidic emissions of sulfur dioxide and nitrogen dioxide are eroding and dissolving the marble monument.[29] In Mumbai there are high levels of sulfur dioxide and an increased prevalence of breathing difficulties, coughs and colds. According to the World Resources Institute, mortality data show a link between dense air pollution in Mumbai and a higher rate of death from respiratory and cardiac conditions.[30]

Airborne lead: a success story

A particularly toxic component in urban air is lead, the heavy metal which for many years was added to gasoline to raise octane levels and help engines run more smoothly. Up until recently, leaded gasoline was the primary source of lead exposure in many developing cities. Lead levels in the air of large African cities such as Cairo, Cape Town and Lagos were up to ten times those of European cities. Lead is one of the most harmful airborne pollutants, causing neurological damage that particularly impacts children. High exposure to lead in children can have more dramatic health impacts than for adults, such as slowed growth, hearing problems, headaches, reduction in IQ and mental retardation. Lead poisoning is one of the most serious environmental health threats to children and is a significant contributor to occupational disease. In 2003 the World Health Organization estimated that 120 million people are over-exposed to lead (approximately three times the number infected by HIV/AIDS) and 99 percent of the most severely affected are in the developing world.[31]

As a result of growing concern over lead poisoning, the World Bank launched the Clean Air Initiative in sub-Saharan Africa and identified the elimination of lead from gasoline as the priority, noting that "switching quickly to unleaded gasoline is one of the most cost-effective steps to protect children's health."[32] Within a few years it had become an international initiative and more than 70 countries agreed to begin phasing out leaded gasoline. By 2009 all countries in Latin American and nearly all sub-Saharan African countries had phased out leaded gasoline and now sell only unleaded gasoline. Only a handful of countries still use leaded gasoline.

The elimination of leaded gasoline has significantly reduced airborne lead—but not all of it. For example, Mexico City still suffers from high levels of lead, both inhaled and ingested (see Box 12.5). Dense traffic and industry contribute to poor air, which still contains high levels of lead even though leaded gasoline was phased out in 2000. Over a decade after the ban of leaded gasoline, the soil and dust are still laden with lead.[33] Causes of lead in the environment include industrial use of lead, from facilities that process lead-acid batteries or produce lead wire or pipes, and metal recycling and foundries. Leaded paint in older homes can also be a source, as can lead-glazed ceramics and pottery, a practice still common in Mexico. Children living near facilities that process lead, such as smelters, have been found to have unusually high lead levels in their blood.

Indoor air pollution

A problem distinct to many cities in the developing world is indoor air pollution that results from the use of wood or animal dung or kerosene for heat and cooking.[34] Indoor air pollution may pose a greater risk in developing cities than ambient air pollution. Studies from Asia, Africa and the Americas show that indoor air pollution levels in households reliant on biomass fuel or coal are extremely high. For example, typical 24-hour mean levels for PM10 in homes using biomass fuels are around 1,000 $\mu g/m^3$, compared to the current limit of 150 $\mu g/m^3$ set by the US EPA.[35]

Dependence on polluting and inefficient household fuels and appliances is both a cause and a result of poverty. Over two billion people around the world use coal and biomass fuels such as wood, cow dung, charcoal and grass for cooking and heating. The pollutants that result from the burning of these fuels in inefficient stoves combined with poor ventilation have severe and sometimes fatal consequences. Often the open fires or poorly functioning stoves prevent complete combustion, which means the fuels do not burn cleanly, and emit many types of toxic gases such as carbon monoxide.

Indoor air pollution disproportionately impacts women and children, who devote a large portion of the day to cooking or other tasks inside the home.[36] Particularly in urban slums, women often perform their daily cooking in small, enclosed areas, relying on some form of biomass (Figure 12.3). It is common for women to be exposed for 3–7 hours each day. Young children, who are often carried on their mother's back or kept close by, are also vulnerable. A World Health Organization report noted that indoor air pollution is the most lethal killer in the developing world after malnutrition, unsafe sex and lack of safe water. Indoor air pollutants are 1,000 times more likely to reach people's lungs than outdoor air pollutants.

Figure 12.3 Indoor air pollution. A woman cooks over an indoor fire while her children look on

Source: Photo by Elizabeth Chacko

BOX 12.5

Exporting lead poisoning to Mexico City?

In the US, recycling of car batteries is high and most states have laws that mandate that stores take back old batteries. Lead batteries come from cars, truck, solar power systems, golf carts and forklifts. Whether deposited at the store where they were purchased or with a local mechanic, used batteries are redirected to recycling plants, where the real goal is not environmental stewardship but extracting the precious lead. Along the way, however, used batteries can be sold to a middleman, who then ships the batteries out of the US to take advantage of lower costs and less regulation and enforcement.

Approximately 12 percent of used lead batteries generated in the United States are exported to Mexico. This is due to the tightening of US air quality standards in 2008 which made the recycling process more complex and more expensive. A 2011 report prepared by two environmental organizations, Occupational Knowledge International, a US organization, and Mexico's Fronteras Comunes, shows that there are unanticipated consequences to this practice.

In Mexico lead is often extracted by crude methods that are illegal in the US, exposing plant workers and local residents to dangerous levels of the toxic metal. Spent batteries house up to 40 pounds of lead, and when broken for recycling the lead is released as dust and, during melting, as lead-laced emissions. Many of the largest lead battery recyclers are in Mexico City.

In the US, lead battery recycling plants operate under strict controls, with monitored smokestacks and lead-monitoring devices. But according to *New York Times* reporter Elisabeth Rosenthal, in Mexico City, "batteries have been dismantled by men wielding hammers, and their lead melted in furnaces whose smokestacks vent to the air outside, where lead particles can settle everywhere from schoolyards to food carts." One of the largest battery recyclers in the city is on the same street as an elementary school. While Mexico does have some regulation for smelting and recycling lead, the laws are poorly enforced. Airborne lead emissions reported by battery recycling plants in Mexico are 20 times higher than comparable plants in the US.

"It is remarkable that both governments allow US companies to export batteries to Mexico where there is neither the regulatory capacity nor the technology in place to recycle them safely," said Perry Gottesfeld, executive director of Occupational Knowledge International.

Chronic lead poisoning in children is hard to diagnose because the symptoms are fairly common, among them low IQ and attention issues. The only definitive diagnosis is a blood test, but that costs about $100—often beyond the reach of poor families.

Americans may think they are doing a good thing by recycling their lead batteries, but the whole picture is more complex. This type of "trade" is creating a potentially more hazardous waste stream and one that has received very little attention or scrutiny. While Mexico City has done much to eliminate airborne lead by outlawing leaded gasoline, it appears a new source of lead may be exposing city residents to the poison once again.

Sources: Occupational Knowledge International and Mexico's Fronteras Comunes (2011) "Exporting hazards: U.S. shipments of used lead batteries to Mexico take advantage of lax environmental and worker health regulations." http://www.okinternational.org/docs/Exporting%20 Hazards_Study_100611v5.pdf (accessed July 2012). Rosenthal, E. (2011, December 8) "Lead from old U.S. batteries sent to Mexico raises risks." *New York Times*. http://www.nytimes.com/2011/12/09/science/earth/recycled-battery-lead-puts-mexicans-in-danger.html?pagewanted=all (accessed July 2012).

In India, 80 percent of households use bio-fuels and the estimated child mortality rate from indoor air pollution is 130,000 children per year, mainly from acute respiratory disease. According to the World Resources Institute the exposure of pregnant women to indoor smoke results in a 50 percent increase in stillbirths.[37] Women using biomass for cooking are up to 75 times more likely to contract chronic lung disease.[38] Between 3.5 and 7 percent of the national burden of disease in developing countries is attributable to respiratory and other problems related to indoor air pollution, a much larger percentage than is ascribed to urban and industrial pollution.

On a positive note, a wide range of interventions are available to reduce indoor air pollution and associated health effects. Interventions can include: (1) interventions on the source of pollution; (2) interventions to the living environment; and (3) interventions to user behavior. For example, large reductions in emissions can be accomplished by switching from solid fuels (biomass, coal) to cleaner and more efficient fuels and energy technologies such as liquid petroleum gas (LPD), solar stoves or electricity. Changes to the residence such as installing chimneys or smoke hoods, or even enlarging windows, can improve ventilation of the cooking and living area, which can contribute significantly to reducing exposure to smoke. Finally, changes in user behavior can also play a role in reducing pollution and exposure levels. For example, drying fuel wood completely before use improves combustion and decreases smoke production. Keeping young children away from smoke reduces exposure of this most vulnerable age group to health-damaging pollutants.[39]

Air quality trends in developing cities

In the past decade, many developing countries have adopted National Air Quality Standards, using either World Health Organization Air Quality Guidelines (AQG) or US EPA National Ambient Air Quality Standards (NAAQS). By 2010 Thailand had updated its standards four times since they first established them in 1981. India also recently revised its standards, establishing stricter PM2.5 and O_3 standards. Recent developments were also observed in Vietnam; Hong Kong Special Administrative Region is also currently undergoing review of its Air Quality Objectives (AQO).[40] Nonetheless, there are a number of Asian countries still without any NAAQS—Afghanistan, Bhutan, Lao People's Democratic Republic (PDR) and Pakistan. In addition, many Asian countries have yet to set standards for particulate matter.

BOX 12.6

Beijing: cloaked in smog

Since the mid-1990s, China has experienced rapid economic development and high rates of urbanization. In the past decade, annual GDP has increased 8–11 percent. The joke is that you can smell China's GDP in the air. This economic development has been accompanied by the rapid growth of energy consumption. An emerging middle class has meant increasing vehicle ownership; statistics show vehicles have increased 10 percent annually for the last decade. Not surprisingly, air pollution in Chinese cities such as Beijing is significant. In some cases the smog has been so bad, high-rise office buildings disappear into the mist. It is estimated that two-thirds of China's cities have not attained air quality standards. Beijing's case reflects the problems that other rapidly developing cities in China must face. While economic growth is enviable, informed urban residents demand a better quality of air (see Figure 12.4).

In the 1990s and first part of the twenty-first century, much of the air pollution came from sulfur dioxide (SO_2). China has abundant sources of coal and relied on this for much of its

Figure 12.4 Air pollution in Shanghai. Some residents wear masks to prevent inhaling particulate matter and other forms of air pollution

Source: Photo by John Rennie Short

energy. However, in recent years SO_2 emissions have declined, but particulate matter has increased, reflecting the development of mixed-source pollution and increased automobile exhaust. Today, coal burning, automobile emissions and dust are the major sources of air pollution in Beijing. Research by Zhang and colleagues found that the economic costs to health during a five-year period were estimated to lie between $1,670 million and $3,655 million annually, accounting for about 6.5 percent of Beijing's GDP each year.

At approximately 19 million people, Beijing is China's second largest city after Shanghai and is the country's political, cultural and educational center. Air pollution was a high-profile issue to the local and national government and to the people living in the city, especially after it was selected to host the 2008 Olympics. In response, the government implemented a series of measures to control air pollution in Beijing. These included: the use of low-sulfur coal; the partial replacement of coal with natural gas or liquefied petroleum gas; the phase-out of leaded gasoline; and the relocation of high-polluting industry from the city. Government officials also implemented a number of air improvement schemes for the duration of the Games, including stopping work on all construction sites, closing many factories in and around Beijing, closing some gas stations and reducing motor traffic by half by limiting drivers to odd or even days (based on their license plate numbers). Significant money went to build two new subway lines and thousands of old high-emissions taxis and buses were replaced to encourage residents to use public transport. Beijing added 3,800 natural gas buses, one of the largest fleets in the world. The city also planted hundreds of thousands of trees and increased green space in an effort to make the city more livable. Nearly $17 billion was spent to reduce pollution in the city.

A study by Hao and Wang, however, suggested that controlling only local sources in Beijing is not sufficient to attain the air quality goal set for the Beijing Olympics. Prevailing winds blow high levels of particulate matter and ozone into Beijing from neighboring Hebei and Shandgon Provinces and Tiajin city. On average, 35–60 percent of the ozone can be traced to sources *outside* the city.

The government claims pollution levels have decreased, and the number of so-called "blue sky days" has increased in the past 13 years as cities have relocated factories, reduced coal burning and adopted stricter vehicle emission standards. A World Bank analysis of the government's data found that average concentrations of particulates measuring ten microns or less—a group that includes both fine and coarser particulates—fell 31 percent from 2003 to 2009 in 113 major cities.

However, a 2012 *New York Times* article reported that the Chinese government has monitored exposure levels in 20 cities and 14 other sites, reportedly for as long as five years, but has kept the data secret. The article questions whether government statistics are reliable, noting other studies have shown PM2.5 levels increasing by 3–4 percent annually, and rising levels of ozone. Some of the discrepancy between reported pollution levels is due to omission. The Beijing government monitors but does not report fine particulate (PM2.5) and ozone, both of which are pollutants linked to lung disease and premature death. PM2.5 is smaller in diameter than PM10, and therefore more dangerous to human health as these particles can lodge more deeply in the lungs. The Chinese government, however, only measures PM10, not PM2.5.

In 2011 Beijing experienced its worst air pollution since 2008. The American consulate began issuing its own air quality readings, reporting that particulates are significantly worse than government statistics indicate. The Chinese government responded that American officials had insulted the Chinese government by posting readings from the PM2.5 monitor atop the embassy on their Twitter account, Beijing Air. A Foreign Ministry official warned that the embassy's data could lead to "social consequences" in China and asked the embassy to restrict access to it. The embassy refused. Within weeks of the diplomatic kerfuffle, and in a rare bow to public pressure, the Beijing local government began releasing data for PM2.5. Currently the US embassies in Beijing, and consulates in Shanghai and several other Chinese cities, monitor PM10 and PM2.5 and ozone, and offer free apps for smartphones. In 2012 Chinese entrepreneurs that run the Chinese equivalent of Twitter, Weibo, began offering air quality assessment that relied on both Chinese and US embassy data.

This brief overview of air pollution in Beijing reveals important trends. In many developing cities where economic growth is robust and urbanization is rapid, air pollution may get worse before it gets better. Second, public access to reliable information is important in creating meaningful environmental reform: in non-democratic countries access to reliable data is not always guaranteed. However, social media and the power of information technology may find a way around governments that limit access to information.

Sources: Andrews, S. (2011, December 5) "Beijing's hazardous blue sky." *Chinadialog.* http://www.chinadialogue.net/article/show/single/en/4661-Beijing-s-hazardous-blue-sky (accessed May 2012). Chan, C. and Yao, X. (2008) "Air pollution in mega cities in China." *Atmospheric Environment* 42 (1):1–42. Hao, J. and Wang, L. (2005) "Improving urban air quality in China: Beijing case study." *Air & Waste Management Association* 55: 1298–1305. LaFraniere, S. (2012, January 27) "Activists crack China's wall of denial about air pollution." *New York Times.* http://www.nytimes.com/2012/01/28/world/asia/internet-criticism-pushes-china-to-act-on-air-pollution.html?pagewanted=all (accessed May 2012); Zhang, M., Song, Y. and Cai, X. (2007) "A health-based assessment of particulate air pollution in urban areas of Beijing in 2000–2004." *Science of the Total Environment* 376 (1–3): pp. 100–8.

In many African cities, air quality is worsening due to rapid urbanization, the importing of older vehicles and an absence of laws that require catalytic converters or vehicle inspection and maintenance. The higher use of diesel fuel has also been problematic, and international organizations continue to work with African countries to adopt lower sulfur emissions standards. Another common problem is that industrial manufacturing is often located close to residential areas. International studies have noted that African cities are at a relatively early stage in developing air pollution abatement policies, which means that policymakers can seize opportunities for integrated programs and regional cooperation.

Egypt, for example, started its National Air Quality Standard network in 1997 with 44 sites. By 2010 it had expanded to 78 sites (16 in Cairo),

Table 12.6 Global Atmospheric Pollution Forum: partnership of international organizations and regional air pollution networks

United Nations Environment Programme (UNEP)
UNEP-sponsored networks and programs in East Asia and South Asia
UN Economic Commission for Europe (UNECE)
Long-range Transboundary Air Pollution (LRTAP)
Air Pollution Information Network for Africa (APINA)
Inter-American Network for Atmospheric and Biospheric Studies (IANABIS)
Clean Air Initiative—Asian Cities; Latin America; Africa
Sahara and Sahel Observatory (OSS)
International Union of Air Pollution Prevention Associations (IUAPPA)

measuring the six criteria pollutants. While SO_2 emissions decreased, other pollutants had increased, primarily due to increased vehicles.

The development of new regional air pollution networks in the developing world promises a way to promote regional coordination on air pollution abatement and to support integrated pollution/climate strategies. The Clean Skies Initiative in sub-Saharan Africa and the Clean Air Initiative—Asia (CAI-Asia) and the Clean Air Initiative—Latin America (CAI-LA) are examples. These are networks of institutions and individuals committed to improving air quality management. A primary mission is to promote and demonstrate innovative ways to improve the air quality of cities by sharing experiences and building partnerships. These networks aim to harmonize technical systems and information/data gathering between regions. Table 12.6 shows several global air pollution partnerships.

Conclusions

During the twentieth century air pollution, once a localized problem, became a global one. Nowhere is immune from toxic fallout and yet the most intense effects on both ecosystems and human health are local—and urban. Unfortunately, many see development and air quality regulation as incompatible, believing that air quality controls are barriers to development. There is the potential, however, to see air pollution as being directly related to a significant public health burden. By reducing emissions, there are often immediate improvements in public health and economic productivity.

Progress on air pollution has been mixed. On the one hand, there has been a reduction in many point sources of pollution, in both developing and developing cities. However, non-point sources are now the primary causes of pollution and this is a challenge because many non-point sources include automobiles and other vehicles. Trends in both developed and developed cities show mobile sources of emissions to be increasing, as more people purchase cars, and/or drive more miles. This is happening more rapidly in developing cities. In the developed world, cities have improved air quality for nearly all of the criteria pollutants. The picture is more complex in the developing world, because some countries have not yet set standards for ozone or particulate matter. Lastly, there remain many air pollutants for which much progress remains to occur: toxic and hazardous pollutants, and indoor sources of air pollution remain problems in cities around the world.

Guide to further reading

American Lung Association. (2012) "Key findings, State of the Air 2012." http://www.stateoftheair.org/2012/key-findings (accessed June 2012).

Brimblecombe, P. (ed.) (2003) *The Effects of Air Pollution on the Built Environment.* Singapore and River Edge, NJ: Imperial College Press.

Bulkeley, H. and Bestill, M. (2005) *Cities and Climate Change.* New York: Routledge.

Elsom, D. (1996) *Smog Alert: Managing Urban Air Quality.* London and Sterling, VA: Earthscan.

Gonzalez, G. A. (2005) *The Politics of Air Pollution: Urban Growth, Ecological Modernization, and Symbolic Inclusion.* Albany, NY: State University of New York Press.

National Academy of Engineering (2008) *Energy Futures and Urban Air Pollution: Challenges for China and the United States.* Washington, DC: National Academy Press.

Rock, M. T. (2002) *Pollution Control in East Asia: Lessons from the Newly Industrializing Economies.* Washington, DC: Resources for the Future.

Schwela, D., Haq, G., Huizenga, C., Han, W., Fabian, H. and Ajero, M. (2006) *Urban Air Pollution in Asian Cities: Status, Challenges and Management.* Sterling, VA: Earthscan.

Uekoetter, F. (2009) *The Age of Smoke: Environmental Policy in Germany and the United States, 1880–1970.* Pittsburgh, PA: University of Pittsburgh Press.

Watt, J., Tidbald, J., Kucera, V. and Hamilton, R. (2009) *The Effect of Air Pollution on Cultural Heritage.* New York: Springer.

For up-to-date information on ozone/smog and interactive maps, go to: www.epa.gov/air/ozonepollution. Also, www.scorecard.org allows you to profile the air, land and water pollution in your state, city or community in the US.

13 **Climate change**

Cities today consume 75 percent of the world's energy and emit 80 percent of the world's greenhouse gases (GHGs). Cities are both the major generators of GHGs and the key players in reduction. There is growing attention on the relationship between cities and climate change. In this chapter we begin with a discussion of the potential impacts of climate change on cities. We then examine the two ways that cities are dealing with the issue: mitigation and adaptation. Mitigation refers to measures that alter the behavior of the citizens to reduce anthropogenic effects on climate change, mainly efforts that reduce emissions of GHGs. Adaptation, on the other hand, involves pursuing projects that work to accommodate citizens to the inevitable future of global environmental changes. While mitigation and adaptation were often seen as competing processes, today there is a general consensus that there is a need for both approaches to be pursued simultaneously. This view holds that while climate change is inevitable and will need to be adapted to, it is still possible to mitigate the severity of this change by altering human behaviors. Cities have the ability to experiment and innovate, and this is the hope and core of resilience around climate change.

Climate science: the basic facts

Climate change is linked to GHGs. There are four principle GHGs: carbon dioxide, methane, nitrous oxide and fluorinated gases. Three of the four gases are naturally occurring, while fluorinated gases are man-made. Carbon dioxide, methane and nitrous oxides are continuously emitted into and removed from the atmosphere by natural processes on earth. Anthropogenic (man-made) activities, however, can cause additional quantities of these and other GHGs to be emitted or sequestered, thereby changing their global average atmospheric concentrations. Most anthropogenic emissions come from the combustion of carbon-based fuels, principally wood, coal, oil and natural gas. According to the EPA, the primary sources of these gases are as follows:

- Carbon dioxide (CO_2). Carbon dioxide enters the atmosphere through burning fossil fuels (coal, natural gas and oil), solid waste, trees and wood products, and also as a result of certain chemical reactions (e.g., manufacture of cement). Carbon dioxide is removed from the atmosphere (or "sequestered") when it is absorbed by plants as part of the biological carbon cycle. In 2010 CO_2 accounted for about 84 percent of all US GHG emissions from human activities (Figure 13.1).
- Methane (CH_4). Methane is emitted during the production and transport of coal, natural gas and oil. Methane emissions also result from livestock and other agricultural practices and from the decay of organic waste in solid-waste landfills.
- Nitrous oxide (N_2O). Nitrous oxide is emitted during agricultural and industrial activities, as well as during combustion of fossil fuels and solid waste.
- Fluorinated gases. Hydrofluorocarbons, perfluorocarbons and sulfur hexafluoride are three synthetic, powerful GHGs that are emitted from a variety of industrial processes.

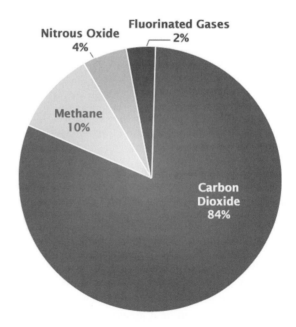

Figure 13.1 Greenhouse gases in the atmosphere by percentage volume
Source: US EPA

Carbon dioxide emissions in the US increased by about 12 percent between 1990 and 2010.[1] Since the combustion of fossil fuel is the largest source of GHG emissions in the US, changes in emissions from fossil fuel combustion have historically been the dominant factor affecting total emission trends. Methane is the second most prevalent GHG emitted in the US from human activities. In 2010, methane accounted for about 10 percent of all US GHG emissions from human activities.[2] According to the US EPA, methane's lifetime in the atmosphere is much shorter than carbon dioxide but methane is more efficient at trapping radiation than carbon dioxide. Pound for pound, the comparative impact of methane on climate change is over 20 times greater than carbon dioxide over a 100-year period. Globally, about 40 percent of total nitrogen oxide emissions come from human activities. Nitrous oxide molecules stay in the atmosphere for an average of 120 years before being removed by a sink or destroyed through chemical reactions. The impact of one pound of nitrous oxide on warming the atmosphere is over 300 times that of one pound of carbon dioxide. Finally, fluorinated gases, unlike many other GHGs, have no natural sources and only come from human-related activities. These gases represent the smallest amount of GHGs in the atmosphere, but they are quite potent. Small atmospheric concentrations can have large effects on global temperatures. They can also have long atmospheric lifetimes—in some cases lasting thousands of years. Like other long-lived GHGs, fluorinated gases are widely diffused in the atmosphere, spreading globally after they are emitted. Each of these gases can remain in the atmosphere for different amounts of time, ranging from a few years to thousands of years. GHGs remain in the atmosphere long enough to become well mixed, meaning that the amount that is measured in the atmosphere is roughly the same all over the world, regardless of the source of the emissions. These four gases absorb some of the energy being radiated from the surface of the earth and trap it in the atmosphere, essentially acting like a blanket that makes the earth's surface warmer than it would be otherwise (see Figure 13.2).

GHGs are necessary to life as we know it, because without them the planet's surface would be about 60°F cooler than present. But, as the concentrations of these gases continue to increase in the atmosphere, the Earth's temperature is climbing above past levels. According to NOAA and NASA data, the Earth's average surface temperature has increased by about 1.2–1.4°F since 1900. The ten warmest years on record (since 1850) have all occurred in the past 13 years.[3] Most of the warming in recent decades is very likely the result of human activities. Other aspects of the climate are also changing, such as rainfall patterns, snow and ice cover and sea level. The United

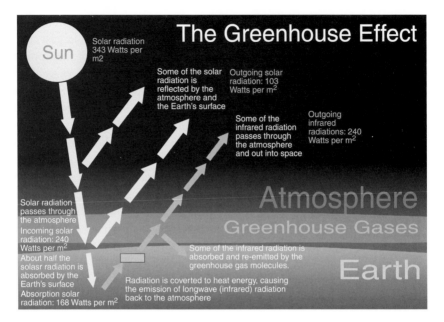

Figure 13.2 The greenhouse effect
Source: US EPA

BOX 13.1

Contribution to GHG emissions, São Paulo, Brazil

The São Paulo Metropolitan Region has a population of 18 million and is the largest urban area in Brazil. The city is a major driving force for the national economy, with a GDP of $102 billion in 2012. The service industry is the main driver, accounting for 62.5 percent of GDP. This is followed by the industrial sector, which accounts for 20.6 percent. A comprehensive GHG emissions inventory was conducted in 2005. It shows that energy use accounts for more than three-quarters of the city's emissions. Approximately two-thirds of this was associated with diesel and gasoline, and 11 percent with electricity generation. However, the contribution of urban transportation to GHG emissions is still relatively low as a result of the mandatory blend of ethanol (23 percent) and gasoline (77 percent) used in most of the private fleet. Similarly, the contribution of the electricity generation sector is low as the city relies heavily on hydro-electric generation. Solid-waste disposal accounted for almost one-quarter of the city's emissions, or 3.7 million tons of CO_2eq. However, Clean Development Mechanism (CDM) projects at the Bandeirantes and São João landfills will prevent the generation of 11 million tons of CO_2 by 2012—almost removing the contribution of solid waste to the city's emissions.

Per capita emissions from the city are low, at about 1.4 metric tons CO_2 per year (in 2011), compared to a national average for Brazil of 8.2 tons; compare that also to Berlin, with an average of 5.8 metric tons per capita, Seoul at 4.7, New York at 6.6 and Tokyo at 7.6. Despite this, the growing importance of reducing global GHG emissions means that cities in middle-income countries will increasingly need to identify their emissions reduction potential and act on this. It is important to note that although the city of São Paulo accounts for 6.8 percent of the population of Brazil, its GHG emissions are relatively small. This is because Brazil is a large emitter of GHGs from agriculture, land-use change and forestry. In the case of deforestation, due to high rates, emissions account for 63.1 percent of total national emissions of CO_2 and methane. The agriculture sector as a whole is responsible for 16.5 percent of the same gases, mainly because of the size of the national cattle herd. In the case of the extremely urbanized city of São Paulo, these emissions are insignificant.

Source: UN Habitat (2011) *Cities and Climate Change: Case Studies.* http://www.unhabitat.org/downloads/docs/GRHS2011/SomeCaseStudies.pdf (accessed April 22, 2012).

Nations International Panel on Climate Change notes that if GHGs continue to increase, climate models predict that the average temperature at the Earth's surface could increase from 2.0 to 11.5°F above 1990 levels by the end of this century. Warmer temperatures influence the patterns and amounts of precipitation, reduce ice and snow cover as well as permafrost, raise sea levels and increase the acidity of the oceans.

Since 1990, the EPA has conducted a yearly national GHG inventory and submits this to the United Nations in accordance with the Framework Convention on Climate Change. In 2011 the EPA reported that since 1990, US GHG emissions have increased by 10.5 percent.[4] Globally, climate records have shown several important trends, as shown in Table 13.1.

The EPA also reports that by 2100, the average US temperature is projected to increase by about 4–11°F. An increase in average temperatures worldwide implies more frequent and intense extreme heat events, or heat waves. The number of days with high temperatures above 90°F is expected to increase throughout the US, especially in areas that already experience heat waves. For example, cities in the southeast and southwest currently experience an average of 60 days per year with a high temperature above 90°F. These areas are projected to experience 150 or more days per year above 90°F by the end of the century, under a higher emissions scenario. In addition to occurring more frequently, these very hot days are projected to be about 10°F hotter at the end of this century than they are today, under a higher emissions scenario.[5]

So is there a genuine "climate debate?" Yes and no. Scientists are certain that human activities are changing the composition of the atmosphere, and that increasing the concentration of GHGs will change the planet's climate. But they are not sure by how much it will change, at what rate it will change, or what (and where) the exact effects will be. This level of uncertainty about impact is what climate change deniers have focused on in their effort to prevent regulation of energy use or the signing of important international treaties to limit emissions. Extreme climate change deniers also question whether anthropogenic emissions are significant, and point to the earth's history of fluctuating climate patterns. However, climate scientists and experts have proven that anthropogenic emissions are significant and that these concentrations do affect global average temperatures. Climate experts agree that man-made emissions of carbon dioxide and methane have rapidly accelerated since the nineteenth century as industrialization evolved and expanded.

Table 13.1 Climate change observations

0.1-0.2 meter (one-third to two-thirds of a foot) rise of global average sea levels
10 percent decrease in snow cover since the 1960s
More frequent, persistent and intense El Niños since the 1970s
More frequent and severe droughts in Africa and Asia

Source: Bulkeley, H. and Betsill, M. (2005) *Cities and Climate Change: Urban Sustainability and Global Environmental Governance*. New York: Routledge.

Impacts of climate change

Climate change is impacting society and ecosystems in a broad variety of ways. For example, climate change can increase or decrease rainfall, influence agricultural crop yields, affect human health, cause changes to forests and other ecosystems or even impact energy supply.[6] Table 13.2 lists just some climate change impacts. Climate change could have significant effects

Table 13.2 Potential climate impacts

Water resources	Shift in the timing of spring snowmelt to earlier in the spring
	Lower summer streamflows, particularly in snowmelt-dependent water systems in the western US
	Increased risk of drought
	Increased risk of flooding
	Increased competition for water
	Warmer water temperature in lakes and rivers
	Changes in water quality (varies by water quality parameter)
Recreation	Increased opportunities for warm season activities in milder regions of the US
	Decreased opportunities for warm season activities during the hottest part of the year, particularly in the southern US (e.g., from heat, forest fires, low water levels, reduced urban air quality)
	Reduced opportunities for cold season recreation due to decreased snowpack and/or reduced snow or ice quality
	Increased reliance on snow-making at ski areas
	Shifts in tourism dollars within a community from one recreation sector to another, or from communities losing recreational opportunities to communities gaining opportunities
Energy	Reduced heating demand during winter months
	Increased cooling demand during summer months
	Increased or decreased hydroelectric generating capacity due to potential for higher or lower streamflows
Infrastructure	Need for new or upgraded flood control and erosion control structures
	More frequent landslides, road washouts and flooding
	Increased demands on stormwater management systems with the potential for more combined stormwater and sewer overflows
	Reduced effectiveness of sea walls with sea-level rise

Continued

Table 13.2 Continued

Public health	More heat-related stress, particularly among the elderly, the poor and other vulnerable populations
	Fewer extreme cold-related health risks
	Increase in vector-born illnesses (e.g., West Nile virus)
	Reduced summer air quality in urban areas due to increased production of ground-level ozone
Business	Price volatility in energy and raw product markets due to more extreme weather events
	Increased insurance premiums due to more extreme weather events
	Fewer shipping disruptions associated with snow and ice
	Impacts on business infrastructure located in floodplains or coastal areas
	Shifts in business opportunities
Coastal areas and aquatic ecosystems	Increased erosion or damage to coastal infrastructure, dunes, beaches and other natural features due to sea-level rise and storm surge
	Loss of coastal wetlands and other coastal habitats due to sea-level rise and erosion
	Increased costs for maintenance and expansion of coastal erosion control (natural or man-made)
	Saltwater intrusion into coastal aquifers due to sea-level rise
	Increased risk of pollution from coastal hazardous waste sites due to sea-level rise
	Loss of cultural and historical sites on coastlines due to sea-level rise and related impacts

Source: adapted from Center for Science in the Earth System (The Climate Impacts Group) Joint Institute for the Study of the Atmosphere and Ocean University of Washington and King County, Washington (2007) *Preparing for Climate Change: A Guidebook for Local, Regional and State Governments.* http://www.iclei.org/fileadmin/user_upload/documents/Global/Progams/CCP/Adaptation/ICLEI-Guidebook-Adaptation.pdf, pp. 40–41 (accessed September 2012).

BOX 13.2

Climate models

A global climate model is a mathematical representation of the interactions between and within the ocean, land, ice and atmosphere. Each of these components is distinct within the model. To construct climate models, scientists divide each of the Earth's components into a set of boxes. Simple models may have only a few boxes. The most complex models may have more than 100,000.

The movement of energy, air and water are represented as horizontal and vertical exchanges between the boxes. In this way, models represent interactions between different parts of the climate system and the world. Models attempt to capture the very complex interactions. Climate models use mathematical equations based on well-understood principles to depict the behavior of processes in each box. As models incorporate more factors or greater spatial detail, more computer power is needed. Scientists use some of the largest supercomputers in the world to run climate models.

To test model accuracy, scientists simulate past climate conditions. They then compare the model results to observed conditions. The observed average global change in temperature over the past century is relatively well replicated by models. Temperature is one of the variables that is most accurately simulated by models. However, models do not agree on many small-scale details.

Scientists continually refine models as knowledge increases, trying to improve representations of features that are not well simulated. Models that account only for the effects of natural processes (such as volcanic eruptions) are not able to explain the recent warming of the climate, but models that also account for the GHGs emitted by humans are able to explain this warming. Most of the observed increase in global average temperature since the mid-twentieth century is very likely due to the increase in human-generated GHG concentrations.

Source: US EPA (2012) "Future climate change: climate models."http://epa.gov/climatechange/science/future. html (accessed August 2012).

on food production around the world. Heat stress, droughts and floods may lead to reductions in crop yields and livestock productivity. The risks of climate-sensitive diseases and health impacts can be high in poor countries. For example, the spread of meningococcal meningitis is often linked to climate change, especially drought. Areas of sub-Saharan and West Africa are particularly sensitive to the spread of meningitis. In addition, mosquito-borne diseases such as malaria may increase in areas that receive more precipitation and flooding. Climate change also has the potential to exacerbate national security issues and increase the number of international conflicts over water and food.

In April 2011, the US experienced record-breaking floods, tornadoes, drought and wildfires, all within a single month. NOAA's National Climatic Data Center had already reported ten weather events from 2011 for which damages and/or costs reached or exceeded $1 billion each, exceeding the previous annual record of nine events recorded over the entire year of 2008. NOAA estimated the total damage of property and economic impacts for all weather-related disasters during the spring and summer of 2011 at more than $45 billion. In the summer of 2012, record heat waves and drought affected large areas of the US Midwest and east. In Washington, DC, for example, July set an all-time temperature record 3°F higher than any other July on record. The severe and costly losses suffered during recent extreme weather events demonstrate the importance of increasing the resilience of the US to climate variability and change in order to reduce economic damages and prevent loss of life.

There is a geography to the potential impacts of climate change. Although climate change is an inherently global issue, the impacts will not be felt equally across the planet. There is a profound unevenness in terms of those countries that are causing climate change and those that will be most affected. Predictions are that many of those countries and cities that are among the low-producing GHG countries—most of which are developing countries—are the most vulnerable to climate impacts such as sea-level rise and storm surges.

Until recently, much of the debate about climate change has been at the global scale or national scale. The urban scale was often ignored, but in the last ten years or so, this area of study has expanded significantly. Two major effects of climate change will be rising sea levels and rising global average temperatures. The major danger to human populations in cities will probably occur from extreme events such as increased storm surges (related to increasing mean sea levels) and temperature extremes (related to increasing average temperatures). In other words, climate change is not just about warming up, it also involves increases in and increased frequency of weather-related events such as rainfall patterns, snow and ice cover and sea levels.

At the urban level, cities also face different vulnerabilities and geographies to climate change. For example, in the US, inland cities such as Las Vegas and Denver may see an increase in soil erosion and a loss of water availability. Summer heat waves and droughts may affect the type of yield of crops. Underground water sources may become stressed and overused and competition for water may intensify. Cities in the Midwest and northeast may also experience wetter winters with heavier snowfall. Cities located on a river, such as Kansas City, Cincinnati, Memphis and Sacramento, may see their rivers experience flooding in the spring, as more snowpack melts, but reduced flows late in the summer; the potential for summer water shortages could affect the supply of water for drinking, agriculture, electricity production and ecosystems.

Many of the world's cities are close to the sea, so the most vulnerable cities are those in coastal locations. Alex de Sherbinin and colleagues examined the vulnerability of global cities to climate hazards. They looked at Mumbai, Rio and Shanghai and found that recent floods and monsoons had disproportionately impacted urban slums.[7] As the chapter on hazards showed, there is also an unevenness to those most at risk within the city. Those in slums and without adequate housing are at high risk, as are cities that have poorly developed drainage systems. Infants and elderly are more likely to be unable to cope with extreme heat and evacuation in areas with floods or mudslides. Finally, the urban poor may not be able to deal with illness, injury or loss of income as easily as the more affluent.

Climate change and urban water

A major emerging issue is the connection between climate and water. Climate change will have a big impact on water in terms of sea-level rise and flooding rivers. More than half of the countries of the world are at risk of water problems due to climate change. Changing climate patterns will affect the availability of fresh water both locally and regionally.

There are several factors that will impact cities and water. First is sea-level rise. Global average sea levels rose at an average rate of around 1.7 ± 0.3 mm per year from 1950 to 2009. Two main factors contribute to sea-level rise. First is thermal expansion, which means as ocean water warms, it expands in volume. The second is from the contribution of land-based ice (from glaciers and ice sheets) due to increased melting. Scientists report increased melting of ice at rapid rates in the Arctic Ocean. In 2007 the Intergovernmental Panel on Climate Change (IPCC) projected that during the twenty-first century, sea levels will rise another 7–23 inches (18–59 cm). But at the climate conference in Copenhagen in 2011, scientists suggested that the 2007 report was a drastic underestimation of the problem, and that oceans were likely to rise twice as fast. Recent satellite and other data show seas are rising by more than one-eighth of an inch (3 mm) each year—more than 50 percent faster than the average for the twentieth century. There is a widespread consensus that substantial long-term sea-level rise will continue for decades to come.[8] Rapidly melting ice sheets in Greenland and Antarctica are likely to push up sea levels by 3.5 feet (1 meter) or more by 2100.

Rising sea levels threaten already vulnerable salt marshes and other coastal habitats that provide storm protection to cities; heavy precipitation caused by more extreme weather events will increase sewer overflows, degrade water quality and increase the likelihood of waterborne diseases.[9] From more severe and frequent droughts to unprecedented flooding, many of the most profound and immediate impacts of climate change will relate to water.

Figure 13.3 Predicted flooding of Charleston, South Carolina

Source: http://www.csc.noaa.gov/digitalcoast/_/pdf/chsflood.pdf

Impacts include sea-level rise, saltwater intrusion, harm to fisheries and more frequent and intense storm events. Some climate modelers have even predicted large areas of coastal cities such as Miami, London, Amsterdam and New York will be flooded by the increase in sea levels.

In addition to rising sea levels, storms and storm surge may cause widespread flooding in coastal cities. Many climate experts predict that climate change will increase the number and intensity of storms and hurricanes. Today, some 600 million people live along the coastlines, referred to as lower elevation coastal zones (LECZ). Deltas and small island states are particularly vulnerable to sea-level rise. For example, experts estimate that a 3.5-foot (1-meter) sea-level rise will flood 17 percent of Bangladesh. In Britain lower-lying areas along the east coast, from Lincolnshire to the Thames estuary, could experience a greater risk of catastrophic storm surges. Cities may also face problems from impacts that include increased coastal erosion, higher storm-surge flooding, inhibition of primary production processes, more extensive coastal inundation, changes in surface water quality and groundwater characteristics, increased loss of property and coastal habitats, increased flood risk and potential loss of life, loss of non-monetary cultural resources and values, impacts on agriculture and aquaculture through decline in soil and water quality, and loss of tourism, recreation and transportation functions.

Cities in the developing world are equally vulnerable, but in some cases less able to spend billions of dollars on adaptation measures. A 2012 Asian Development Bank report notes that megacities in Asia are at great risk from climate change. While many of the low-lying islands in the Indian and Pacific Oceans—such as Kiribati and the Maldives—have received a lot of attention, it is the megacities where climate migrants are expected to move to. Yet many of these cities are also highly vulnerable to flooding and sinking. Cities such as Guangzhou, Seoul, Nagoya, Dhaka and Mumbai, Bangkok, Manila and Ho Chi Mihn City are already low-lying areas, but they have not yet created plans to deal with the threat of increased flooding. In 2011, for example, the Caho Phraya river system in Thailand was overwhelmed by heavy rainfall and Bangkok saw large areas flooded; the city only just avoided a catastrophe. The World Bank estimated the floods costs Thailand $46 billion in economic damage. The Thai government is urgently reviewing its policies and considering relocating factories currently located on low-lying land.

The city of Jakarta in Indonesia is also challenged by flooding that accompanies the annual monsoons. The capital lies in a low, flat basin less than 30 feet (10 meters) above the Java Sea and is prone to flooding. At the peak of the rainy season, flooding is a common problem. In 2007 floods inundated nearly three-fifths of the city, killing 52 people and costing nearly $1 billion. Five years before that, floodwaters killed about 60 residents and forced 360,000 from their homes.[10] A plan developed in cooperation with the World Bank is to dredge the city's 13 rivers and four reservoirs. Much of the city's garbage is tossed directly into rivers and canals. Sludge and silt has accumulated, exacerbating the flooding. The urgency to complete the clean-up and dredging is compounded by the fact that Jakarta is sinking. A recent World Bank report says that land subsidence from compaction from new skyscrapers and increased groundwater extraction for a growing population has caused the city to sink ten times faster than the Java Sea is rising because of climate change.

Responding to climate change: mitigation and adaptation

Throughout the 1990s and early 2000s, much of the literature on climate change viewed mitigation and adaptation as two competing methods. An early emphasis on mitigation was the result of the United Nations Framework Convention on Climate Change (UNFCCC) in 1992, one of the first international treaties that promoted policies to reduce GHG emissions, and the subsequent, more binding 1997 Kyoto Protocol. Mitigation focuses on the reduction of GHGs in order to meet the ultimate objective of the UNFCCC, which is to achieve "stabilization of greenhouse gas concentrations in the atmosphere at a level that would prevent dangerous anthropogenic interference with the climate system." The International Panel on Climate Change (IPCC), for example, defines mitigation as "an anthropogenic intervention to reduce the sources or enhance the sinks of greenhouse gases."

While mitigation tackles the causes of climate change, adaptation, on the other hand, tackles the effects of the phenomenon. Adaptation refers to the ability of a system to adjust to climate change (including climate variability and extremes), to moderate potential damage, to take advantage of opportunities or to cope with the consequences. The IPCC defines adaptation as follows:

> Adaptation to climate change refers to adjustment in natural or human systems in response to actual or expected climatic stimuli or their effects, which moderates harm or exploits beneficial opportunities. Various types of adaptation can be distinguished, including anticipatory and reactive adaptation, private and public adaptation, and autonomous and planned adaptation.[11]

For many years, mitigation was a more popular focus, in part because adaptation carried with it a sense of fatality, and also an implicit critique of those nations more responsible for high emissions. For example, it was only in 2006 that the Cities for Climate Protection Programme added adaptation to its overall Strategic Plan. In recent years, however, there is a movement toward a blend of both mitigation and adaptation agendas. This has occurred for several reasons. One is that there are limits to the ability to adapt, so actions to mitigate climate change must continue at the same time cities plan for adaptation. For example, the relocation of communities or infrastructure may not be feasible in many locations, especially in the short term. Over the long term, adaptation alone may not be sufficient to cope with all the projected impacts of climate change. Second, while climate change is inevitable and will need to be adapted to, it is still possible to mitigate the severity of this change by altering human behaviors. Recent research in the last several years has seen more of a blend of agendas that focus on both reducing GHG emissions and preparing for inevitable changes.

Governments at all levels are implementing policies geared toward both mitigation and adaptation. But there still remain those who resist making changes, despite the overwhelming evidence. Some resist climate mitigation or adaptation actions because they believe it will be a significant economic drain. In response to arguments that we cannot financially afford to enact climate change policies, Ted Nordhaus and Michael Schellenberger have framed the issue as an opportunity for new economic strategies. After the failure of climate legislation in the US Senate (for the third time), and the inability to implement a system of carbon trading, Nordhaus and Schellenberger contend that climate policy should focus not on making fossil fuels expensive through regulation but rather on making clean energy cheap. They say that climate change is not an "environmental problem" like acid rain or local water pollution that could be solved through regulation—in this case, cap-and-trade programs like the Kyoto Protocol.[12] Rather, it is an all-encompassing threat that will demand changes to our global energy system far more revolutionary than anything that could realistically be achieved by regulation. They write enthusiastically about environmentally sustainable green technology and green jobs as opportunities to counterbalance the costs of GHG reduction.[13]

Mitigation

Many cities have confronted the challenge of climate change with a focus on mitigation. Mitigation focuses on reducing the concentrations of GHGs either by reducing their sources or increasing their sinks. There are several broad categories, including reduction of fossil fuel use (by using alternative energy sources), energy efficiency and conservation, and the promotion of carbon sinks. There are many ways to approach these through specific strategies and policies; a few are highlighted in Table 13.3. Most mitigation efforts are in broad categories that include low-carbon energy, carbon storage, carbon science and carbon policy. Stephen Pacala and Robert Socolow's discussion on stabilization wedges, for example, provided numerous ways in which GHG emissions could be mitigated through basic changes to human activities, including transportation usage, fuel management and nuclear fission efforts.[14] Despite a thorough analysis on how these efforts would show serious results in reduction by 2050, some mitigation strategies are seen as controversial, especially those that promote nuclear energy. Recently there is greater emphasis on mitigation that is seen to have a multiplier effect. For example, green buildings are a fixture of sustainability and climate change discussions as they are linked to energy conservation and stormwater run-off. Also, because climate mitigation is essentially about reducing GHG emissions, mitigation efforts can also have positive side-effects on public health and ecosystems.

Table 13.3 Selected climate mitigation efforts

Use alternative energy sources
- renewable energy such as solar, hydroelectric, biofuels
- nuclear power
- carbon dioxide capture and carbon sequestration from power plants or industrial sources

Improve energy efficiency and conservation
- transportation
  - improve vehicle efficiency (including use of hybrid or zero-emission vehicles)
  - reduce the number of miles traveled
  - encourage bicycle use
  - promote public transportation
- urban planning and building design
  - encourage green buildings and green roofs
  - increase the energy efficiency of buildings

Sinks and conservation
- plant trees and promote reforestation

Adaptation

At one level adaptation is about ensuring that the infrastructure of a city, from buildings and industrial facilities to roads and sewerage systems, has as low a vulnerability as possible to likely future climate changes, and that they are as flexible as possible, providing an ability to cope with the high level of uncertainty over the actual nature of future changes in climate. At another level it involves putting institutional frameworks in place that are "adaptation friendly"—for example, urban planning offices that build adaptation requirements into a wide range of ongoing planning activities.[15] Adaptation involves responding to changes and to building the capacity to be resilient. Adaptation strategies including the ability to:

- make decisions and plan in the face of complexity and uncertainty;
- identify, assess, prioritize and manage risks related to climate change;
- engage communities in risk-management processes;
- ensure transparency in making and communicating decisions on risk treatment options and implementation plans;
- foster leadership and culture change in councils to ensure the development of a strategic approach to managing high-priority risks/opportunities.[16]

Recent research highlights a need to improve our understanding of climate vulnerability and adaptation in urban areas. This is particularly important in developing cities with large numbers of urban poor. Thomas Tanner and colleagues assessed climate change resilience in ten Asian cities.[17] They found that the involvement of poor and marginalized groups in decision making, monitoring and evaluation is a key characteristic of a city intent on improving conditions and building climate resilience.

National governments now recognize that most climate adaptation occurs at the local level. A 2011 National Research Council's report, *America's Climate Choices*, noted that communities across the US are already experiencing a range of climatic changes, including more frequent and extreme precipitation events, longer wildfire seasons, reduced snowpack, extreme heat events, increasing ocean temperatures and rising sea levels. A 2011 survey of 396 mayors from all 50 states found that over 30 percent are already taking climate impacts into account within their planning and improvement programs, demonstrating growing local concern about climate risks.[18] For example, the City of Chicago, anticipating a hotter and wetter future, is already taking steps to adapt, such as repaving alleyways with permeable materials to handle greater rainfall and reduce flood risks, and planting trees that can tolerate warmer conditions.[19] Even smaller cities are

BOX 13.3

Cities for Climate Protection (CCP)

Since its inception in 1993, the Cities for Climate Protection (CCP) Campaign has become the centerpiece of the climate action project of the Council for Local Environmental Initiatives (ICLEI—now called the Local Governments for Sustainability). The five-milestone process is considered as one of the unique features of the CCP. Within this framework, following a political commitment made to the representative of their local governments, participating cities are expected to:

1   measure their emissions of GHGs, generated through the actions of their local government administration (government emissions) and through the actions of the community they serve (community emissions);
2   commit to an emissions reduction target with respect to a base year and a target year;
3   plan their actions (e.g., energy efficiency in buildings and transport, introduction of renewable energy, sustainable waste management) at the government and community level to reach this committed reduction target;
4   implement their local climate action plan;
5   monitor emissions reductions achieved by their mitigation actions.

Following the rapid dissemination in the US, Australia and Europe in early 1990s, the CCP campaign was used to initiate urban GHG mitigation actions in many developing countries in South Asia, Latin America and Africa in the late 1990s. The availability of software tools for accounting and monitoring of urban GHG emissions, capacity building and training activities, city-to-city and network learning support were vital elements of success that enabled the CCP to influence the climate mitigation action of cities.

The experience with urban GHG emissions globally helped ICLEI to develop the concept of a need for a standardization of local government emissions inventory. Thus, in 2009, the first version of the International Local Government GHG Emissions Analysis Protocol (IEAP) was developed that followed principles that were previously adapted by the WRI/WBCSD GHG Protocol.

Source: International Council for Local Environmental Initiatives (n.d.) "Cities for Climate Protection programme." http://www.iclei.org/index.php?id=10828 (accessed September 2012).

making adaptation plans. Greg Heartwell, mayor of Grand Rapids, Michigan says:

> The City of Grand Rapids is addressing various climate-related threats such as extreme heat and more intense precipitation events. We see these climate strategies as an extension of responsible governance and an imperative investment in the future prosperity of our city. As an inland watershed city, we have focused on restoring and maintaining a high quality of water in the Grand River, with over $240 million in combined sewer separation investment. This prepares us for ever-increasing precipitation levels now and into the future.[20]

Cities are now planning for climate adaptation, and these plans involve a range of examples and strategies, some of which are highlighted in Table 13.4.

Although cities are taking the lead in climate adaptation, they will need federal governments to support the costs of many types of projects. For example, cities on the coast are developing plans to relocate water and sewer infrastructure to higher elevations. For smaller, lower-wealth cities, this could be a significant financial burden without grants from the federal government.

In 2009, President Barack Obama established the Interagency Climate Change Adaptation Task Force. The Task Force was charged to assess key steps needed to help the federal government understand and adapt to climate change. All federal agencies were asked to assess how climate change may impact agency missions and operations, to identify necessary adjustments to reduce risk, avoid unnecessary costs and take advantage of opportunities. Agencies with emergency management and health missions will likely focus

Table 13.4 Examples of adaptation approaches

Coastal city adaptation	Identify and improve evacuation routes and evacuation plans for low-lying areas to prepare for increased storm surge and flooding.
	Shore protection techniques and open space preserves that allow beaches and coastal wetlands to gradually move inland as sea level rises.
Ecosystems adaptation	Protect and increase migration corridors to allow species to migrate as the climate changes.
	Promote land and wildlife management practices that enhance ecosystem resilience.
Energy	Increase energy efficiency to help offset increases in energy consumption.
	Harden energy production facilities to withstand increased flood, wind, lightning, and other storm-related stresses.
Human health	Implement early-warning systems and emergency response plans to prepare for changes in the frequency, duration and intensity of extreme weather events.
	Plant trees and expand green spaces in urban settings to moderate heat increases.
Water resources	Improve water use efficiency and build additional water storage capacity.
	Protect and restore stream and river banks to ensure good water quality and safeguard water quantity.
Infrastructure	Relocate vulnerable water and sewer infrastructure.

Source: US EPA (2012) "Adaptation overview." http://www.epa.gov/climatechange/impacts-adaptation/adapt-overview.html (accessed August 2012).

on planning that reduces climate change risks to communities; those with infrastructure responsibilities will emphasize planning that enhances resilience and minimizes disruption; and agencies that support particular sectors (e.g., agriculture, energy) will focus on climate risks to production and security. For example, in 2011 the EPA announced it would complete a Climate Change Adaptation Plan in 2012. Every program and regional office within the EPA will develop an Implementation Plan outlining how each considers the impacts of climate change in its mission, operations and programs, and carry out the work called for in the agency-wide plan.

There are signs that federal agencies are working with cities on climate adaptation. Southeast Florida cities such as Miami and Palm Beach are already experiencing the impacts of extreme weather and sea-level rise, compromising drainage systems and sea walls during high-tide events. With continued sea-level rise and the prospect of more intense hurricanes and heavy downpours, the region faces greater risks of flooding, safe water supply shortages, infrastructure damage and natural resource degradation. In response, Broward, Miami-Dade, Palm Beach and Monroe Counties entered into the Southeast Florida Regional Climate Change Compact in 2010 to address these threats collaboratively. Local and regional offices of Federal agencies—including NOAA, USGS and the EPA—have supported these counties with regional adaptation planning. For example, the US Army Corps of Engineers and NOAA provided technical assistance to evaluate threats of future sea-level rise. USGS applied advanced hydrologic models and provided financial resources to support projects related to saltwater intrusion of groundwater supplies and flood risks. The EPA provided coordination support, helping connect the Compact partners with critical technical, planning and programmatic resources.[21]

Climate plans

Climate scholars acknowledge that while national and international policies have an important role in reducing GHG emissions, it may be a half-century or more before these policies lead to any substantive reduction in atmospheric concentrations of GHGs and global average temperature. More importantly, the impacts of climate change will be felt most acutely at the local scale. Managing these impacts will require developing locally based strategies.[22]

Harriet Bulkeley and Michele Betsill argue that cities are a significant arena through which to address climate change because they are sites of high consumption of energy and the production of waste; local authorities have been engaging with issues of sustainability; local authorities can facilitate action by others in response to climate change by developing small-scale demonstration projects; and local authorities have expertise in addressing a range of environmental impacts within the fields of energy management, transportation and planning.[23] Bulkeley and Betsill note that there is evidence of a "new wave" of urban climate change response with a greater emphasis on adaptation and an effort to include a broader range of cities.

Many cities already have strong local policies and programs in place to reduce global warming pollution, but more action is needed at the local, state and federal levels to meet the challenge. On February 16, 2005 the Kyoto Protocol, the international agreement to address climate disruption, became law for the 141 countries that have ratified it to date. In response, mayors in cities such as Seattle and London called for more radical local action to reduce global warming pollution.

In March 2005 Seattle mayor Greg Nickles and nine other US mayors representing more than three million Americans joined together to invite cities from across the country to take additional actions to significantly reduce global warming pollution. Three months later in June, the Mayors Climate Protection Agreement was passed unanimously by the US Conference of Mayors. The Mayors Agreement on Climate Change is remarkable in that it occurred primarily because the Bush administration refused to ratify the Kyoto Protocol. In the absence of national leadership, cities have committed themselves to the agreement. The agreement tasks participating cities to take the following three actions:

1 Strive to meet or beat the Kyoto Protocol targets in their own communities, through actions ranging from anti-sprawl land-use policies to urban forest restoration projects to public information campaigns.
2 Urge their state governments, and the federal government, to enact policies and programs to meet or beat the GHG emission reduction target suggested for the US in the Kyoto Protocol—7 percent reduction from 1990 levels by 2012.
3 Urge the US Congress to pass GHG reduction legislation, which would establish a national emission trading system.

By 2012, the Mayors Agreement on Climate Change included 1,054 mayors from the 50 states, the District of Columbia and Puerto Rico, representing a total population of over 88 million citizens. Many US cities have also initiated local environmental legislation that exceeds EPA standards, or, as is the case with the Mayors Agreement on Climate Change, occurs in the absence of national leadership. In some cases national environmental organizations are also providing leadership. One example is the Sierra Club's "Cool Cities" Program (Box 13.4).

Mayoral leadership on climate change goes beyond the US example. In October 2005, representatives of some 20 cities met in London at the World Cities Leadership Climate Change Summit, which was organized by the Mayor of London, Ken Livingstone. The conference launched an international collaboration on the pressing issue of climate change and brought together city leaders who are taking the most productive and radical steps to adapt to and mitigate climate change, to share ideas and provide leadership to the rest of the world. One result of the conference was the C40 group.

BOX 13.4

Sierra Club's Cool Cities

The Sierra Club, one of the largest grassroots environmental organizations in the US, launched the "Cool Cities
Program" in 2005. The Cool Cities Program is a collaboration between community members, organizations, businesses and local leaders to implement clean energy solutions
that save money, create jobs and help curb global warming.
Since the start of the program, more than 1,000 cities have
made the commitment to curb their carbon footprint.

The Sierra Club certainly recognizes the continued inability of
the national government to take the lead on climate change,
noting that

> at a time when the federal government is failing to take
> action to solve global warming, Cool Cities offers an
> opportunity and a road map by which to work toward
> realizing our shared vision of a safe and clean energy
> future in the cities where we live.

The program has two important goals. The first goal is to
encourage cities (and eventually states and the federal
government) to take action with smart energy solutions
that reduce global warming emissions. Cool Cities focuses
on reducing energy consumption of buildings, and works
in partnership with the US Green Building Council. The
program also focuses on energy and transportation. Efforts
can include cleaner transportation fleets, energy-efficient
buildings and use of renewable energy. A second goal is
to energize, support and build local volunteer activism and
influence, both inside and outside the Sierra Club. The Cool
Cities Program has established five milestones. Cities should
aim to:

1 establish a Cool Cities Campaign;
2 engage the community;
3 sign a municipality commitment agreement;

4   implement initial solution steps;

5   implement advanced smart energy solutions.

One example of a Cool City is the town of Blacksburg, Virginia. Blacksburg is home to approximately 43,000 residents, about half of whom are Virginia Tech students. Since beginning a formal campaign through the organization Sustainable Blacksburg in 2006, the town has accomplished four out of five of the Cool Cities milestones. Some of their notable initiatives have been replacing traffic lights with LEDs, conducting an energy audit of municipal buildings and making needed changes to reduce energy use, purchasing five hybrid Ford Escape vehicles for the town fleet, collecting leaves from residential neighborhoods and making them available to gardeners and increasing recycling for residents and businesses.

Sources: Sierra Club (2012) "About Cool Cities" http://coolcities.us/about.php?sid=595c47f7394e0a1ea35b5ed95c3d5960 (accessed April 2012). Sierra Club (2012) "City profile: Blacksburg, VA." http://coolcities.us/cityProfiles.php?city=70&state=VA (accessed April 2012).

The C40 group is a group of the world's largest cities committed to tackling climate change to reduce carbon emissions and increase energy efficiency in large cities across the world. C40 members include: Bangkok, Berlin, Bogotá, Buenos Aires, Cairo, Caracas, Chicago, Delhi NCT, Dhaka, Houston, Istanbul, Jakarta, Johannesburg, London, Los Angeles, Madrid, Melbourne, Mexico City, Moscow, New York, Paris, Philadelphia, Rome, São Paulo, Seoul, Tokyo and Toronto. Smaller, affiliated cities include: Austin, Barcelona, Copenhagen, Curitiba, Heidelberg, New Orleans, Portland, Rotterdam, Salt Lake City, San Francisco, Seattle and Stockholm.

One outcome from these networks and agreements is the creation of city climate plans. Most climate plans address both mitigation and adaptation. For example, in 2007 London adopted its own Climate Change Action Plan that outlines numerous measures that aim to save London 20 million tons of carbon by the year 2025.[24] The plan centers around four programs: green homes, green organizations, green energy and green transport.

Ken Livingstone, London's mayor, announced that £78 million will be reprioritized over three years within existing Greater London Authority finances this year to launch these programs. Said Livingstone:

> The actions set out in this plan are radical—the most comprehensive for any city I know. But they will need to be accompanied by further action from government. It is completely inadequate to simply talk about climate change or make purely token actions. This Plan sets out the beginning of a comprehensive programme to tackle climate change in London in the next twenty years.[25]

London has implemented many innovative policies. It is the only major city in the world to achieve a shift away from private car usage to public transport, cycling and walking, thus stabilizing emissions from road traffic. In part, this has been driven by the pioneering move to charge vehicles entering central London, which has cut carbon emissions by 16 percent within this zone. The London congestion charge is a fee charged on most motor vehicles operating within the Congestion Charge Zone in central London between 07:00 and 18:00 (Monday–Friday only). The charge, which was introduced in 2003, remains one of the largest congestion zones in the world and aims to reduce congestion and to raise money for public transport. Automobiles are charged £10 (approximately $15 dollars) each day. Private auto use is down and air quality monitoring over the last several years shows that nitrogen dioxide, particulate matter and carbon dioxide levels have been reduced from pre-2003 levels. In addition, London has ambitious goals to reduce carbon dioxide to levels far below those of the UK and other nations (Figure 13.4).

Many European cities have been in the forefront of climate action plans, including Amsterdam, which is profiled in Box 13.5.

In its 2007 climate plan New York City established the goal of reducing city-wide GHG emissions by 30 percent below 2005 levels by 2030. In 2009 it reported that city-wide GHG emissions were lower than in 2008, the first year a decrease in GHG emissions has been recorded. As with many cities, New York's GHG emissions are dominated by energy consumed in transportation and energy consumed in buildings. Roughly 78 percent of New York City's GHG emissions are related to heating, cooling, powering and lighting buildings, and 20 percent are related to transportation. Not surprisingly, its climate plan focuses on ways to reduce energy use. Similarly, Chicago adopted its Chicago Climate Action Plan in 2008. The broad goal is to eliminate GHG emissions by 80 percent of their 1990 level. Divided into five main strategies, the plan mostly deals with ways to mitigate climate change with less attention on adaptation. Chicago proposes to reduce GHG

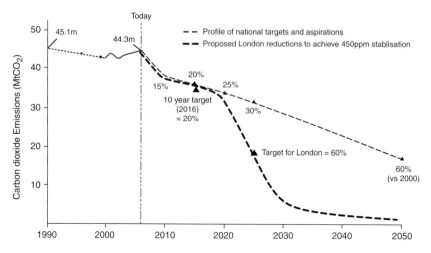

Figure 13.4 London's Climate Action Plan calls for a reduction in carbon dioxide emissions that exceeds UK targets

Source: London's Climate Action Plan at: http://www.london.gov.uk/thelondonplan/climate

BOX 13.5

Amsterdam: balancing mitigation and adaptation

Located less than ten feet above sea level along the North Sea, Amsterdam is highly vulnerable to the potential impacts of climate change, primarily sea-level rise. In order to deal with this growing threat, the city has implemented a balanced approach of both mitigation and adaptation. As a result, Amsterdam has become a world leader in many of these initiatives.

In the mid-2000s, Amsterdam Innovation Motor initiated *Amsterdam Smart City*, a program geared toward reducing GHG emissions. Fuel cell usage is being promoted to lower energy consumption. Such technology is being implemented for a trial run in several areas in the city center. The aim is to lower CO_2 emissions by 50 percent. Amsterdam has also implemented programs to improve public building energy consumption, calling for municipal buildings to be energy-neutral by 2015. This initiative will involve measuring energy usage through an online portal, with the goal of raising greater awareness of daily consumption. Finally, the city can build on an already existing bicycle culture. Some 75 percent of Amsterdamers own at least one bicycle.

In addition to mitigation efforts, Amsterdam is also preparing for inevitable changes in the near future through adaptation. For example, in 2011 the Nieuw-West City District of Amsterdam submitted an Adaptation Action Plan. As winters are expected to get wetter, summers are expected to get drier, and year-round temperatures are expected to increase, the plan proposes several programs to cope with these changes. One proposal involves producing a comprehensive climate change impact map to highlight the most vulnerable areas in the district. Emphasis is then placed on ensuring that every spatial-related policy plan contains some reference to climate change adaptation, such as reducing local flooding and restructuring buildings whose old infrastructure would be exposed due to possible extended droughts. As the numerous documents submitted throughout the city's institutions show, Amsterdam is not taking the effects of climate change lightly and plans on being as prepared as possible for the future.

Source: (2012) "Amsterdam Smart City." *Amsterdam Innovation Motor and Liander*.http://www.amsterdamsmart-city.nl (accessed March 2012).

emissions by retrofitting older buildings with more energy-efficient technology, investing in clean and renewable energy sources, reducing waste and industrial pollution. In order to improve transportation, the Chicago Transit Authority has introduced over 200 hybrid buses that now comprise nearly 13 percent of the city's fleet. The plan also considers adaptation focusing on what the city sees as the two main effects of climate change: extreme heat and decreased precipitation. To combat increased urban heat islands, the city has pledged to plant more than 10,000 trees. Since 2008 the city has also added 55 acres of permeable surfaces.[26]

Conclusions

The development of city climate plans is relatively recent, and many of these are initial blueprints for action. Over the next several years, it is likely that cities will further develop their climate plans to include a better balance of mitigation and adaptation strategies. Climate change is inextricably linked to cities, and this is an area of research and policy that will continue to be of critical importance in the years to come.

Guide to further reading

ActionAid (2006) *Unjust Waters: Climate Change, Flooding and the Protection of Poor Urban Communities—Experiences from Six African Cities.* London: ActionAid.

Aerts, J., Botzen, W., Bowman, M., Ward, P. and Dircke, P. (eds.) (2011) *Climate Adaptation and Flood Risk in Coastal Cities.* London and New York: Routledge.

Bicknell, J., Dodman, D. and Sattherwaite, D. (eds.) (2009) *Adapting Cities to Climate Change, Understanding and Addressing the Development Challenges.* London: Earthscan.

Bulkeley, H. (2012) *Climate Change and the City.* London and New York: Routledge.

Bulkeley, H. and Betsill, M. (2005) *Cities and Climate Change: Urban Sustainability and Global Environmental Governance.* New York: Routledge.

Girardet, H. (2008) *Cities, People, Planet: Urban Development and Climate Change.* 2nd edition. Chichester: Wiley Academic Press.

Nordhuas, T. and Schellenberger M. (2007) *Break Through: From the Death of Environmentalism to the Politics of Possibility.* Boston, MA: Houghton Mifflin Company.

Rosenzweig, C., Solecki, W. D., Hammer, S. A. and Mehrotra, S. (eds.) (2011) *Climate Change and Cities: First Assessment Report of the Urban Climate Change Research Network.* Cambridge: Cambridge University Press.

Sherbinin, A., Schiller, A. and Pulsipher, A. (2007) "The vulnerability of global cities to climate hazards." *Environment and Urbanization* 19 (1): 39–64.

Stone Jr., B. (2012) *The City and the Coming Climate: Climate Change in the Places we Live.* Cambridge: Cambridge University Press.

14 Garbage

Garbage is a fundamental urban problem. Problems of collection and disposal and the environmental hazards associated with refuse continue to challenge many cities. Underlying social and cultural values have made reform in refuse lag far behind water and air pollution reform. In this chapter we explore how cities are confronting waste, the acceleration in the consumption of goods and resources per person which increases the generation of solid waste and how new technologies, particularly those developed in the chemical industry such as plastics, radically changed the composition of garbage, presenting new challenges and problems for the environment. In the developed world, cities generate very high per capita rates of waste, and recycling rates are low and part of the formal economy. In addition, many cities have sought to "export" their garbage problems to developing countries. In the second part of the chapter we examine cities in the developing world. Many developing cities face similar issues with waste, although they generate less per capita waste. Many cities have fewer resources to ensure that collection and disposal will not create environmental and health hazards. Interestingly, in most developing cities, recycling rates are high because the urban poor separate recyclables as part of the informal economy.

Waste trends in the US

In this section we focus on US cities because America generates more refuse per person than any other country in the world. It produces some 30 percent of the world's garbage (with only 5 percent of the world's population). Today the average individual in the US generates about 4.43 pounds of waste per day, and recycles and composts only about 1.51 pounds. These rates are the highest in the developed world: for comparison, waste generation rates (in pounds per person per day) are 3.1 in Sweden, 3.5 in Germany and 3.4 in the UK. The garbage problem in US cities is more acute than anywhere else.

Solid waste is defined as "material that has no apparent, obvious or significant economic or beneficial value to humans that is intentionally thrown away for disposal." Unlike air and water pollution, solid waste must be collected to be disposed of. The classification of municipal solid waste (MSW) includes waste generated from two main streams of waste: residential and commercial/institutional. Residential waste (including waste from apartment houses) is about 60–65 percent of total MSW generation. Commercial waste from businesses, schools, hospitals and construction accounts for about 35–40 percent. Table 14.1 lists different sources of waste.

In the US postwar period, economic prosperity created a "throwaway" culture that generated vastly increased volumes of waste. New materials such as plastics, other synthetic products and toxic chemicals made their way to landfills. In addition, new changes in the packaging industry created innumerable goods with very short lives. Between 1955 and 1965 per capita refuse increased by 78 percent in New York City; between 1958 and 1968 it increased 51 percent in Los Angeles.[1] By the 1960s solid waste had become

Table 14.1 Sources and types of municipal solid wastes

Source	Typical waste generators	Types of solid waste
Residential	Single and multi-family dwellings	Food wastes, paper, plastics, textiles, leather, yard wastes, glass, metals, ashes, bulky items like refrigerators, tires, hazardous wastes
Industrial	Manufacturing, construction sites, power and chemical plants	Packaging, food wastes, construction materials, hazardous wastes, ashes, slag, scrap materials, tailings, special wastes
Commercial	Stores, hotels, restaurants, markets, office buildings	Paper, cardboard, plastics, wood, food wastes, glass, metals, special wastes
Institutional	Schools, hospitals, prisons	Same as commercial plus medical wastes
Construction and demolition	New construction sites, road repair, renovation sites, demolition of buildings	Wood, steel, concrete, dirt
Municipal services	Street cleaning, landscaping, parks, beaches, wastewater treatment plants	Sludge from wastewater treatment plants, landscape and tree trimmings, general wastes from parks

Source: US EPA.

a critical environmental issue and land pollution joined air and water pollution as a triad of blights deserving federal attention.[2] In 1969 the US collected 30 million tons of paper and paper products, 4 million tons of plastic, 100 million tires, 30 billion bottles, 60 billion cans and millions of tons of grass, tree trimmings, food waste and sewage sludge.[3] In 2010 Americans generated about 250 million tons of trash, and recycled or composted over 85 million tons of that,[4] and about 136 million tons were discarded in landfills. That tonnage represents a 60 percent increase from the 1960s. In 2011 New York City alone collected 1,200 tons of trash and recycling each day, for a total of 3.2 billion tons.[5] Figure 14.1 shows the rising trends in municipal solid waste from 1960 to 2010; the good news is that material discarded in landfills peaked around 2005 and has declined as cities have implemented more effective recycling and composting programs. In 1960, US cities recycled about 5 percent of waste; in 2010 they were recycling about 34 percent of the waste. According to a 2010 EPA report, there are about 9,000 curbside recycling programs and about 3,100 community composting programs.[6]

For the first half of the twentieth century, most refuse was simply placed in "open" dumps. Open dumps were natural or human-made depressions in the landscape. The hazards of open dumps included fires and the leaching of

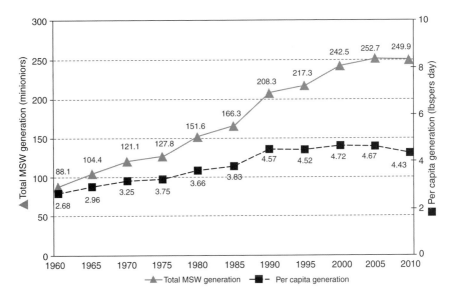

Figure 14.1 Municipal solid waste trends, 1960–2010

Source: US EPA (2010) "Municipal solid waste generation, recycling and disposal in the United States: facts and figures for 2010"

numerous chemicals into the soil or groundwater. They also posed other health and safety issues. Human fecal matter is commonly found in municipal waste. Insects and rodents attracted to the waste can spread diseases such as cholera and dengue fever. The EPA has identified 22 human diseases that are linked to improper solid waste management.[7] Until very recently, open dumps were not regulated.

In the postwar years, advances in sanitary engineering, notably by the Army Corps of Engineers, improved upon the open dump and established standards for the proper use of the sanitary landfill. Contemporary landfills are huge engineering feats that include layers of non-permeable clay or synthetic lining and a network of pipes to collect leachate and methane as an energy source. Landfills were sited according to drainage patterns, wind and distance from the city, rainfall, soil types and the depth of the water table. Since the 1950s most US cities have instituted service charges for the collection and disposal of household refuse. The most popular form of disposal is land disposal. By the 1960s most cities outlawed the open dump, by then considered a danger and a health menace, and looked to the landfill as the solution to the problem of refuse.

Additional technological developments of the late twentieth century aided in both collection and disposal. The introduction of compaction vehicles, which compress wastes by 30 percent, allow more volume to be collected curbside. Similar compaction systems can also be found in private residences. In-home garbage disposals that grind food wastes were once promoted as a technology that would completely eliminate household garbage; in reality disposals just transfer the pollution problem from curbside collection to wastewater treatment plants. The development and use of transfer stations—collection points where trucks unload into larger vehicles or temporary storage facilities—increase efficiency and assist in sorting.[8] Technological "fixes" often provide partial solutions, or in some cases alleviate the problem, but rarely solve the underlying issue of the production of waste.

The garbage crisis?

Despite the improvements in refuse technology, solid waste remains a growing problem. In the last 30 years, many cities have experienced a "garbage crisis." Many landfills have or will soon close as they reach their capacity. In 1988 there were 7,924 landfills in the US, in 2000 there were 2,216, although the average landfill size did increase.[9] In 2010 there were 1,908 landfills. Although the number of landfills has declined with time, national capacity has not changed significantly because older MSW landfills tended to be smaller and more numerous. Many of these landfills closed due to the cost of meeting new federal and state regulations. Older landfills were replaced by newer, larger landfills. While the EPA believes there is enough capacity in existing landfills, there are regional disparities and some cities do not have local access to a landfill.

There is a geography to the garbage crisis. Cities in the northeast, New England and mid-Atlantic are running out of space to dump their refuse, while inland cities such as Dallas, Phoenix and Santa Fe have landfill capacity for more than 40 more years. Arkansas reports enough capacity to go more than 600 years without opening another facility. Massachusetts and Rhode Island, on the other hand, have just 12 years of capacity remaining. New Jersey has to ship 50 percent of its solid wastes, or 11 million tons per year, to nearby states. New York State ships most of its trash across state lines, and has only 25 years of capacity left. Today, New York City exports 20 percent of its trash to other parts of New York, Pennsylvania, Virginia and other states. Canadian cities are not immune either: Toronto outsources up to 40 percent of its refuse, some of it to the US. The garbage crisis is more regional than national.

For those cities facing landfill closures there are much higher tipping fees for both collection and disposal. The average landfill tip fee is around $42 per ton. A "tip fee" is the price paid, usually on a per ton basis, to dispose of trash at a landfill. The average tipping fee in the US has risen consistently since 1985, when it was $8.20 per ton. But regional variations are found here, too: the northeast (CT, ME, MA, NH, NY, RI and VT) has the highest tipping fees at $96 per ton, while the south central (AZ, AR, LA, NM, OK and TX) and west central (CO, KS, MT, NE, ND, SD, UT and WY) states have the lowest tipping fees at $36 and $24 per ton, respectively.[10] A recent EPA study found that areas with high tipping fees tended to recycle and compost more of the waste stream, while cities in the south and west tended to send a higher percentage of their waste stream to landfills. Thus, higher fees for refuse collection and disposal may provide an economic incentive for cities to improve their recycling and composting programs. Cities that can divert more refuse to recycling will save money.

At the same time as older landfills have reached capacity and are closing, it has become more difficult to open new landfills as many environmental organizations have effectively opposed them on the basis of potential environmental hazards. Since 1990 only a handful of new landfills have been approved and opened. In the midst of increasing refuse and decreasing landfills, many

BOX 14.1

Transforming New York City's Fresh Kills landfill

One of the more infamous landfills that has closed is Fresh Kills. Fresh Kills (originally from the Dutch "kill"=water) covers some 3,000 acres and is located on the western shore of Staten Island in New York. From 1948 to 2001, Fresh Kills was New York City's sole landfill. At its height, Fresh Kills received some 17,000 tons of trash per day. It was a putrid mountain of waste, the largest human-made structure in the history of the world. In 1996, officials announced that Fresh Kills would close. The city would expand its recycling program and search for alternative landfill destinations. The Fresh Kills landfill was officially closed in March 2001. By order of Governor Pataki it was reopened on September 12, 2001 to accept material from the World Trade Center. Following the end of the WTC Recovery Project in 2002, it was prepared for final closure. Today, two of the four sections of the landfill have been capped; a third mound is nearing completion. Estimates are that the city has already spent nearly $1 billion to close the landfill, and may spend up to another $1 billion for leachate and methane management. In recent years, the sale of methane gas from the landfill has generated $4.5 million in revenue and eliminated the equivalent of 600,000 metric tons of CO_2 and associated greenhouse gases.

Since its closure the mounds at Fresh Kills have been covered with a foot of dirt, topped with a foot of sand, followed by a plastic liner and two more feet of soil. The plan is to "reclaim" the landfill as a "postmodern forest." Landscape architects envision it as a thriving park and bird sanctuary, where in another 30 years or so people might walk on natural trails or picnic among the freshly made wilderness. Recreational facilities are also planned—tennis courts, soccer fields, golf courses and even boat docks. When Fresh Kills Park is completed at 2,200 acres, it will be more than three times the size of Central Park. In 2010, New York City Mayor Bloomburg broke ground on Schmul Park. New

York State has awarded grants to begin work on some of the recreation areas and smaller parks proposed, but it will be many years before the landfill is transformed and open to public use.

Today, trash from the Bronx, Brooklyn and Manhattan is now exported out of New York State. However, the exporting of trash costs the city some $300 million each year and there have been unpredicted costs as well. For those who live along city truck corridors, there has been a "remapping" of the flow of garbage. Floors and windows of homes and apartments vibrate when the trucks roll by; the smell of rotting garbage and truck exhaust fills the air outside. Proposed solutions to the refuse problem include transporting the trash by rail or barge to landfills in New Jersey, Virginia and states in the Midwest and South.

Source: Department of Sanitation New York (2011) *Annual Report, 2011.*

cities have turned to one of two solutions. First, many cities now truck their garbage out of the urban region to other parts of the state, and in some cases even across state lines. Philadelphia, for example, sends some of its trash to Delaware landfills. Ironically, the state of Pennsylvania imports garbage from cities in other states. Many cities in the northeast transport their refuse to landfills in the Midwest, where large areas of land are still available. The problem is not limited to the US. Toronto, for example, saw one of its major landfills close in 2002 and began shipping 400,000 tons of waste each year to Michigan. However, the city set a deadline to stop shipping waste to Michigan and is looking for ways to divert waste from landfills.

Second, and perhaps more ethically problematic, many cities have attempted to make arrangements to ship their municipal waste to developing countries. But critics call this practice "waste imperialism" and argue that governments are endangering their own citizens by accepting refuse and hazardous wastes from rich countries. With no solution to the garbage problem forthcoming, cities desperately search for new landfills and disposal sites.

Consider the very famous case of the city of Philadelphia and the cargo ship, *Khian Sea*. In September 1986, the *Khian Sea* left Philadelphia with 14 tons of ash from refuse that had been incinerated. Initially, Philadelphia had contracted with a firm that found a shipping company willing to take the ash to a man-made island in the Bahamas. Alerted by environmental organizations such as Greenpeace about the potential environmental hazard of the ash, the Bahamian government denied permission for the *Khian Sea* to dock. Over the next two years, the *Khian Sea* attempted to make arrangements to unload its ash with 11 different countries including Honduras, Haiti, the Dominican Republic, Senegal, Sri Lanka, Indonesia and the Philippines. Finally, in November 1998, the ship arrived in Singapore—without its ash. Somewhere along the way, the ship had been sold, and twice renamed. The ash was most likely dumped illegally into the ocean.

Ethical questions of waste disposal in places far beyond the source of origin have become a recurring theme as air, water and now refuse cross national boundaries. Is it equitable for cities to intentionally export their pollution problems to other, often poorer areas, states or countries?

BOX 14.2

China's paper recycling industry

For many years there has been a growing and massive trade imbalance between the US and China. In 2010 the trade imbalance was $273 billion. Economists believe the trade imbalance will likely increase as China begins to export more high-end manufactured goods such as cars, trucks and planes.

The richest woman in China, Zhang Yin, is worth $3.4 billion. But unlike other Chinese entrepreneurs who have made their money by exporting to the West, Zhang built her fortune another way: she imports scrap paper from the West. She is the "queen of waste paper," China's largest importer of scrap paper. Ms. Zhang started Nine Dragons Paper a few years ago to become the world's richest self-made woman, surpassing US talk-show host Oprah Winfrey and J. K. Rowling, the *Harry Potter* author. The good news is her business recycles paper into products like cardboard boxes. The strange news is that this paper comes from the US.

According to the US International Trade Commission, "the United States shipped 7.7 million metric tons of waste paper to China in 2005." Between 1995 and 2005, the USITC reports, "Chinese imports of wood pulp and waste paper from the United States increased by 500 percent over the same period, while imports of finished paper declined by 12 percent." According to the *Economist*, China's number one export to the US is computer equipment (nearly $50 billion) while the US's number one export to China is waste paper and scrap metal (approximately $8 billion). Other big buyers of waste paper for recycling include Mexico, India and Vietnam.

A 2010 *Economist* article noted that the huge and growing Chinese market for US scrap paper created by Zhang and others has important landfill implications in the United States. The waste paper shipped to China each year for recycling into paper, cartons and other products represents nearly 8.5 million tons of paper that will not be deposited in US land-fills. On the other hand, it means that the US is exporting paper for recycling elsewhere, at the expense of growing domestic recycling facilities.

Sources: *China Daily* (2006, November 20) "China's richest woman: from waste to wealth." http://www.china-daily.com.cn/china/2006-10/20/content_713250.htm (accessed July 2012); US International Trade Commission (2006) "The effects of increasing Chinese demand on global commodity markets," pp. 1–14. http://www.usitc.gov/publications/332/working_papers/pub3864-200606.pdf (accessed June 1, 2012); *The Economist* (2010) "The number one U.S. export to China: waste paper and scrap metal." http://theeconomiccollapseblog.com/archives/the-number-one-u-s-export-to-china-waste-paper-and-scrap-metal (accessed July 2012).

The growing concern over potential environmental hazards associated with refuse and incinerator ash may play out in the legal system. For example, some states in the US have passed laws that restrict or prohibit trash imports. In New Jersey, state and local leaders blocked a plan to ship New York City's trash to Newark landfills, and one utility company filed suit to block trash deliveries by New York City garbage trucks to a transfer station in Elizabeth, New Jersey. Said Elizabeth's Mayor J. Christian Bollwage: "They [NY] are forcing their will on others because they don't have the creativity to solve their own problems."[11] In Docklands, UK, the siting of a solid-waste transfer station generated local protest (Figure 14.2).

The environmental hazards of landfills

Unlike air and water pollution, garbage and solid waste is regarded as more of an engineering problem than a public health issue.[12] However, there are several potential environmental problems and public health issues associated with landfills. A significant part of the composition of the US waste stream is biodegradable organic material such as food wastes, yard wastes and tree trimmings (Figure 14.3 shows the sources and types of solid wastes in the waste stream).

Figure 14.2 Docklands protest

Source: John Rennie Short

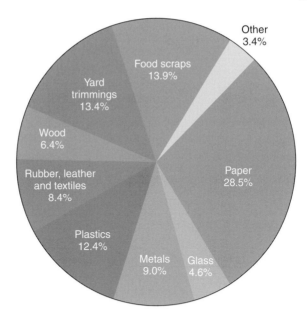

Figure 14.3 Composition of the US waste stream, 2010

Source: US EPA (2010) "Municipal sold waste generation, recycling and disposal in the United States: facts and figures for 2010"

In theory, over time, organic material should biodegrade, creating nutrient-rich soils. In reality, however, landfills are so compacted that there is not enough air for microbes and bacteria to break down organic debris. One of the greatest myths about the US waste stream is that it is primarily composed of plastics and disposal diapers. While it is true that there is a significant amount of packaging waste, in reality most of what people discard could be recycled, reused or composted, but often is not. Landfills, even the most recently engineered ones, still have a potential to contaminate soil and groundwater. Trash bags leak their contents. Water from rain also seeps through refuse, picking up a variety of liquids and poisons such as ammonia, chlorides, zinc, lead and acids (Table 14.2 shows for a list of household hazardous wastes commonly sent to the landfill). Leachate is any liquid that, in passing through matter, extracts solutes, suspended solids or any other component of the material through which it has passed. Leachate and liquid poisons can then leak into aquifers, contaminating fresh water sources.

Modern landfills are designed with a leachate drainage system responsible for the collection and transport of the leachate collected inside the liner. In theory, this prevents leachate from contaminating water sources, but

Table 14.2 Household hazards in landfills

Used motor oil
Auto batteries
Antifreeze
Degreasers
Weed killers
Insect killers
Ant poisons
Bug sprays
Lighter fluid
Gasoline
Oil-based paint
Turpentine
Concrete cleaners
Drain and oven cleaners
Aerosol products
Cleaners with bleach or ammonia
Nail polish remover

Source: US EPA.

leachate collection systems can experience many problems, including clogging with mud or silt.

A consuming mentality

Americans discard more garbage per capita than citizens of other prosperous nations, and far more than those in the developing world. In some regards, the production of garbage reveals much about levels of affluence, household formation, commercial activity and values. Our "garbage crisis" is as much social and political as physical.

Garbage, and particularly household refuse, is one pollutant that is often invisible. We throw away our unwanted food scraps and household items into bags, then into bins, which are then wheeled to the sidewalk or curbside. A large collection truck comes by, usually when we are gone from the home or apartment, and it magically disappears. Garbage is effectively removed from our lives. Few people ever visit landfills and see the vast accumulation of trash. And unlike air and water pollution, which is often visible on a daily basis, garbage is something we rarely see or think of as impacting the environment. Writer Elizabeth Royte noted: "From the moment my trash left my house and entered the public domain … it became terra incognita, forbidden fruit, a mystery that I lacked the talent or credentials to solve."[13]

Landfills can be considered silent "monuments" to our consuming lifestyle. Many landfills are visible from space—Fresh Kills landfill in Staten Island can be seen from orbiting satellites. So too can the Great Pacific Garbage Patch, which is said to be composed of 3.5 million tons of garbage, mostly floating plastics. The 2011 earthquake and tsunami that hit Japan dispersed millions of tons of garbage and plastics into the sea. Much of that garbage is now beginning to wash up on the coasts of Alaska and the Pacific northwest.

In many developed economies, but particularly in the US, we have favored convenience over conservation, short-term needs over long-range resourcefulness. The fast-food of McDonald's is both indicative of a cultural value that embraces convenience and accepts the short-term duration of packaging. Food arrives within minutes. The food wrappers and drink cups have a commercial life span of less than one hour, yet they survive in landfills for years. The fast-food society is a disposable society. This extends beyond the fast-food restaurants into many aspects of our culture. It is often less expensive to purchase a new radio or DVD player rather than repair a broken one. Even iPhones and laptops become obsolete in just a few years. We now purchase purposely "disposable" products such as razors, toothbrushes, paper plates and cups and writing pens.

The urban refuse problem stands for larger issues associated with the production of waste in general. As with water and air pollution, few legislative reforms punish or prohibit the production of waste; rather, reforms often deal with financial incentives to develop new technologies to reduce the pollutants. Cities deal with refuse as an "end of the pipe" issue, rather than challenging the production of refuse. American consumers have few incentives to decrease their own refuse stream: there are virtually no economic costs associated with the volume of trash per household; generally these services are a flat fee. New technological developments in the plastics industry have made packaging thinner and lighter; however, consumers still generate and then discard a synthetic product that does not biodegrade in the landfills. Table 14.3 lists numerous products that are synthetic plastic-based. In 1945 the US produced some 400,000 tons of plastic products. By 1998 the US generated 47 million tons, worth more than $200 billion. In 2012, the US generated 90 million tons, worth some $374 billion. It is estimated that every year, approximately 30 million tons of plastics enter the waste stream, and only about 12 percent is recycled.

Table 14.3 Plastics in the waste stream

Plexiglas (polymethylmethacrylate)

Polyesters

Polyvinyl chloride (PVC)

Teflon (polytetrafluoroethylene)

Polyurethane

Other products such as: golf clubs, bike helmets, backpacks, "fleece" sweaters and jackets, Gortex, toothpaste, toothbrushes, chapstick, zippers, billiard balls, knobs and buttons.

Reducing waste

In the last decade, cities have made more deliberate efforts to reduce the amount of waste destined for landfills. There are several ways to reduce waste: source reduction (Table 14.4) and reusing material. Source reduction is a preferred option since it reduces not only ultimate disposal, but also avoids transportation and management costs associated with recycling. Recycling and composting also work to reduce waste destined for landfills. Finally, some cities see incineration or waste-to-energy facilities as ways to reduce the waste stream. Figure 14.4 shows the EPA's preferred waste "hierarchy." The EPA now encourages cities to prioritize source reduction, recycling and composting before turning to incineration.

As recycling programs strengthen, and manufacturers reduce packaging materials, solid-waste generation in many cities in rich countries may become stable or even decrease. Already there are signs that governments are doing more to reduce materials that end up in landfills, or make the manufacturing side of consumer goods more responsible for the wastes they generate.

Table 14.4 Source reduction activities

Minimize the volume of packaging material required to deliver products by selecting products packaged efficiently or buying in bulk.

Identify opportunities to reuse products and packaging in the home or community rather than disposing of or recycling them.

Encourage companies to implement source reduction programs and purchase products with post-consumer recycled content.

Reduce consumption of disposable goods and purchase products from reuse centers.

Reduce food waste (27 percent of edible food is wasted at the consumer level) through efficient meal planning and composting of scraps.

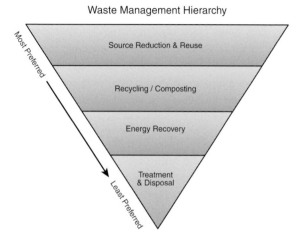

Waste Management Hierarchy

Figure 14.4 The waste management hierarchy

Source: EPA: http://www.epa.gov/osw/nonhaz/municipal/wte/nonhaz.htm

BOX 14.3

Taxing plastic bags

Many cities have begun to levy taxes on plastic, carry-out shopping bags. Toronto was one of the first cities in North America to do so. In 2007, Toronto adopted a goal to divert 70 percent of waste from landfills and to reduce the impact and cost of litter to the city. One of the major strategies for waste reduction was to reduce the number of plastic bags found in Toronto's waste stream, and to increase the number of plastic bags collected and recycled. As part of that commitment, Toronto introduced a 5¢ levy on in-store packaging including carry-out, plastic shopping bags. The introduction of that 5¢ levy had a dramatic and positive effect on plastic bag use and its ultimate presence in Toronto's waste stream. City reports note that measurable environment and cost reductions impacts included a 53 percent reduction in plastic retail bag generation, and a 65 percent reduction of these bags in the residual waste stream.

Bag taxes have also occurred in other cities. One of San Francisco's waste reduction measures, the city's Plastic Bag Reduction Ordinance, also achieved national and international attention. Within a year of its implementation, Boston, Portland and Phoenix were investigating such plastic bag bans, and other cities were looking into the program.

Whether called a levy or a tax, efforts to reduce plastic bags this way can be seen as part of a "polluter pays" principle. Residents can choose the extra levy, or they can chose alternative, reusable options. The tax made many people become aware of the overuse of plastic bags, the unnecessary costs of managing them once in the waste and litter streams, and the ready availability and ease of choosing alternatives.

Source: Recycling Council of Ontario (2012) "Open letter to Mayor Rob Ford." www.rco.on.ca (accessed July 2012).

In many cases, European cities have done far better than their US counterparts in strengthening recycling and composting programs. In 1991 Germany passed a law requiring industry to take back, reuse and/or recycle packaging materials. By making industry take back its packaging, the ordinance shifts the burden of managing packaging waste away from cities toward manufacturers, distributors and retailers.[14] Germany is a true pioneer in the field of recycling: one poll suggests that nine out of ten householders willingly separate their trash. This is demonstrated clearly in the capital, Berlin, where seven different bins exist for all the various recyclable waste—general waste, paper, compost, plastic/metal, amber glass, clear glass and green glass (in contrast, US cities have at the most three bins—trash, recycling and composting). The Netherlands and Sweden also have extended product responsibility frameworks. The Dutch government implemented a new policy that requires a lifecycle assessment at each stage for manufactured products. Manufacturers currently recycle as much as 75 percent of potential recyclable material. Switzerland also recycles at a 75 percent rate; and in Sweden, a new law promotes more efficient use of resources in the production, recovery and reuse of waste. Canada passed a National Protocol on Packaging in order to reduce the amount of packaging that goes to landfills. In 2000 the European Declaration on Paper Recycling was signed by 12 states. This voluntary industry initiative set a target to increase the recycling rate from 49 percent to 56 percent by 2005. Today, the EU recycles about 65 percent of all paper products. A second declaration in 2010 aims to increase paper recycling to 70 percent by 2015. This is far more ambitious than US goals, although some US cities are striving for 75–100 percent recycling by 2020.

The US does not have federal regulation on recycling; rather, these programs are left to municipalities. Some cities are very successful at recycling and reusing. Seattle, Portland, Los Angeles and Minneapolis have achieved recycling rates of over 50 percent of their solid waste stream. In contrast, cities without curbside recycling, such as El Paso and Detroit have recycling rates of less than 10 percent. These success stories have mandatory curbside recycling, and accept a wide variety of materials. The specific laws vary in scope and forcefulness. Some are mandatory, some are not; some only request separation from the remaining trash, others are aimed at only businesses and not residences, finally, some states passed laws to forbid certain types of materials from landfills, like yard wastes.

There are also creative ways to prevent food wastes from reaching landfills. The San Francisco Food Bank collects some 37 tons of edible food a month from wholesalers and distributes it to local service agencies; composting food scraps and leftovers at the University of Massachusetts diverts 48 percent of the materials destined for landfills. In the US, there are more than 10,000 curbside recycling programs. This has led to an overall increase in municipal recycling rates to nearly 35 percent (Figure 14.5). One challenge for successful composting is that there is no universal label for "compostable" materials and packaging. Some products have a "green-colored stripe" to denote compostability. Others have the logo of a certifier or the word "compostable." Some have no marking at all. As a result, composters are having a difficult time identifying products that are actually 100 percent compostable as these products look very similar to traditional plastic products. It is important that facilities that accept food waste have a universally recognized label of compostability. Many hope that recycling and composting reforms will mark the end of the throwaway society.

But getting to zero waste is more than just recycling and composting. It is about reducing packaging and waste in the first place. As we make efforts to reduce wastes going to landfills, compostable packaging is going to be critical to encourage the composting of food waste by consumers and businesses and to ensure that commercial compost piles that accept food waste are not contaminated by plastic or other traditional packaging.

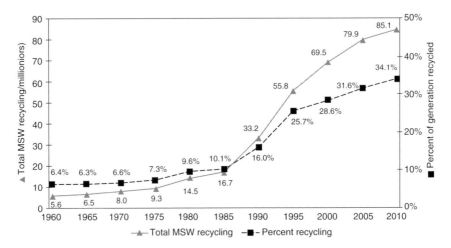

Figure 14.5 Recycling rates since 1960

Source: US EPA (2010) "Municipal sold waste generation, recycling and disposal in the United States: facts and figures for 2010"

One example is getting companies to use compostable packaging to package food. In 2009 Frito-Lay introduced compostable bags made of plant material for its SunChips line. Sadly, a year later, the company took the bag off the market, citing declining sales and complaints from consumers that the bag was too noisy—due to the unusual molecular structure that makes the bag more rigid. Some compared the sound of opening the chip bag to a busy street corner and tests showed it registered around 80–85 decibels, compared to the average 70 decibels for other chip bags. After refining the bag material, Frito-Lay reintroduced the "quieter" bag in 2011. Sales are back and complaints are down.[15] The bag illustrates the sometimes unexpected bumps that can trip up companies trying to do the right thing environmentally.

Incinerators: the solution to garbage?

In 1976 the US Congress passed the Resource Conservation and Recovery Act (RCRA). The RCRA gave the EPA authority to regulate dangerous and hazardous materials, which expanded their ability to track both the generation and the disposal of these dangerous chemicals. The EPA defines a hazardous waste as one which meets one of the following characteristics: ignitability, corrosivity, reactivity or toxicity. The agency has since identified 450 wastes as hazardous. By the late 1990s the US produced some 197 million tons of hazardous waste annually.

BOX 14.4

Composting cities

A significant component of the US waste stream is organic material and food waste. Currently, the US only composts 3 percent of its food waste, which means 97 percent is going into landfills. Some 32 million tons of food waste, or approximately 13 percent of the total trash in the US, is being sent to landfills. Of the nation's 3,400 commercial composting facilities across the country, only 8 percent accept food waste.

One of the major problems with food waste in landfills is that as food waste decomposes, it lets off methane, a GHG emission that is more potent than carbon dioxide. One estimate is that methane from landfills accounts for 34 percent of all human-related methane in the US.

Composting is an alternative to food waste in our landfills. Composting involves decomposing food scraps and other organic waste, like cardboard rolls and cotton rags, into a useful, rich substance called mature compost which can be used as fertilizer. Anyone can compost, and those with yards can set aside a small area as a compost area. Another way to compost is at the city-wide scale: sending food waste to composting facilities. Modern composting facilities grind up biodegradable waste into the desired texture and stuff it into long plastic or fabric tubes. While in the tubes, the compost is aerated, its moisture levels are monitored and the temperature is kept at the ideal warmth to speed up decomposition. These actions help microorganisms to carry out the decomposition process.

Cities that compost benefit by reducing the costs of landfill disposal. In addition, the waste byproduct can be valuable and useful. Once waste is turned into compost, agricultural operations such as local vineyards will pay for the nutrient-dense material.

Curbside composting programs have taken root in several US major cities, including Seattle, San Francisco and Boulder. Residents are given wheeled composting bins, like recycling or garbage bins, that are set curbside on a designated pick-up day. Some are also given composting pails for the kitchen.

In the mid-2000s, San Francisco set the ambitious goal of keeping 75 percent of its trash out of landfills by the year 2010, and reaching the point of zero waste by 2020. By 2007 it had reached 72 percent trash diversion, but the city realized that voluntary participation in recycling and composting would not get to zero waste. In 2009 the city of San Francisco passed the San Francisco Mandatory Recycling and Composting Ordinance (No. 100-09), which requires all residents to separate their recyclables, compostables and landfilled trash and to participate in recycling and composting programs. Since the mandatory ordinance went into effect, composting increased by 45 percent, and the city now sends nearly 600 tons of food scraps, soiled paper and yard trimmings to its compost facilities daily. In 2010 San Francisco mayor Gavin Newsom announced the city's diversion rate had reached 77 percent. Today the city recycles and composts at large public events such as baseball games and conventions.

Getting to zero waste is more than just recycling and composting. It is about reducing packaging and waste in the first place. As we make efforts to reduce wastes going to landfills, compostable packaging is going to be critical to encourage the composting of food waste by consumers and businesses and to ensure that commercial compost piles that accept food waste are not contaminated by plastic or other traditional packaging.

Sources: Toothman, J. (2012) "What US city composts the most waste?" http://home.howstuffworks.com/city-com-posts-the-most1.htm (accessed June 2012).

Related videos about composting are available at: http://sfenvironment.org/video/green-spot-tv; "From Food Scraps to Wine," http://sfenvironment.org/video/test.

Another very important aspect to the RCRA was its implicit solution to the garbage crisis. The RCRA was influenced by three growing concerns: increasing landfill closures; public anxieties about the public health and environmental impacts of landfills; and the growing dependence on foreign oil imports which made Congress look for other potential sources of energy. These three issues coalesced to influence the RCRA legislation in a profound way. The law phased out open dumps and set higher standards for landfill design. More importantly, the law established tax incentives and government grants for cities to build waste-to-energy incinerators. In municipal America's war on trash, incinerators became the strategic weapon of choice. The incineration movement had seized the political moment.

Waste-to-energy incinerators burn garbage at very high temperatures, approximately 1,800°F, in a combustion chamber, superheating water in the boiler, which in turn spins a turbine that then generates electricity which can be sold to the power company. Figure 14.6 shows the incineration process. In theory, because incinerators burn at such high temperatures, the heat destroys any and all toxic material. In reality, early incinerators imported from Europe had problems coping with the high organic content of the waste stream, often causing incomplete combustion and the release of toxic or hazardous chemicals. This solution to garbage reflects much of what we have discussed in previous chapters. Many cities, often because of the encouragement of federal laws, seek the high-tech solution to pollution.

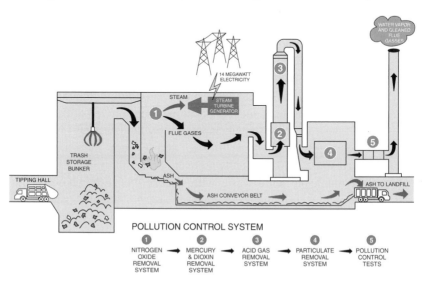

Figure 14.6 Operations at a typical waste to energy facility, or incinerator
Source: US EPA: http://www.epa.gov/osw/nonhaz/municipal/wte/basic.htm

Incinerators cost as much as $300 million to construct and are expensive to maintain. At the same time the new law did little to encourage waste reduction in the first place and did not support recycling and reuse programs with the same level of financial assistance and incentives that it did the incineration option. Ironically, because incinerators operate most efficiently when at capacity, many private incinerator firms contracted with cities to receive as much refuse as possible; in some cases forestalling the development of municipal recycling programs.

Between 1976 and 1990 there were more than 140 incinerators constructed in cities around the US. Gradually, concerns emerged about the potential environmental hazards. Nearly all incinerators produce bottom ash—ash that falls to the bottom of the combustion chamber. Bottom ash, which tends to have high concentrations of heavy metals, is disposed of by sending it to landfills. Airstack emissions (which in theory should emit only water vapor) have been found to contain heavy metals, sulfur dioxide, nitrogen oxides and carcinogens such as dioxins. Dioxins are generated by the incineration process. When pesticides and plastics are burned along with organic material, the result is the creation of dioxins, which are fat soluble in the human body and highly carcinogenic. In Europe, where the incinerator technology was perfected, the waste stream is different from the US (the US has higher organic materials). As a result, many of the incinerators in the US did not efficiently burn refuse due to the "wetter" content of organic materials. The challenge to operating an incinerator power plant is to control the burn temperature so that it is high enough to prevent the creation of dioxins from plastics and low enough so that not too many nitrogen oxides are produced. Many incinerators initially failed this challenge. The health impact of dioxins includes damage to the central nervous system, the immune system, reproductive health system and can impair thyroid function. Because of many of these concerns, local opposition groups proved effective at preventing the construction of new incinerators. By the late 1990s incinerators had become controversial and politicized, and few new incinerators have been approved since 2000. Currently there are 86 facilities in the US for combustion of municipal solid waste with energy recovery. These facilities are located in 25 states, mainly in the northeast. No new plants have been built in the US since 1995, but some plants have expanded to handle additional waste and create more energy. These facilities have the capacity to produce 2,720 megawatts of power per year by processing more than 28 million tons of waste per year.[16] The long-term "solution" to garbage in US cities is unlikely to be incinerators, but for other cities in countries such as Japan, incinerators are the key method for dealing with solid waste.

Waste in the developing world

Developing cities face many of the same general issues of waste as those in the developed world. In recent years many developing cities have significantly improved waste collection, particularly in commercial districts and tourist areas (Figure 14.7). Developing cities are also setting ambitious goals to reduce waste, recycle and compost more. For example, Singapore and Taiwan have begun a zero-waste plan. However, there remain significant challenges in dealing with and providing resources for refuse in cities of the developing world. Some of these challenges are highlighted in Table 14.5.

Figure 14.7 Many developing cities like Hanoi have significantly improved garbage and recycling programs in tourist areas and in public spaces, but still struggle to provide waste pick-up for all residential areas

Source: Photo by Becky Barton

Table 14.5 Comparisons of solid waste management for selected cities

Activity	Lowest income cities: Lima, Kinshasa, Lahore, Jakarta	Middle income: Seoul, Rio de Janeiro, Bangkok	High-income: London, Frankfurt, New York City, Tokyo
Source reduction	No organized programs but low per capita waste generation rates	No organized programs	Education programs, some voluntary source reduction programs
Collection	Sporadic and inefficient. Service limited to high-visibility areas, wealthy areas	Improved services and increased collection from residential areas; small vehicle fleet	Collection rate greater than 90 percent. Highly mechanized vehicles common
Recycling	Recycling through informal sector, waste pickers; localized markets for recyclables	Informal sector still involved; some high-technology sorting	Formalized recycling collection services and high-technology sorting. Increased attention to developing long-term markets
Composting	Rarely undertaken	Some small-scale composting, but not on large scale	Becoming more popular at large-scale facilities
Incineration	Not common; high moisture content in waste	Some limited use of incinerators	Prevalent in areas with high land costs. More prevalent in European cities than in US
Landfills	Mostly open dumps	Mix of open dumps and sanitary landfills	Engineered sanitary landfills with liners, methane and leachate collection systems
Costs	Collection costs are 80–90 percent of solid waste budget	Collection costs represent 50–80 percent of budget	Collection costs represent less than 10 percent of budget; other funds allocated to landfills, incinerators or recycling programs

Source: Urban Development Unit, World Bank (1999) *What a Waste: Solid Waste Management in Asia.* Washington, DC: World Bank, p. 19.

First, in developing countries individuals living in urban areas use nearly twice as many resources per capita than those living in a rural setting.[17] Because they consume more, they also generate more solid waste than their rural counterparts. For example, Delhi now generates about 8,300 tons of waste each day. Second, there are differences in the waste stream. Typically in cities in developed countries more than one-third of waste is paper; plastic accounts for 9 percent; and organic materials such as food wastes and yard trimmings account for 28 percent. In contrast, in the lowest income countries, some 40–85 percent of the waste stream is organic compostable matter, and paper accounts for around 5 percent. Cities in India and China diverge from this trend because they use coal as a household fuel source, thereby generating large quantities of dense ash. The percentage of consumer packaging (plastic, paper, glass and metal) tends to be lower; packaging wastes correlate with the nation's degree of wealth and urbanization. However, as developing countries become richer and more urbanized, they may see a significant increase in paper and packaging—more newspapers and magazines, fast-service restaurants and single-serving beverages.

BOX 14.5

Solid waste management in Dhaka, Bangladesh

Dhaka, the capital of Bangladesh, is one of the fastest-growing megacities in the world. It is a city with a population of over ten million. By 2025, the UN predicts Dhaka will be home to more than 20 million people—larger than Mexico City, Beijing or Shanghai. It is estimated that some 45–55 percent of the population are poor and live in slums and squatter settlements with little or no access to municipal sanitation services. The city generates about 4,600 tons of solid waste each day, of which 67 percent is food scraps, and as a result, the waste stream is far wetter and heavier than in many other cities. Only 45 percent is collected and disposed of by the municipality. Informal sector waste pickers, popularly known as *tokais*, collect many of the recyclable materials, such as aluminum and glass. They account for about 15 percent of recycling efforts. However, they tend to ignore the organic material and so there remains considerable organic material in the solid-waste stream. Often the waste not collected by the municipality or removed by the *tokais* is left to rot in the heat and humidity of the city's open spaces. The resulting stench, rodents and clogged drains pose a serious health risk to Dhaka's residents. In addition, almost half of the waste generated in Dhaka is disposed of in an environmentally unsound way. As the city expands horizontally, it is becoming even more difficult to locate large waste-disposal sites within easy access of the city. As a result, the city pays higher transportation costs to haul solid waste longer distances.

A Bangladesh non-governmental organization, "Waste Concern," initiated a program that promotes the use of organic solid waste for compost to help improve the depleting topsoil fertility in rural areas. The city benefits from organic waste recycling through composting because this reduces disposal costs, prolongs the life span of disposal sites and reduces environmental impacts of landfills. Involving the population in the use of compost programs promotes awareness of garbage while composting activities help create employment and generate income. This innovative partnership includes: the government, which makes small parcels of vacant land available free of charge for composting plants; Waste Concern, which collects, separates and turns solid waste into organic compost; the community; and a private business that markets the organic compost. Currently this program serves 30,000 people in Dhaka and another 100,000 people in 14 other cities in Bangladesh.

One of the interesting aspects to this program is that the composting plants are not highly mechanized, but are small-scale, decentralized, community-based composting areas that use manual labor. The composting facilities have generated employment for about 16,000 urban poor, particularly women. Consider the story of Makusda, a 32-year-old woman who works in one of the community-based composting plants. Previously she worked in a garment factory, but left because she was often forced to work long hours with no holidays and the salary was very low. There were also lull periods where there was no work in the garment factory and she often went for two or three months without a salary, causing hardship for her family. In the composting factory she has a weekly holiday, fixed hours of work with rest time and access to toilet and bathing facilities. She also earns more money than she did at the garment factory. More recently Waste Concern introduced solid-waste composting in two of the city's slums by supplying specially designed barrels for composting on-site. This not only reduces the waste disposal problem at the source, but also generates income for slum dwellers.

Sources: Enayetullah, I. and Sinha, A. H. M. M. (2001) "Public–private community partnerships in urban services (solid waste management & water supply) for the poor: the experience of Dhaka City." Bangladesh country report prepared for the Asian Development Bank, Manila. United Nations Economic and Social Commission for Asia and the Pacific (2002) "Solid waste management in Bangladesh." http://www.unescap.org/rural/bestprac/waste.htm (accessed June 13, 2012). Waste Concern (2009) "2009 waste data base of Bangladesh." http://www.wasteconcern.org/database.html (accessed July 2012). See also Waste Concern: www.wasteconcern.org.

A third difference is that many cities in the developing world lack proper infrastructure (landfills and institutionalized waste diversion programs) for dealing with solid waste. Corrupt bureaucracies and poor planning mean that the few official facilities are inadequate. In addition, lack of resources means most of what a city spends is on collection, rather than the upgrading or maintenance of dumps to sanitary landfills. Few have taken steps to construct, operate or maintain sanitary landfills. In many developing cities, particularly megacities, waste is destined for open dumps. In Brazil, for example, only 10 percent of solid waste is dumped in sanitary landfills; 76 percent is dumped in illegal landfills and another 13 percent in open dumps. Beijing is said to have more than 461 illegal open dumps. Open dumps are not sanitary and lack liners, methane collections systems, etc., all of which are now mandated in the developed world. This presents a pressing problem for developing cities. In most open dumps surface and ground water is being contaminated, explosive methane gas is being created and disease-bearing insects and rats are breeding and thriving. Open dumps, like Payatas in Manila, Dhapa in Kolkata, Bantar Gebang in Jakarta and Matuail just outside of Dhaka, are more than eyesores and smellsores. They are hazards. For example, in Malaysia's capital, Kuala Lumpur, the waste dump Taman Beringin is a smelly, 12-hectare dump for the city's garbage. Nearby residents put up with flies, rats, disease and the reek of rotting garbage.[18] When the dump caught fire in 2004, it burned for more than two weeks. Illegal open dumps create risks for the water supply and the health of the urban population. Few cities have the resources to monitor the environmental effects of waste disposal in open dumps.

Fourth, like many cities in the developed world, many developing cities face the problem of increased waste generation and where to put it. They too have a landfill crisis. In Mexico City, for example, Bordo Poniente, the city's main

Figure 14.8 Informal trash heap. A sacred cow sits atop trash along the roadside to Agra, India. As in many developing cities, lack of sanitary land-fills and collection programs lead to informal dumping of household wastes

Source: Photo by Michele A. Judd

garbage dump, receives 12,000 tons of waste per day. It is one of the largest landfills in the world. Mexico City officials also count some 1,000 illegal dumping sites around the city. At present the city's generation of garbage is increasing at a rate of 5 percent a year and there are likely to be even more "clandestine" fields. The city government was urged to take action. Initially the city considered waste-to-energy incinerators, but many environmental organizations, including Greenpeace, protested. Instead, Mexico City launched an ambitious formal recycling program and ambitiously aims to get to zero waste by 2020. Mexico City has required its residents to separate trash since 2003, but for many years did not enforce the law, nor did it provide necessary recycling equipment. As a result, the city had recycling rates of about 5 percent. In 2009 the city launched a public campaign to try to create a culture of recycling. The city also began distributing recycling bins in households, commercial and service businesses and industrial facilities. As a result, recycling rates in the city are now at about 20 percent. A newly formed Waste Commission is working to build four state-of-the-art processing centers in the next four years to recycle, compost or burn for energy 85 percent of Mexico City's trash. That is the good news.

However, in December 2011 the city closed the Bordo Poniente landfill 12 days ahead of schedule. Apparently no one bothered to inform workers where they could deposit the 12,600 tons of trash that comes out of Mexico City daily. Confusion and poor coordination followed; an interim plan to take refuse to smaller dumps outside the city fell apart as municipalities adjacent to Mexico City refused to take the city's trash. Mounds of garbage began piling up on city streets and garbage trucks queued up for more than six hours to dump loads at transfer stations. Critics say the city was unprepared, and it was not clear why an alternative waste system was not in place after earlier plans to build four new garbage processing plants were abandoned. The story highlights how the absence of a comprehensive policy for urban waste collection, disposal and processing can create a major problem. It also underscores the politics of garbage. The closure of Bordo Poniente sparked a conflict between the national, city and state governments: the national government had been exerting pressure since 2008 to close the site, but the Mexico City government kept it going until 2011, while the state government continues to resist the location of garbage dumps in its territory. City leaders were eager to transform their waste system from one of the dirtiest to one of the greenest, but ran into unexpected issues. The story is informative because many other megacities in Latin America and Asia have dumps that have reached capacity or soon will.

Today, cities in Asia generate about 760,000 tons of municipal solid waste per day; by 2025 this figure could increase to 1.8 million tons and they will likely need to double their current $25 billion per year spending on solid waste management.[19] As a whole, urban populations in the developing world will likely triple their current rate of MSW generation over the next 20 years, while cities in Nepal, Bangladesh, Vietnam, Laos and India might see four to six times the current amount. Such dramatic increases will place enormous stress on already limited financial resources.[20] How will developing countries deal with increasing waste and the need for more landfills or disposal systems? Will rising affluence equate with increased waste generation, as it has in the developed world?

Informal recycling

In many developing cities there is not a long history of dealing with significant amounts of waste, and so they often lack established infrastructure. In many cases municipal waste programs in developing cities are characterized by inconsistent collection, a lack of equipment and open-air dumps. But perhaps one of the most distinguishing features of waste in developing cities is the fact that most of the recycling is in the informal sector.

In many large cities in the Global South, the majority of solid waste goes to open-air dumps (in contrast to formal, highly regulated landfills of the Global North). Slums have emerged on land around dumps. Residential areas adjacent to dumps are exposed to the delivery and smell of the open-air dumps. If they have narrow or unpaved streets, it is unlikely that they have door-to-door collection by garbage trucks. They may also be home to a growing number of collectors and scavengers who fill an important need for waste collection, waste separation and recycling. A major contrast between developed cities and developing cities is that recycling is formally institutionalized as part of municipal services in developed cities. In developing cities, however, recycling is part of the informal economy. Christian Rogerson refers to this as the "waste economy," and notes that in many developing cities this is a source of income for many.[21] The waste economy is not often regulated by the government. David Wilson and colleagues have shown that despite the "informal" nature of the waste economy, it has helped developing countries reach high recycling rates of up to 25–40 percent.[22] For example, in Mexico more than 30 percent of waste is recycled by scavengers.

The informal waste economy has networks and organization. At garbage dumps scavengers sort through the trash as it arrives, recovering materials with value. These various front-line collectors then sell to middlemen who then sell to the recycling industries. Although the informal sector does not follow government regulations or guidelines, it is nonetheless complex and structured. For waste pickers or scavengers, the risks and dangers they face due to the work is sometimes mitigated by collaborating and working together. Informal workers have formed highly organized cooperatives and micro-enterprises. Often, these cooperatives will collect their materials and then sell them in bulk to the recycling industry, earning a better rate by eliminating the middleman. These cooperatives can also provide a support system for workers in what can be a dangerous line of work, or in a society where they are often ostracized.

Phnom Penh in Cambodia has done much to distribute clean water to its residents, as we documented in Chapter 11, but still struggles with waste management. From 1965 to 2009, the city had only one dump. All types of waste were dumped, including hazardous waste. By 2003 some 10,000 people lived around the dump, many of whom worked as waste scavengers. In 2009 the city opened a new landfill, Dorng Kor, which was planned as a new sanitary landfill. However, insufficient funds resulted in only partial operation of the dump.[23] A lack of resources also means that the city does not provide waste collection as a city service. Instead, the city contracted with a private company to provide waste collection and disposal, but the contractor has a poor performance record. Only 64 percent of waste generated is being collected and disposed of by the company; the rest is collected by informal workers.[24] Informal workers are active at several stages in the waste collection process. There are buyers who go to households, there are also waste pickers who collect recyclable material from litter and trash piles in the streets. There are also scavengers at the dump, who pick through the waste looking for any overlooked materials. But because the city has contracted to private firms, many of the informal workers are harassed by authorities or by employees of the company, who see them as competition.

The city of São Paulo is the third largest generator of domestic solid waste behind New York and Tokyo. Almost 14,000 tons of refuse and 5,000 tons of industrial solid waste is collected each day at a cost of $150 million and sent to one of three nearby landfills. The city of Rio de Janeiro is home to Jardim Gramacho, one of the world's largest garbage dumps (Box 14.6). However, Brazil has a higher rate of recycling compared to the US and Japan. While efforts to privatize recycling have largely failed, the brute fact of high unemployment has created a new economic activity in which poor families collect aluminum cans and paperboard. Some 110,000 Brazilians make their living collecting cans on the street and earn an average of

BOX 14.6

The *catadores* of Jardim Gramacho

On the outskirts of Rio de Janeiro is the Jardim Gramacho landfill. It is 300 feet high and spans 14 million square feet (the equivalent of 244 American football fields). It is the largest landfill in Brazil and in all of South America. Built in the late 1970s, it received close to 8,000 tons of trash daily, 70 percent of all the trash in the Rio metro area. The landfill became home to a community of scavengers during the economic crises of the 1970s and 1980s. An estimated 50 percent of the residents of the favela of Jardim Gramacho make a living from recycling. They are known as *catadores*. At Jardim Gramacho, about 5,000 *catadores* spend their days scavenging the landfill for plastic, paper, wood, metal and anything else that can be sold to recycling companies (Figure 14.9). They collect about 200 tons of recyclables each day. The *catadores* have extended the life of the landfill by removing materials that would have otherwise been

Figure 14.9 *Catadores* in Rio's Jardim Gramacho dump

Source: from Wikimedia: http://en.wikipedia.org/wiki/File:LixaoCatador
es20080220MarcelloCasalJrAgenciaBrasil.jpg

buried and have contributed to the landfill having one of the highest recycling rates in the world. In landfills across Brazil it is estimated there are more than one million *catadores*.

Catadores at Jardim Gramacho generally earn about the Brazilian minimum wage, roughly $268 per month. But in many cases, they can bring home twice as much. This is because they have formed a cooperative, the Association of Recycling Pickers of Jardim Gramacho. The cooperative led the way in community development, creating a decentralized system of recycling collection in neighboring municipalities, building a recycling center, securing professional recognition of the *catador*, enabling *catadores* to be contracted for their services, establishing a 24-hour medical clinic, and constructing a daycare center and skills training center. In addition to their community initiatives, the cooperative leads a national movement for greater professional recognition for the *catador*. The *catadores* and the cooperative were catapulted to fame in Vik Muniz's Oscar-nominated 2010 documentary *Waste Land*. Vik Muniz, an artist, photographed the pickers individually at the landfill, and then projected the images onto the floor of a massive warehouse nearby. Together with the *catadores* the artist gathered recyclable items from the dump and used these discarded things to recreate the images on the floor.

In 2012 the city closed the landfill. The closure occurred less than three weeks before the United Nations RIO + 20 summit, where world representatives gathered to discuss sustainable solutions for the Earth's environment. The closure also occurred in advance of Rio's preparations to host the 2014 World Cup and the 2016 Summer Olympics. The landfill closure was seen as an example of substituting an untreated open facility for a modern waste treatment plant. In addition, the closure of the landfill will include methane capture, which will reduce by 1.5 percent the amount of methane released into the atmosphere by Brazil. The closure was a purposeful act to show Brazil's commitment to being more sustainable.

With the closure, however, came concern for the future of the *catadores*. The cooperatives were able to negotiate monetary compensation for the *catadores*. According to a *Rio on Watch* article, the greatest challenge in this process, according to Tião Santos, president of the Association of Recycling Pickers of Jardim Gramacho, is getting pickers to register their identity and open a bank account, a requirement of the compensation. The closure of the landfill marks an important point for the city of Rio. The cooperative and its members know this is a step forward in rebuilding and conserving Rio's environment, but for many this also means their traditional source of income and identity is gone.

This may not be limited to Rio or even Brazil. There are plans to close landfills in Delhi and Beijing, which may affect the livelihoods of thousands of waste pickers.

Sources: Brocchetto, M. and Ansari, A. (2012) "Landfill's closure changing lives in Rio." *CNN Online*. http://www.cnn.com/2012/06/05/world/americas/brazil-landfill-closure/index.html (accessed July 2012). *Rio On Watch* (2012) "Waste land pickers struggle from landfill closure." http://rioonwatch.org/?p=4032 (accessed July 2012); for more information on the film *Waste Land*, go to: http://www.wastelandmovie.com.

$200 dollars per month. These are sold to the industries, generally through middlemen, and are then recycled. Brazil recycles more than 64 percent of its aluminum cans, 35 percent of its glass, 37 percent of its paper and 12 percent of its plastic.[25] Alcana, a major aluminum manufacturer, operates its largest recycling facility in South America in the city of São Paulo. Hence recycling programs in cities such as São Paulo or Rio de Janeiro take place through the informal economy; and, ironically, the success of formal recycling programs could eliminate this form of income for the urban poor. Waste picking means higher rates of recycling and provides opportunities for the urban poor to generate some income. However, waste pickers are exposed to severe hazards and can pose a safety threat to themselves and to landfill employees by interfering with operations at the tipping face and accidentally starting fires.

Informal recovery has become a common practice in many developing cities. In Mexico City some 15,000 "scavengers" live in dumps and entire families work to recover cardboard, glass, metals and plastics. In Oaxaca, Mexico, Sarah Moore has shown that garbage scavengers and residents of communities adjacent to garbage dumps are further marginalized because they are portrayed as "dirty, defiled, dangerous."[26] She examined a series of political protests in which people that work or live near the dump blockaded access to the landfill. As garbage piled up in the streets, the city had no way of disposing of its trash and was forced to negotiate. The result is that some of these neighborhoods surrounding the dump now have a basketball court, a medical center, a meeting center and some electricity. Moore concludes that these blockades are acts of transgression that force the rest of the city to "see" the material effects (garbage) of the process of urban development and environmental justice.

The work of Anne Scheinberg and Justine Anshutz has looked at government attitude toward the waste economy.[27] In some cases, they note, the government has tried to limit access to landfills or to fine waste pickers for scavenging. In other cases the government has tried to integrate the informal sector into a more formal specific role. Such is the case in Lagos, Nigeria.

Lagos has many of the problems with waste management that are common in developing cities. The government lacks resources to conduct formal trash collection, which means residents either pay for private services or pay an informal scavenger to collect waste. In Lagos, most of the scavengers are young men, called *barro'* boys, a named derived from the wheel barrows or carts used to collect the waste. *Barro'* boys go door to door, collecting trash for a fee based on bargaining at the time of service.[28] However, the *barro'* boys tend to create more illegal, open dumps in open spaces and on vacant plots of land. A study by Abel Omoniyi Afon recommended that Lagos create transfer stations where the *barro'* boys can bring their waste and operations could be registered.[29] This reflects an emerging trend where municipal governments are trying to formalize the informal sector.

Conclusion

In this chapter we have explored how cities are coping with increased levels of municipal solid waste. This form of pollution, perhaps more than any other, is directly linked to levels of affluence and to cultural values that avoid challenging the production of such vast volumes of consumer goods in the first place. For cities in the developing world, many of which cannot cope with either the collection or disposal of garbage, the future may prove even more challenging should these societies achieve the levels of consumption found in the developed world.

Guide to further reading

Engler, M. (2004) *Designing America's Waste Landscape*. Baltimore, MD and London: Johns Hopkins University Press.

Gandy, M. (1994) *Recycling and the Politics of Urban Waste*. New York: St. Martin's Press.

Hawkins, G. (2005) *The Ethics of Waste: How We Relate to Rubbish*. Lanham, MD: Rowman and Littlefield.

Ludwig, C., Hellweg, S. and Stucki, S. (2003) *Municipal Solid Waste Management: Strategies for Sustainable Solutions*. Berlin: Springer.

Mancini, C. (2010) *Garbage and Recycling*. Farmington Hills, MI: Greenhaven Press.

Myers, G. A. (2005) *Disposable Cities: Garbage, Governance and Sustainable Development in Urban Africa*. Burlington, VT: Ashgate.

Rathje, W. and Murphy, C. (1992) *Rubbish! The Archeology of Garbage*. New York: HarperCollins.

Rogers, H. (2006) *Gone Tomorrow: The Hidden Life of Garbage*. New York: The New Press.

Royte, E. (2005) *Garbage Land: On the Secret Trail of Trash*. New York: Little, Brown.

Strasser, S. (2000) *Waste and Want: A Social History of Trash*. New York: Owl Books.

Vaughn, J. (2008) *Waste Management: A Reference Handbook*. Santa Barbara, CA: ABC-CLIO, Inc.

Williams, P. T. (2005) *Waste Treatment and Disposal*. 2nd edition. Chichester: Wiley.

Young, G. C. (2010) *Municipal Solid Waste to Energy Conversion Process: Economic, Technical, and Renewable Comparisons*. Hoboken, NJ: Wiley.

Part V
(Re)aligning urban–nature relations

15 Race, class and environmental justice

The city is both an environmental and a social construct. It is predicated upon ecological processes; indeed, it is a complex ecological system in its own right, yet also a social artifact that embodies and reflects power relations and social differences. The city is at the center of a social–environment dialectic that connects the environmental and the political. In this chapter we will explore how issues of class, race and gender interconnect with environmental issues.

Urban environments of inequality

It is a consistent finding that toxic facilities are predominately concentrated in lower-income and minority-dominated areas of the city and major infrastructure projects with negative environmental impacts such as urban motorways are more commonly found in poorer and more minority neighborhoods. Study after study reveals a correlation between negative environmental impacts and the presence of racial/ethnic minorities. A 1987 study in the US revealed that race was the most significant variable associated with the location of hazardous waste sites and that the greatest number of commercial hazardous facilities is located in areas with the highest composition of racial and ethnic minorities. The study also showed that three out of every five black and Hispanic Americans lived in communities with one or more toxic waste sites. Although socio-economic status was also an important variable in the location of these sites, race was the most significant.[1] Paul Mohai and Robin Saha, in a more recent detailed study of hazardous waste facilities in the US, found that the magnitude of racial disparity was much greater than previous studies suggested.[2]

In cities around the world, the poor and the marginal more often than not live in the areas of the worst environmental quality. Social inequalities are expressed and embodied in urban environmental conditions. Marco Martuzzi and colleagues, for example, found that in Europe waste facilities were disproportionately located in lower-income areas and in places where ethnic minorities lived.[3] Urban environmental inequality is large and pervasive.

Take the case of Chester, Pennsylvania: a typical industrial town outside the city of Philadelphia (Figure 15.1). It grew as a manufacturing center with steel mills, shipyards, aircraft engine plants and a Ford Motor Company plant. By the 1970s, however, deindustrialization began to erode much of the manufacturing base. As firms closed, workers left and the town's population became increasingly poor, older and black. According to the 2010 Census the population of around 34,000 was 74 percent black, compared to the state average of 10 percent. One-third of the town's population lived below the poverty line and the median income in the town was half the state median income. Cancer rates in the town are 2.5 times greater than the state average.

Eager to lure tax-generating facilities, the city government in the 1980s sought to redevelop old factory sites, attract business and generate jobs and tax revenue. Chester provided an ideal opportunity for certain industries as it

Figure 15.1 Chester, Pennsylvania. This former industrial city, with a predominant African American population, has been the site for the location of several waste incinerators since the 1980s

Source: Photo by John Rennie Short

was poor, desperate and had land that was cheap and available. Local community groups, where they existed, had limited political power. Other cities and areas would have provided stronger resistance to polluting enterprises. A large real-estate developer bought up the land rights in the old industrial area and leased space to other businesses. In 1987 the State Department of Environmental Protection granted permits for three waste facilities in the city of Chester and two just outside the city. By the mid-1990s the city housed the nation's largest concentration of waste facilities, including a trash transfer business, an incinerator, a medical waste sterilizing facility, a contaminated soil burning facility, a rock crushing plant and wastewater plant that handled effluent from factories and a refinery. More than two million tons of waste was processed in the city each year. A local citizens' group, The Chester Residents Concerned for Quality Living, claimed that the toxic emissions led to low birth weight babies and local cancer clusters. They also claimed environmental racism, taking their case against the State Department of Environmental Protection (DEP) to the courts. They argued that between 1987 and 1996 the DEP approved permits for 2.1 million tons of landfill in black areas of the city, but approved only 1,400 tons in white areas. A federal judge threw out the lawsuit in 1996 stating that there may have been a discriminatory *effect*, but the residents could not prove a discriminatory *intent*. A court of appeals reversed the judge's decision but the Supreme Court dismissed the case in 1998. However, by even agreeing to hear the case at all the Supreme Court signaled the viability of environmental racism as a legal argument.

In Chester there was a cluster of noxious, polluting facilities in predominantly poor, black residential areas. This is an extreme case, but one that highlights the character of environmental racism. It is less a formal legal issue, since the permits for the facilities were issued correctly and formal procedures were followed, and more a moral issue. Vulnerable and poor residents were dumped on both metaphorically and literally. Since the court ruling, improvements have been made. Publicity about the case and the actions of local community activist groups, socially concerned corporations and *pro bono* law group advocates led to a clean-up of waste sites and a $60 million renovation of an old power plant that now gives public access to the waterfront. While there is an improvement in the urban environment, the city remains poor, black and declining. Between 2000 and 2010 the city lost almost 8 percent of its population.

It is often difficult to untangle the intent and effects of environmental racism. The term "environment racism" implies both, but in the case of Chester, while the effect was obvious the intent was more difficult to prove.

BOX 15.1

Racialized topographies

Richmond, Virginia, whose residents in 2010 numbered 204,214 of whom 50 percent were African American, exhibited a pattern of lower-lying, inner-city black areas and surrounding white-dominated hills. The correlations between the percentage of black population and altitude were –0.41 in 1990 and –0.47 in 2000. Richmond has a long history of segregating races by elevation. In the 1880s African American communities formed in the Duval's Addition and Madison Ward communities, located in a valley bottom and centered on a stream. Fulton Bottom was a white, working-class neighborhood that evolved into a black one as whites moved to nearby Fulton Hill, above the smoke and direct pollution of the factories located on the James River. Richmond was one of the first southern cities to embrace racial zoning in 1910. Richmond's local elites used the state to control the expansion of black communities, including annexation of dispersed, low-density areas dominated by rural whites. African Americans were crowded into the dilapidated houses in the Jackson Ward area and East End and Church Hill absorbed a large influx of blacks after World War II. Residential discrimination extended into the public arena and into public policy decisions in the placement of public housing. Sixty-four percent of public housing units built prior to 1970 were placed in the Church Hill area and Jackson Ward. Not only did these decisions further galvanize the racial topography of Richmond, they created a polarized class within black neighborhoods themselves, with poor black areas located in east Richmond and middle-class in the northeast. Although the black community is frequently viewed as a "homogenous whole," this spatial differentiation indicates an economic stratification within black communities.

Source: Ueland, J. and Warf, B. (2006) "Racialized topographies: altitude and race in southern cities." *Geographical Review* 96 (1): 50–78.

In counterarguments it was argued that the city was so poor that it needed to attract tax-paying operations to fund city services. The rationale for this cost–benefit analysis may be unfair; the health of residents should outweigh the contribution to the city's coffers. However, the simple causal processes implied by the cavalier use of the term racism needs some caution.

Just because a hazardous site is situated in a minority community does not necessarily mean that environmental racism is at work. Consider the case of trash transfer in Washington, DC. The city needed sites to consolidate its garbage collection and disposal system, the narrow city streets meant that only relatively small trucks could collect garbage in the city, but larger trucks were needed to move the trash to remote incinerators and landfills. A plan emerged from independent committee in the 1990s to build a new transfer station in Ward 8 in the city. The site was ideal; it was flat, the city owned the property, it had good accessibility for the truck traffic and was large enough to satisfy a city ordinance requiring a 500-foot buffer zone between transfer stations and residences. The new proposed site also allowed the closure of older, noisier and more polluting transfer stations in other minority neighborhoods. But Ward 8 houses a predominantly low-income minority population. In the lively debate that followed, the terms "environmental racism" and "environmental injustice" were often used. A closer reading of the facts, however, show that there were few other options. Local activists could easily employ the widely known terms, and in the polarized racial politics of Washington, DC it was an obvious rhetorical point to make. The terms environmental racism and environmental justice can be powerful words of rhetoric even if they provide little explanatory purchase.

In a careful study of environmental inequity, Christopher Boone and Ali Modarres examined the case of the city of Commerce. This predominantly Latino city east of Los Angeles has a high concentration of polluting manufacturing plants. They show that the business located in the city because it provided vacant land and accessibility. The bulk of industry located in the city when it was predominantly white and before the demographic changes that made it a Latino city. In other words, there were toxic neighborhoods because of factors of accessibility and land availability more than because they were Mexican neighborhoods. The Boone and Modarres study suggests that immigrant, migrant and minority communities may develop around toxic areas because of the operation of the housing market, where the poorest people get the least choice and end up in the worst neighborhoods, rather than the case that toxic areas are knowingly located in minority neighborhoods.[4]

This is not to dispute the connection between environmental quality and minority residential areas. In the US among counties that have three or more pollutants 12 percent are majority white, 20 percent are majority African American and 31 percent are majority Hispanic. Race and ethnicity intertwine with issues of power and access to power to produce an uneven experience of environmental quality at home and in the workplace.

While the connections between intent and effect are sometimes difficult to disentangle, race and ethnicity can play a role in mobilization. Robert Bullard provides case studies of community disputes in cities in the US, ranging from conflict over solid-waste landfill in Houston to a lead smelter in Dallas and a solid waste incinerator in Los Angeles. Race and class can become sites of mobilization, a shared experience on which to build resistance and to fight against environmental injustice.[5]

Socio-economic status also plays a major role in the environmental quality of urban living. Poorer communities have less pleasant urban environments and often bear the brunt of negative externalities. It is through their neighborhoods that motorways are constructed; it is in their neighborhoods where heavy vehicular traffic can cause elevated lead levels in the local soil and water. There is a direct correlation between socio-economic status and the quality of the urban environment.[6]

The causal web is sometimes complex but often simple. Poor people get dumped on because they are poor, and they are poor because they lack the wealth to generate political power and bargaining strength. The city is a space where the best areas go to those with the most money and those with the least get what is left. On top of this historical relationship there is the current trend in the siting of noxious facilities that tends to skew their location to those with least power to resist. The poor are the least powerful and experience the worst urban environments. In more racially homogenous cities the deciding factor is socio-economic status.

In a counter-example to the prevailing US literature, Francisco Lara-Valencia and colleagues examined the relationship between hazardous waste sites and socio-economic status in the Mexican border city of Nogales. They found that polluting industries were not located primarily with reference to low socio-economic status neighborhoods; rather, the determining factor was location and accessibility to urban infrastructure and good transport connections. Because this infrastructure correlates with higher income neighborhoods, the more affluent are located closer to the hazardous sites. The study reminds us that different types of cities can produce different patterns of inequality. As the authors conclude, "the construction of hazard and equity spaces is highly dependent on local urbanization trajectories and processes."[7]

In some cases the correlation between race and class on the one hand and environmental quality on the other was, and in some cases still is, reinforced by political movements and economic forces that tended to discount environmental quality in favor of economic growth and employment opportunities. Labor movements were slow to realize that environmental issues were social justice issues, not just the superstructural concerns of the affluent. And the brute economic forces of the industrial city often forced a false divide between environmental qualities versus jobs. Matthew Crenson, for example, tells the story of the lack of an environmental movement in many industrial cities in the US because of the supposed linkage between pollution and employment. Smoking chimneys signified good, well-paying jobs. We have also highlighted this issue in our discussion of Onondaga Lake in Syracuse. For years the business elite promoted the issue in terms of "Does Syracuse want its people employed, or do they want the lake cleaned up?" It was only in the 1970s that a cleaned-up lake was reimagined as a vital part of a new postindustrial city. But as long as it was an industrial city, lake clean-up came a distant second to economic growth. Even today in many cities, especially in the developing world, the dictates of business expansion and economic growth often outweigh the environmental concerns of and the quality of urban life for the majority of the people.[8]

We can now see the false dichotomy of jobs versus environment. Low-grade environments have their most negative health impacts on working people. It is not a case of jobs versus the quality of the urban environment, but jobs *and* the quality of the urban environment. As we move further into a greener economy there are direct connections between improving the urban environment and employment opportunities. Recycling, green technologies and greening the city are all ways to create jobs in hard-pressed cities. Karin Martinson and colleagues explore the rise of green jobs associated with improvements in energy efficiency, the promotion of alternative energy sources and the reduction of pollution.[9] There are green-collar jobs programs with a focus on connecting to lower-income groups in poorer communities as part of the wider environmental justice movement. In New York State the Green Jobs/Green Homes program, initiated in 2009, is a revolving loan program to enable property owners, non-profits and businesses to retrofit their building for better energy efficiency. The program calls for job training and hiring of low-income groups. Green development zones were also established in the state which combine "green affordable housing construction, community based renewable energy sources, housing weatherization, green jobs training and urban agriculture," all with the goal of improving housing conditions, creating jobs and improving energy efficiency.[10] This green development zone concept is a possible model for a more neighborhood-based, ecological sustainable economy.

Urban environmental inequalities occur not only through the presence of hazardous sites; they are also found in absences and lack of access. Two studies highlight this theme. Nik Heynen and colleagues found an inequitable distribution of urban trees in the US city of Milwaukee. The more affluent white areas had more extensive urban tree canopy than the poorer and blacker areas of the city. Because trees can positively affect the quality of neighborhoods, through aesthetic and pollution-diminishing properties, the distribution of uneven green space reproduced uneven social space. The city's shift from a public to a more private funding of the urban tree canopy, part of a more general urban neoliberalization, thus marks a regressive social policy, as the wealthier areas can afford the funds more than the poorer areas.[11] Christopher Boone and colleagues examined the distribution of parks in Baltimore, Maryland. They found that if you simply look at the distribution of parks, then African Americans were located closer to parks, defined as within 400 meters or less. Whites had closer access to a larger acreage of parks. The complex finding is a result of decades of discrimination against African Americans, long denied access to neighborhoods and housing in the city. They were shunted to areas with few parks. The predominantly white areas of the city were provided with more parks, but after white middle-class flight in the 1960, 1970s and 1980s, the incoming blacks now had greater access to the parks.[12] This example shows that equitable results may arise from the historic pattern of unequal treatment.

Environmental justice

The existence of urban environmental inequalities, whether in terms of siting of waste plants or the absence of green spaces, raises the issue of environmental justice. The US EPA defines environmental justice as the "fair treatment and meaningful involvement of all people regardless of race, color, national origin, or income with respect to the development, implementation, and enforcement of environmental laws, regulations, and policies."[13]

This is a narrow definition that restricts the issues to one of a relationship with public policies. "Fair treatment" implies that no single group should bear a disproportionate share of negative environmental consequences or a

BOX 15.2

No child left inside

The numbers of young people and children with asthma is rising. In the US, for children aged 5–14 years the death rates from asthma almost doubled between 1980 and 1993. The disease is more common in blacks and in city-dwellers than in whites and those who reside in suburban and rural areas.

The cause of this recent rise in asthma is under debate. Some believe that the increasing use of antibacterial soap, antibacterial cleaners and antibiotics has created a sterile, clean environment, and as a result immune systems of many children today cannot fight off bacterial infections as well as previous generations. A lack of exposure to the natural world may be exacerbating asthma and a cultural fixation on cleanliness has led to an avoidance of dirt and the outside.

Richard Louv's 2008 book *Last Child in the Woods: Saving Our Children from Nature-Deficit Disorder* generated an increased interest in children's environmental awareness. Louv argues that children have become disconnected from nature and are suffering from "nature deficit disorder." Instead of spending time outside riding bikes, climbing trees and exploring parks, kids spend more time playing video games, surfing the web, and watching TV.

The "No Child Left Inside" Movement has grown since then. Its name is a play on the 2001 educational legislation, *No Child Left Behind*. A federal bill, the *No Child Left Inside Act of 2009*, was introduced in the House and Senate. The bill passed in the House, but was never voted on in the Senate. The federal bill proposed that government money be provided to train teachers for instruction on addressing environmental literacy, and provide innovative technology in elementary and secondary school curriculums. Critics of the federal bill have claimed that it is intended to spread a political agenda to children.

Despite the debate at the federal level, several states, including Connecticut, Colorado, Illinois, Massachusetts and

Wisconsin, have established programs in local parks and schools that address children's disconnect with nature. For example, in Connecticut, the No Child Left Inside pledge is a promise to introduce children to the wonder of nature.

Connecticut's *No Child Left Inside*® pledge:

> I pledge to defend the right of all children and every family to play in a safe outdoor environment. I will encourage and support opportunities for them to exercise their right by:

- Splashing in clean water and breathing clean air
- Digging and planting seeds in healthy soil and watching what grows
- Climbing a tree and rolling down a grassy hill
- Skipping a stone across a pond and learning to swim
- Following a trail and camping under the stars
- Catching a fish, listening to songbirds and watching an eagle in flight
- Discovering wildlife in their backyard
- Soaking in the beauty of a sunrise and sunset
- Finding a sense of place and wonder in this ecosystem we call Earth
- Becoming part of the next generation of environmental stewards.

Sources: Augustyn, H. (2011) "Asthma rates on the rise." *Get Healthy, nwtimes.com.* http://www.nwitimes.com/niche/get-healthy/health-care/article_7c2058db-546f-513a-8d62-2a4a44cefd8e.html (accessed May 2012). "No Child Left Inside, Connecticut." http://www.ct.gov/ncli/cwp/view.asp?a=4005&q=471154&ncliNav_GID=2004 (accessed August 16, 2012); Keirns, Carla C. (2009). "Asthma mitigation strategies: professional, charitable, and community coalitions." *American Journal of Preventive Medicine* 37: S244–50. See also Louv, R. (2008) *Last Child in the Woods: Saving Our Children from Nature-Deficit Disorder.* Chapel Hill, NC: Algonquin Books.

disproportionate lack of positive environment consequences. "Meaningful involvement" implies that people should have the opportunity to participate in environment decisions and affect regulatory bodies.

Environmental justice was given legislative existence in the US in 1994 when President Clinton signed Executive Order 12898 that initiated an environmental justice program within the EPA. The aim was to raise awareness of environmental justice issues, identify and assess inequitable environmental impacts and provide assistance to local areas and community groups. The promotion of environmental justice essentially disappeared under the Bush administration, but under Obama's presidency it made something of a comeback. Under the Environmental Justice 2014 Program, the EPA devised strategies to protect health in polluted neighborhoods, empower communities to take action and establish partnerships with local and state governments to promote healthy sustainable communities. The program is small and modestly funded.

The EPA definition restricts the gaze of environmental justice to the operation of public policies. Environmental justice issues arise from the obvious fact that there is a correlation between the siting of hazardous facilities and low-income communities and/or minority communities. However, as the case of Chester shows, legal redress is difficult if procedures were correctly followed. In many cases it is the everyday operation of the market or the civic society wherein environmental injustices are created and maintained. Environmental justice, like social justice, is not possible in the absence of more interventionist methods. The normal workings of a racist, classist society, for example, will produce racist, classist outcomes in the normal course of events even without the aid of illegality or the help of corruption. If the aim is to produce more equitable outcomes we need more positive interventions. Environmental impact statements, for example, need a more explicit assessment of equity and justice issues. Low-income communities facing environmental challenges should also receive greater resources from government, not simply equal treatment under the law. Unless more positive outcomes are engineered, the system will tend to produce inequitable results.

Urban environmental inequities may be the result of long-sustained processes not amenable to easy fixes. The problems of Chester, for example, were decades in the making and arose from prevailing economic conditions. A more expansive notion of environmental justice should involve deep-seated concerns that move beyond simply measuring spatial correlations to examining the social processes that create persistent patterns of environmental injustice.

Moving out from a specific connection with race in the US, environmental justice is now a worldwide concern. There is now a globalization of the environmental justice movement.[14] It now figures in environmental discussions and debates in cities across the world, as urban environmental issues are connected to issues of justice, equity and fairness. As a scholarly endeavor environmental justice links issues of the production of race, class, gender and ethnicity with the creation of the urban environments.[15]

There is also a widening of the discussion of environmental justice. Brendan Gleeson, for example, provides a scathing critique of equity and justice in contemporary Australian cities. He identifies a new landscape of inequality consisting of new sinkholes of urban poverty, pockets of cultural despair inhabited by indigenous Australians, fortified camps of affluence that amount to acts of secession from the public realm and the problematic non-places of suburbia.[16] Julian Agyeman seeks to link environmental justice with concerns of sustainability, through the notion of just sustainabilities, an idea that encompasses issues of spatial justice, food justice and sovereignty.

Environmental justice is now discussed as part of an urban political ecology. Anne Spirn's book, *The Granite Garden*, was an early example.[17] She explored the city of Boston's air, earth, water, plantlife and wildlife. It was the city understood as an ecosystem in an analysis that linked society, nature and political power. More recent studies include Dawn Day Biehler's discussion of the political ecology of insects and pesticides in public housing. Her study combines issues of domestic space and gender relations with environmental justice.[18]

There is also a body of work concerned with the environmental justice of indigenous land claims in urban areas. In many cities around the world the indigenous people who laid claim to the original land were displaced and dispossessed. Many cities around the world are built on acts of dispossession. In both Australia and Canada, countries with similar colonial histories and legal systems, there are interesting parallels of indigenous land claims on urban sites. Lawrence Berg discusses examples from British Columbia and highlights the role of naming of places.[19] Consider also the case of the small city in the middle of Australia, known as Alice Springs. The town sits on land occupied by the Arrernte people, who moved into the region around 35,000 years ago. In 1870 the South Australian government agreed to build an overland telegraph line to Darwin. Two years later a repeater telegraph station was established just outside the present-day town. In 1874 an area of 25 square miles around the telegraph station was simply annexed. The local people were dispossessed of their land.

In 1928 the town was declared a prohibited area for Aborigines: they required a pass before entering the town. In the 1970s there was major change in policy as the rights of indigenous people, long suppressed, were recognized. The 1976 Aboriginal Land Rights Act recognized land claims of indigenous communities for the first time, and the 1993 Native Title Act created a working legal framework for land claims in urban areas. An application was made in 1994 for native title to lands and waters in and around the town. The applicants were representatives of the Arrernte people. In May 2000 the Federal Court decided that Arrernte native titleholders should retain their rights for most of the reserve, park and vacant public land in the town. The ruling gave the land to traditional titleholders, along with a set of new rights: the right to be acknowledged as traditional owners, the right to use the natural resources, the right to make decisions about the land, the right to protect places, the right to manage the spiritual forces and safeguard their cultural knowledge associated with the land. The ruling also created a changed urban identity. The city's occupancy of Arrernte territory is now signaled at official ceremonies, as well as by urban signage. The town is no longer just Alice Springs and just the expression of white explorations; it is Alice Springs/Mparntwe and it encompasses and honors reminders of Arrernte creation stories (Figure 15.2).[20]

Urban environments and social difference

The experience of living in the urban environment varies across the dimensions of social difference. Socio-economic status, gender, age and level of physical ability/disability are all sources of social difference that are embodied and reflected in urban environments.

Take the case of gender. Women have played an important role in bringing environmental issues to the broader public. Lois Gibbs, for example, was a typical suburban housewife with two children living in a suburb in Western New York. When her son Michael developed epilepsy, her daughter Melissa contracted a rare blood disease and after her neighbor's children also got sick she began to look into possible causes. She and others soon discovered that the area where she lived, Love Canal, was built on a toxic waste site, with more than a dozen known carcinogens, including the deadly chemical dioxin. The local soil and water was poisoning the community. Lois Gibbs became one of the community organizers mobilizing public opinion and promoting state and federal involvement. The area was evacuated in 1980, the area's name joining places like Bhopal as a byword for environmental pollution. The publicity given to Love Canal strengthened the case for the 1981

Superfund legislation, in which chemical clean-up of major toxic sites was federally mandated.[21]

Urban toxicity is often experienced by women in households and neighborhoods where women play a strong role in the reproduction of the households. They tend to children's illnesses, connect with local neighbors and may be closer to local issues. It is often women who are on the front line of local environmental issues.

The case of Love Canal, in which children became very sick, is also an example of the effect of age on the experience of urban environmental

Figure 15.2 Sacred site in Alice Springs/Mparntwe

Source: Photo by John Rennie Short

conditions. Embryos in the womb and very young children are especially vulnerable to adverse environmental conditions. They are the human equivalent of the canary in the mine, providing early and tragic warnings of the state of our environment. Older people are also especially vulnerable to air quality and other environmental conditions. Yet in many cities the very young and very old have the least voice but are often the most impacted. Most cities are designed by and for affluent males, and the further from this group's characteristics the population, the more marginal to most discourses of power and influence they are. The marginal, the poor, the young and the old often live in the worst urban environments.[22]

These settings can also become sites of resistance and progressive social change. In Medellin, Colombia, there are now nine library parks—part library, part park and part community center—scattered throughout the slum areas of this 3.4 million population city. The new library buildings are impressive, architect designed, open and accessible. The initiative was promoted by Sergio Fajardo, a political independent who became mayor of the city in 2004. He spent almost half the city's $900 million budget on education, much of it allocated to the library parks for the poorest communities. In the libraries, adults and children receive job training, have access to computers and can connect with social welfare networks. The library parks, a result of community participation, combine green space and social space. The library parks of Medellin improve environmental quality and reduce inequalities by spending most on the poorest areas of the city.

Healthy cities

In 1984 the World Health Organization sponsored a conference in Toronto entitled Healthy Cities. Two years later, the Ottawa Charter for Health Promotion outlined the basic requirement of improving overall health by improving the physical and social environments of cities. Today, there are almost 3,000 healthy city initiatives around the world. New initiatives have been promoted along with existing public health programs under the term "healthy cities." In Europe, where there are 30 national healthy cities networks, the emphasis is on integrating health and urban planning. In much of the rest of the world, healthy cities are also associated with the provision of basic infrastructural services of water and sewerage. The emphasis is on improving the health of citizens, especially in the poorest areas, upgrading environmental health concerns and facilitating networking between different stakeholders. Early assessments of the healthy cities program are encouraging. A study of 20 participating communities in California found that new programs were created, new policies were adopted and new financial resources were leveraged. The identification of assets and needs and the creation of inter-organization linkages led to positive effects on health status.[23] Similarly positive results were found in an assessment of the initiative in European cities.[24]

There has been a more explicit acceptance that healthy cities must connect to issues of environmental justice. In 2001 member countries of the EU ratified the Aarhus Convention, stating that every person has the right to live in an environment adequate to his or her health and well-being, and to achieve this end citizens must be involved in decision making and have access to information. The Convention promoted early and effective participation and various evaluation criteria. Nicola Hartley and Christopher Wood examined public participation in the environmental impact assessments in four UK waste disposal case studies in the wake of the Aarhus

---

BOX 15.3

Pop-up parks

Parks are not only top-down initiatives. Pop-up parks are now part of an interesting and exciting example of rapid, bottom-up, urban revitalization. In the US the Build a Better Block initiative is a form of rapid urban intervention: in 2010, for example, Dallas community organizers with $1,000 of borrowed material and lots of sweat equity transformed a street overnight by painting a bike lane, planting saplings and narrowing the vehicular lanes from three to one. It is a guerilla form of urban park making. Pop-up parks have also appeared in Cleveland, Tulsa and Philadelphia. In Fort Worth the initiative painted bike lanes and crosswalks on South Main Street, long slated to become a highway. The action persuaded the state, with prompting from local people now made aware of the alternatives, to reroute the proposed highway.

The website for the Build a Better Block is http://better-block.org.

BOX 15.4

A healthy city initiative

One obvious healthy city initiative is to encourage more vegetation growth. A careful tree-planting program, for example, can lower summertime temperatures, minimize the urban heat island and reduce air pollution. Some cities actively encourage tree planting. In Sacramento, California, since 1990 over 375,000 shade trees have been given away to city residents with plans for four million more trees to be planted throughout the city. In other US cities, in contrast, tree-planting programs have been cut back. In 24 cities in the US there was a 25 percent decline in tree canopy in the past 30 years. In cities such as Milwaukee the tree-planting program was re-oriented toward the downtown to make it more attractive to investors. The Sacramento experience, however, shows that tree planting not only makes for a healthier city, it also saves money. The city estimates that for every dollar spent on trees it recoups $2.80 in energy savings, pollution reduction, stormwater management and increased property values.

There has been resistance to tree planting from some utility companies who argue that it costs more to secure their power lines though vegetation. In contrast, in states such as Iowa that mandate tree planting by utility companies, the experience has been more positive. In a time of rapid energy costs, utility companies are often grateful for the public relations bonus they receive from partnering in tree planting. In Sacramento the local power company PG&E pledged $25,000 in 2012 to meet the goal of planting another 30,000 trees in the city.

Convention. They report that the Aarhus Convention led to a strengthening of participation procedures but that the level of improvement secured will depend upon how its ideals are interpreted and incorporated into specific legislation. In other words, it is the legislative and administrative details that will structure the implementation.[25]

Conclusions

Although the pronouncements of the WHO Healthy Cities and the Aarhus Convention are easy to make, rhetoric being easier than real action, they do hint at the future connections between issues of social justice, participatory democracy and urban environmental quality. They also indicate that environmental rights will join the list of citizen rights and government obligations that mobilize communities and motivate governments. Securing better urban environmental conditions, especially for the poorest, is not incidental to wider social struggles and longer-term economic objectives, it is central and pivotal.

Guide to further reading

Adamson, J., Evans, M. M. and Stein, R. (eds.) (2002) *The Environmental Justice Reader*. Tucson, AZ: University of Arizona Press.

Agyeman, J. (2013) *Introducing Just Sustainabilities: Policy, Planning and Practice*. London: Zed Books.

Bickerstaff, K., Buckeley, H. and Painter, J. (2009) "Justice, nature and the city." *International Journal of Urban and Regional Research* 33: 591–600.

Bullard, R. D. (ed.) (2005) *The Quest for Environmental Justice: Human Rights and the Politics of Pollution*. San Francisco, CA: Sierra Club.

Bullard, R. D. (ed.) (2007) *Growing Smarter: Achieving Livable Communities, Environmental Justice, and Regional Equity*. Cambridge, MA: MIT Press.

Holifield, R., Porter, M. and Walker, G. (2010) *Spaces of Environmental Justice*. Malden, MA: Wiley-Blackwell.

Madrid, J. and Alvarez, B. (2011) *Growing Green Jobs in America's Urban Centers*. Washington, DC: Center for American Progress.

Mohai, P., Pellow, D. and Roberts, T. (2009) "Environmental justice." *Annual Review of Environmental Resources* 34: 405–30.

Pellow, D. N. and Brulle, R. J. (2005) *Power, Justice and the Environment: A Critical Appraisal of the Environmental Justice Movement*. Cambridge, MA: MIT Press.

Redwood, Y., Schultz, A. J., Israel, B. A., Yoshihama, M., Wang, C. and Kreuter, M. (2010) "Social, economic, and political processes that create built environment inequities." *Family Community Health* 33: 53–67.

Stein, R. (ed.) (2004) *New Perspectives on Environmental Justice: Gender Sexuality and Activism.* New Brunswick, NJ: Rutgers University Press.

Swyngedouw, E. and Cook, I. (2012) "Cities, social cohesion and the environment: towards a future research agenda." *Urban Studies* 49: 1959–79.

Washington, S. H. (2005) *Packing Them In: An Archaeology of Environmental Racism in Chicago.* Washington, DC and Covelo, CA: Rowman and Littlefield.

Webster, P. and Sanderson, D. (2012) "Healthy cities indicators: a suitable instrument to measure health." *Journal of Urban Health.* doi: 10.1007/s11524-011-9643-9.

The environmental justice website of the USA EPA is http://www.epa.gov/environmentaljustice/index.html.

16 Urban sustainability

Cities are now considerably unsustainable in that they cannot continue in the same way—consuming resources and generating waste. But what does urban sustainability mean? There are a variety of terms, adjectives and meanings associated with "sustainability" from ecological, environmental, social, political, cultural and economic perspectives. The concept in part stems from an appreciation that the environment is of limited capacity. It questions the principal Enlightenment project of the mastery of nature and its unlimited exploitation. In this chapter we first examine the intellectual history behind sustainability. Although we concentrate on the US, many countries have a similar experience. We then explore ways in which sustainability is defined and theorized. In the second part of the chapter, we turn to ways cities are practicing sustainability.

Intellectual roots of sustainability

There is a long history of reconceptualizing society–nature relationships. Henry David Thoreau's *Walden* remains a classic read about the transformative power of nature. Frederick Law Olmsted's legacy was to view urban parks as a way to rebalance the industrial city, while Ebenézer Howard's garden cities created green buffer zones around city nuclei. These planners and visionaries had a strong social agenda as well, and saw the natural world as a way to promote democracy and social equity.

Another important intellectual legacy is found in the works of George Perkins Marsh, who published *Man and Nature* in 1864. The book argued that human impact on the environment could have significant effects. He drew attention to the unforeseen consequences of human actions—floods and landslides, for example, were often the result of overgrazing. Marsh emphasized the food chain, the importance of forests to soil conservation and the need for sound ecological principles. These are now such taken-for-granted notions that we often forget they had to be developed and presented. George Perkins Marsh was a major influence on the American conservation movement. Gifford Pinchot (1865–1946) referred to *Man and Nature* as epoch-making.

Gifford Pinchot studied forest management and put many of his ideas into practice at the Vanderbilt's Biltmore estate in North Carolina. On the 5,000-acre estate Pinchot developed his three main principles: profitable production, constant annual yield and improvement of the forest through selective logging. Together, these principles form the concept of "wise use" of resources and became the core of the conservation movement. Pinchot's influence extended to President Theodore Roosevelt, who asked Pinchot to become Chief Forester of the US. In this capacity, Pinchot introduced the term "sustained yield" as a way to practice resource management. Sustained yield and wise use were concepts that limited overgrazing, and set resource use based on the maximization of benefit for the greatest number of people rather than for the profit of a few. He promoted long-term interest over short-term concerns. Pinchot's ideas percolated into legislation, forestry practice and an environmental approach that would dominate for the next 100 years.

During the 1960s and early 1970s numerous writers and thinkers would expand on and add to new discussions about nature–society relations. In 1962 Rachel Carson published *Silent Spring*. It was a pivotal moment in the US environmental movement. *Silent Spring* begins with the image of a birdless spring, a result of pesticides such as DDT destroying birdlife. Her book chronicled the growing use of chemicals like DDT, deildrin, endrin and parathion, detailing their deleterious effects on humans, plants and animals. The book drew upon a wide body of scientific writing; her bibliography ran over 50 pages and included reports, memos and papers in science journals. Today her message seems eminently sensible. At the time, however, there was great controversy. Chemical manufacturers mounted a vicious campaign against her and tried to block publication of the book. Despite the campaign against her and her arguments, *Silent Spring* was a commercial success. In less than six months, half a million copies were sold. In 1963 Carson appeared before a Congressional committee hearing to discuss unregulated pesticides. Just as she was reaching a huge audience, she succumbed to cancer. But her legacy is monumental. She deepened the American consciousness of the ecological web, and the horrendous dangers of toxic pollution from unseen chemicals; she influenced policy and shaped public opinion. *Silent Spring* marked a new shift in environmental awareness. The earlier conservation and preservation movements had focused on the danger of overexploitation of resources. Carson shifted concern to the threat of human and animal extinction.

Jane Jacobs' most influential book was *The Death and Life of Great American Cities*, published in 1961. Jacobs was reacting against urban renewal of the 1950s in the US and what she saw as the destruction of vibrant urban neighborhoods. Hers was a powerful critique of urban renewal, but it also reached beyond planning issues to influence the spirit of the times. Her activism helped create the historic preservation movement in the US and grassroots efforts to block urban renewal projects that would have destroyed local neighborhoods. Her legacy was to offer radically new principles for rebuilding cities that involved the public in planning, and ensuring that cities were ethnically and racially diverse places.

Other important thinkers include Andre Gunder Frank, who wrote widely on the economic, social and political history and contemporary development of the world system, and especially the developing world. His 1967 book, *Capitalism and Underdevelopment in Latin America*, explored why poverty persisted and examined ways in which underdevelopment was linked to capitalism. Donella Meadows led the Club of Rome, a group of scholars who, in 1972, published *Limits to Growth*. The book used computer models to understand how unchecked economic and population growth impacts finite resources. The authors also explored the possibility of a sustainable feedback pattern that would be achieved by altering growth trends among different variables.

Ecological economist Herman Daly's 1973 book, *Steady State Economy*, has been widely read. Daly noted that economic activity degrades ecosystems and interferes with natural processes that are critical to various life support services. In the past, the amount of economic activity was small enough that the degree of interference with ecosystems was negligible. However, unprecedented growth of twentieth-century economic activity significantly shifted the balance, with potentially disastrous consequences. The book combined a limits-to-growth argument, theories of welfare economics, ecological principles and the philosophy of sustainable development into a model Daly called steady-state economics. Together with Robert Costanza, AnnMari Jansson, Joan Martinez-Alier and others, Daly helped to develop the field of ecological economics.

All of these writers and thinkers—and many others—share a rethinking of society–nature dialectics, albeit from different angles and perspectives. Inherent in many of their works is a concern about the environment, efforts to promote equity and a sense that there are limits to economic growth and resource use. Three themes—environment, equity and economy—have been gestating for more than a century.

The term and concept of sustainability did not simply happen. It is less a linear process, and more a convergence of intellectual perspectives. Gradually, from the 1970s to the 1980s, the idea of "sustainability" coalesced around the three issues of environment, economic development and advancing social equity. It was the outcome of a long and rich intellectual evolution of ideas that redefine nature–society relationships. The intellectual roots include writers and thinkers focused on ecology, economics and social equity, as highlighted in Table 16.1.

Defining sustainability

In 1987 the United Nations World Commission on Environment and Development issued a report called *Our Common Future*. The report was the product of a commission of foreign ministers, finance and planning officials, economists, policymakers in agriculture, science and technology. The report is often referred to as the Brundtland Report, after Gro Harlem Brundtland, Chair of the Commission (Box 16.1).

The report defined sustainable development as "development that meets the needs of the present without compromising the ability of future generations to meet their own needs."[1] Seen as the guiding principle for long-term global development, sustainability consists of three pillars: economic development, social development and environmental protection. These are often referred to as the "three Es"—economy, ecology and equity (Table 16.2). Although this definition has become the most widely accepted, ideas about what sustainability is and how to achieve it differ. Despite several decades of discussion, no single definition of sustainable development has emerged.[2] Other definitions include "living within the carrying capacity of supporting ecosystems" or "preservation of ecological and social capital" and "that which improves the long-term health of human and ecological systems."[3] Despite the fact that the 1987 definition is vague, it is not meaningless.

Sustainability has been applied in the urban context. Herbert Girardet, for example, defines a sustainable city as "a city that works so well that all its citizens are able to meet their own needs without endangering the well-being of the natural world or the living conditions of other people, now or in the future."[4] There can be no sustainable city without social justice, political participation, economic vitality and ecological regeneration, themes we have touched upon in many chapters of this book.

Table 16.1 Intellectual contributions on urban sustainable development

Environmentalists	Economists	Urban studies/urban planning	Equity advocates	Spiritual writers and ethicists
John Muir (preservation) George Perkins Marsh	Kenneth Boulding	Patrick Geddes, Ebenezer Howard, Lewis Mumford (comprehensive planning)	Murray Bookchin (social ecologists)	Gary Snyder, Thomas Berry, Dalai Lama
Rachel Carson	Herman Daly (steady-state economics)	Jane Jacobs (preserving neighborhoods and diversity), Norman and Susan Fainstein (critique of rational planning)	Edward Goldsmith, Nicholas Hildyard, Frances Moore Lappe, Arturo Escobar, Vandana Shiva and Martin Khor (development critics)	Charlene Spretnak, Petra Kelly, Carolyn Merchant (ecofeminism)
Donella Meadows	Michael Redclift and David Pearce (environmental economics)	John Logan, Harvey Molotch, Brian Stoker, Peter Calthorpe (urban growth coalitions)	Robert Bullard, Carl Anthony (environmental justice)	Baird Callicott, (environmental ethics)
Brundtland Report	Robert Costanza, Richard Norgaard (ecological economics)	David Gordon, Timothy Beatley (green urbanism)	Eric Swyngedouw, Matthew Gandy (political ecology)	Theodore Roszak (ecopsychology)
EarthSummit/Agenda 21	Paul Hawken (restorative economics)	Michael Hough, Rutherford Platt (urban ecology)		Yi-Fu Tuan (topofilia)
President's Council on Sustainable Development (USA)	William Rees (ecological footprint analysis)	James Kunstler, Edward Relph (placelessness)	Bill Devall/George Sessions (deep ecology)	Kirkpatrick Sale (bioregionalism)

Source: Wheeler, S. (2004) *Planning for Sustainability: Creating Livable, Equitable and Ecological Communities*. London and New York: Routledge, p. 28.

BOX 16.1

Our common future

"The present decade has been marked by a retreat from social concerns. Scientists bring to our attention urgent but complex problems bearing on our very survival: a warming globe, threats to the Earth's ozone layer, deserts consuming agricultural land. We respond by demanding more details, and by assigning the problems to institutions ill-equipped to cope with them. Environmental degradation, first seen as mainly a problem of the rich nations and a side-effect of industrial wealth, has become a survival issue for developing nations. It is part of the downward spiral of linked ecological and economic decline in which many of the poorest nations are trapped. Despite official hope expressed on all sides, no trends identifiable today, no programmes or policies, offer any real hope of narrowing the growing gap between rich and poor nations. And as part of our 'development,' we have amassed weapons arsenals capable of diverting the paths that evolution has followed for millions of years and of creating a planet our ancestors would not recognize.

When the terms of reference of our Commission were origi-nally being discussed in 1982, there were those who wanted its considerations to be limited to 'environmental issues' only. This would have been a grave mistake. The environment does not exist as a sphere separate from human actions, ambitions, and needs, and attempts to defend it in isolation from human concerns have given the very word 'environ-ment' a connotation of naivety in some political circles.

… the links between poverty, inequality, and environmental degradation formed a major theme in our analysis and recommendations. What is needed now is a new era of economic growth—growth that is forceful and at the same time socially and environmentally sustainable."

United Nations World Commission on Environment and Development (1987) *Our Common Future*, pp. 6–7. http://www.un-documents.net/our-common-future.pdf (accessed July 2012).

Table 16.2 The Three E's of sustainability

Ecological principles of sustainability
- Prevention is better than cure
- Nothing stands alone (all things are connected)
- Minimize waste
- Maximize use of renewable and recyclable materials
- Maintain and enhance diversity (biodiversity, cultural diversity)
- Enhance environmental understanding through research

Economic principles of sustainability
- Use appropriate technology, materials and design
- Small is beautiful
- Create new indicators for economic and environmental wealth and stop using GDP measurements
- Create new indicators for economic and environmental productivity
- Establish acceptable minimum standards through regulatory control
- Take action to internalize environmental costs into the market (polluter pays)
- Ensure social acceptability of environmental policies
- Encourage widespread public participation

Equity principles of sustainability
- Intergenerational equity (not compromising the planet for future generations)
- Intergenerational equity: basic needs of people must be met (social justice)
- Eliminate poverty
- Geographical equity: all people and communities are entitled to equal protection of environment, health, employment, housing, transportation and civil rights
- Equity in governance
- Minimal use of non-renewable resources
- Human well-being
- Interspecies equity: other life does not have the moral equivalence of humans but we must preserve ecosystem integrity

---

Source: Dresner, S. (2008) *The Principles of Sustainability*. 2nd edition. New York: Routledge; Rees, William E. (1999) "Achieving sustainability: reform or transformation?" in *The Earthscan Reader in Sustainable Cities*, David Sattherwaite (ed.). London: Earthscan. pp 22–54.

Since *Our Common Future*, international discussions and debates about sustainability continued with Agenda 21, the primary outcome of the 1992 Rio Earth Summit, where 178 governments voted to adopt the program. The final text was the result of drafting, consultation and negotiation, beginning in 1989 and culminating at the two-week conference. The number 21 refers to an agenda for the twenty-first century. The preamble to Agenda 21 notes:

> Humanity stands at a defining moment in history. We are confronted with a perpetuation of disparities between and within nations, a worsening of poverty, hunger, ill health and illiteracy, and the continuing deterioration of the ecosystems on which we depend for our well-being. However, integration of environment and development concerns and greater attention to them will lead to the fulfillment of basic needs, improved living standards for all, better protected and managed ecosystems and a safer, more prosperous future. No nation can achieve this on its own; but together we can—in a global partnership for sustainable development.

Since the 1992 Rio Earth Summit, the international community has met every five years. In 2012, Rio + 20 was held and participants reaffirmed their commitment to Agenda 21 in their outcome document, "The Future We Want." Sustainability is now the guiding agenda for the protection of both the local and the global commons—the biosphere and the atmosphere.

Urban sustainability

Part of the evolving discourse on sustainability is that sustainability takes place at a variety of geographical scales, including—but not limited to—the city. For cities, one of the important aspects of Agenda 21 is known as Local Agenda 21. Chapter 28 of the report focused on the role of local governments and authorities to take steps to implement sustainability. From this mandate emerged the group, Council for Local Environmental Initiatives (ICLEI)—now called the Local Governments for Sustainability—which formed in 1990. ICLEI is widely regarded as an example of a vehicle that effectively promotes the implementation of Agenda 21. Today, more than 1,200 cities, towns, counties and their associations in 70 countries are members of the organization. There are now networks of cities on a range of regional and global issues. For example, the Aalborg Charter of 1994 outlined sustainability objectives for European cities (Box 16.2). Subsequent international conferences such as the Istanbul UN City Summit of 1996 and the Local Government Declaration to the 2002 UN Johannesburg Earth Summit expressed similar ideas about directing urban growth in more sustainable ways.

Defining urban sustainability involves multiple dimensions and can include everything from environmental protection, social cohesion, economic growth, neighborhood design, alternative energy and green building design.

This means that for cities to embark on sustainable urban development, they need to address not only local environmental impacts, but the urban impacts on the global climate, biodiversity and energy use.

Cities and metropolitan regions are an important venue in tackling sustainability and in advancing a green agenda. It is at this scale that many things are possible, that committed citizens and organizations can exert pressure and make a difference.

BOX 16.2

European cities and sustainability

In 1994 the European Sustainable Cities & Towns Campaign was launched in Aalborg, Denmark. The participants at this first European conference discussed and adopted the Charter of European Cities and Towns Towards Sustainability (the Aalborg Charter). To date, more than 2,000 European local and regional authorities (metropolitan areas, cities, towns, counties, etc.) from numerous European countries have signed up to the Aalborg Charter. Below are several principles in the Aalborg Charter.

The role of European cities and towns

"We have learnt that present levels of resource consumption in the industrialised countries cannot be achieved by all people currently living, much less by future generations, without destroying the natural capital.

We are convinced that sustainable human life on this globe cannot be achieved without sustainable local communities. Local government is close to where environmental problems are perceived and closest to the citizens and shares responsibility with governments at all levels for the well-being of humankind and nature. Therefore, cities and towns are key players in the process of changing lifestyles, production, consumption and spatial patterns."

Urban economy towards sustainability

"We, cities & towns, understand that the limiting factor for economic development of our cities and towns has become natural capital, such as atmosphere, soil, water and forests. We must therefore invest in this capital. In order of priority this requires:

1. investments in conserving the remaining natural capital, such as groundwater stocks, soil, habitats for rare species;
2. encouraging the growth of natural capital by reducing our level of current exploitation, such as of non-renewable energy;
3. investments to relieve pressure on natural capital stocks by expanding cultivated natural capital, such as parks for inner-city recreation to relieve pressure on natural forests); and
4. increasing the end-use efficiency of products, such as energy-efficient buildings, environmentally friendly urban transport."

Social equity for urban sustainability

"We, cities and towns, are aware that the poor are worst affected by environmental problems (such as noise and air pollution from traffic, lack of amenities, unhealthy housing, lack of open space) and are least able to solve them. Inequitable distribution of wealth both causes unsustainable behaviour and makes it harder to change. We intend to integrate people's basic social needs as well as healthcare, employment and housing programmes with environmental protection. We wish to learn from initial experiences of sustainable lifestyles, so that we can work towards improving the quality of citizens' lifestyles rather than simply maximising consumption.

We will try to create jobs which contribute to the sustainability of the community and thereby reduce unemployment. When seeking to attract or create jobs we will assess the

> effects of any business opportunity in terms of sustainability in order to encourage the creation of long-term jobs and long-life products in accordance with the principles of sustainability."
>
> Source: http://www.aalborgplus10.dk/media/key_documents_2001_english_final_09-1-2003.doc (accessed March 2007).

Theories of sustainability

Sustainability is now a global buzzword. Nations have pledged to work toward a more sustainable future. Cities have joined initiatives and are developing sustainability plans. While the concept has grown in currency, scholars have attempted to better define and elaborate on the concept. Some scholars equate sustainability with concepts like freedom, justice or quality of life: elusive to define, but you know it when you see it. Ideas about sustainability have evolved in many directions over the past 25 years. These debates give impetus to re-evaluate the ethical underpinnings of policy and analysis. Today there are many approaches to defining and implementing sustainability. Theories about sustainability range from "all we need is a bit of regulation and reform" to "we need a total transformation of our economic system." We can characterize a spectrum of theories, from light green to dark green. Discussing these theoretical approaches to sustainability allows us to see the complexity of the concept, and the challenges for its realization.

Light green

Those theories characterized as "light green" would hold that institutions can be reformed in order to avoid an ecological crisis. There is no need to leave the path of modernization that has dominated the last 200 years. For example, neoliberals argue that sustainability should develop with the efficacy of the market. Government investment or involvement should be minimal. Neoliberals privilege private ownership over collective ownership, private consumption over public consumption, private automobiles over public transportation. This ideology posits that regulation is a barrier and a burden and that it clashes with individual liberty and freedom. Neoliberals see personal responsibility as more important than social responsibility toward others. They often see current environmental regulation as excessive and they have a strong belief that the earth's carrying capacity can be overcome by technological innovations. This is not to say that neoliberals have rejected the concept of sustainability. They assume free and open markets will maximize health and ensure sustainability. They do not worry about resource scarcity because they place great confidence in price as an indicator of scarcity and on the mechanics of the marketplace. They also see environmental degradation as the result of poverty, and they believe the way to repair the environment is to eliminate poverty through economic growth. Neoliberals have embraced notions of the "quality of life," walkable neighborhoods, safe communities. However, they are skeptical of efforts to implement environmental or social regulation.

Progressives are also on the light green spectrum. Progressives link environmental degradation and economic inequality and believe that a systematic problem requires systematic solutions. They believe government can do good works for the public good. Democratically elected government can and should ensure environmental sustainability and social equity, and the government must intervene, for example, by creating affordable housing, or by passing environmental laws. Unlike neoliberals, progressives see a critical role for government and regulation, but they also believe that individuals need to change personal behavior as well. Many progressive activists and professionals are motivated by a deep moral or ethical concern for the environment and social justice. They do not believe that economic systems must be overthrown to achieve a sustainable future.

Dark green

Criticisms of the underlying purpose and principles behind *Our Common Future* have emerged. Dark green theorists believe that the wider economic structure itself must be transformed to achieve environmental and social justice. Some argue that the Brundtland Commission began with an assumption that the priority was economic growth. Simon Dresner notes the starting point of sustainable development was a meeting point for environmentalists and developers, and that this was "deliberately conceived as being somewhat more palatable than the hard line environmentalist message."[5] Tim O'Riordan also argues early concepts of sustainable development were growth-centered, not environmentally centered.[6] He notes that this represents the status quo, not a radical rethinking of society–nature relationships. William

Rees argues that mainstream sustainability ideas have failed and that so far sustainability is "shallow" as opposed to "deep." Rees notes that environmental assessment, pollution control and legislation, and growth strategies can produce positive local effects but they are more cosmetic in that they have not significantly changed fundamentally unsustainable environment–economy relationships.[7]

The emergence of political ecology and urban political ecology approaches are highly critical of mainstream efforts at sustainability. A neo-Marxian political ecologist builds on a more radical political perspective that critiques modern capitalism and cities. A political ecology approach urges us to look at the wider ideological assumptions about growth, development and the environment to understand how and why decisions are made. This approach deconstructs the ideologies that undergird and inform political debates about sustainability and urban–nature relations. Political ecologists and urban political ecologists tease out who gains from and who pays for socio-environmental change.

For example, urban political ecologist Roger Keil views *Our Common Future* and early efforts at sustainability as "a neoliberal approach to greening capitalism."[8] He notes that the Commission had revolutionary potential as it was an implicit critique of capitalism, but it did not follow through. The result was "light green" sustainability, a recipe for the survival of capitalism. Many urban political ecologists believe that the politics of sustainability must include an agenda that redirects accumulation into products and services that help sustain human and natural metabolism.

Graham Haughton also argues that *Our Common Future* does not reject market-driven capitalism, but tries to reform it to be more environmentally compatible. Haughton argues that we need change at two levels: first we need to improve political, economic, regulatory and legal systems; second, we need to devise systems that ensure responsibility.[9] Haughton argues that sustainable cities should be:

- self-reliant and reduce negative external impacts of the city beyond its own bio-region;
- redesigned to bring nature back into the city and reduce sprawl;
- fair by ensuring that environmental assets are traded on a fair basis.

Urban political ecologists critique neoliberals, who believe that the allocation of urban space, as a scarce resource, should be determined by market forces. Because they also see an inherent problem in capitalism, they also critique progressives for believing that reform is possible. Urban political ecologists Eric Swyngedouw and Nik Heynan have written that the material conditions that comprise urban environments are controlled and manipulated and serve the interests of the elite at the expense of marginalized populations. The result is—and always will be—highly uneven urban environments. A sustainable future that promises equity is not possible without social, economic and political transformations.

The theoretical approach to sustainability may also inform how sustainability is implemented. For example, dark greens see light greens as moving toward sustainability with surface appearances. Land reclamation and tree planting is merely "green washing"; such modest efforts will result in weaker forms of sustainability. The counterargument is that sometimes modest reforms create a momentum for more radical and meaningful measures. Dark greens believe that to really create meaningful sustainability, society needs to grapple with the impacts of production and consumption, both socially and at an individual level.

Practicing sustainability

In this section we look at a range of sustainability practices. We first consider the ways in which cities deal with growth. We then examine how cities create and implement sustainability plans.

Dealing with growth

Urban growth is a pervasive feature of the past 100 years. In cities large and small and in countries rich and poor, growth has been a dominant feature. Cities that once occupied only a few square miles now cover hundreds. Urban residents that once walked most places now depend on the automobile. Suburban sprawl has increased as metropolitan areas extend their reach further out into the countryside. The metropolitan sphere of influence lengthens its shadow across the landscape as farmland turns into tract housing, woods into subdivisions and the prairie into gated communities. We can characterize five main responses to urban growth as *resistance, smart growth, new urbanism, slow growth* and *historic preservation.*

Resistance

Resistance occurs at all scales. It occurs especially in growth areas with more affluent households. In high-growth areas, rapid increase in traffic, overcrowded schools and local property taxes can all stimulate local resistance. In Howard County in Maryland, for example, population grew 34 percent from 2000 to 2010. Some residents have begun to resist the latest round of development proposals. The powerful development-house building lobby marshaled against such resistance. When residents in Howard County signed a petition to create a referendum to challenge rezoning proposals, a group of landowners sued the council to cancel the referendum.

Resistance is particularly strong when the local residents are wealthy and organized. Mark Singer describes a struggle in a wealthy Connecticut town as one between the haves and the haves. The town of Norfolk in Connecticut is a place of old money and obvious displays of affluence are frowned upon. Almost 80 percent of the town's 30,000 acres is designated as forest, agriculture or park. An area of 780 acres, known as Yale Farms, came on the market in 1998. A plan to develop a luxury golf course and 100 homes, each on four acres, generated intense resistance from groups that called themselves the Canaan Conservation Coalition and the Coalition for Sound Growth. The fight, according to Singer, has given people a chance to affirm their shared values. Behind the debates about loss of green space and loss of community was fear of change and a distaste for the incoming *nouveaux riches* who had enough money to disrupt the traditional moral code with their ostentatious wealth.[10]

Resistance takes many forms from "not here," "not this here," to "not this here now." The success of the resistance depends upon the wealth, organizational skill and effective links to political power of the pressure groups. But the battle is uphill and in many cases unsuccessful. The resistance movement has to compete with powerful development, real estate and property investment interests. Successful resistance is not a global phenomenon. In some parts of the world poorer citizens lack access to power because political leverage is linked to income and status. In China, for example, which has experienced arguably the highest rates of recent urban growth, rural peasants and urban dwellers have little recourse in the face of state-prompted urban development.

Smart growth

One planning response to sprawl is the *smart growth* initiative that stresses mixed land uses and compact building designs that create high densities with lower environmental impact. Smart growth has emerged as a strategy to deal with the constant pull of development toward greenfield sites on the city's edge; it focuses on existing developments in order to utilize their infrastructures and to preserve open space and farmland. It is a framework for municipalities facing heavy development pressure and looking for principles and policies to halt the abandonment of urban infrastructure and the building of greenfield sites. In 1996 the Smart Growth Network enunciated its principles:

• mix land uses;
• design more compact buildings;
• construct walkable communities;
• create a sense of place;
• preserve open space;
• direct development toward existing communities;
• provide a variety of transport choices;
• make fair, predictable and cost-effective decisions;
• encourage community involvement in development decisions.

Smart Growth argues the remedy for sprawl is to create housing development dense enough to encourage public transportation and to reduce the need for the automobile.

In 1997 the state of Maryland in the US established a set of smart growth policies with three main goals: to invest state resources in areas where infrastructure was already in place; to preserve farmland and natural resources; and to resist public investment in building infrastructure that promoted sprawl. A set of priority funding areas was identified for state investment in

Table 16.3 Sprawl vs. Smart Growth

Sprawl	Smart growth
Car dependent	Walkable
Scattered subdivisions of single-family homes	Diversity of housing types in many neighborhoods
Cul de sacs and wide roads that funnel traffic into a few highways choked with traffic	Connected street network that distributes traffic throughout the system
Low density	Higher average densities around commercial centers
Little public open space	Networks of parks, greenways and natural areas
Spread out	Compact centers
Single-use office parks and shopping centers surrounded by parking lots	Mixed-use centers (shops, offices, housing, restaurants, schools) served by transit
Limited or no public transit service	Frequent and convenient transit service

Source: adapted from the Chattanooga Climate Action Plan http://www.chattanooga.gov/chattanoogagreen (accessed May 12, 2012).

transportation, water and sewage. In effect, the policy guided higher-density development in areas already served by public infrastructure. The Maryland smart growth initiative under the leadership of Governor Parris Glendenning was a model for other states. The state-wide scheme was abandoned in 2002, with the election of Republican Governor Robert Ehrlich, but at the county level a variety of smart growth strategies continue. In Montgomery County, Maryland, for example, planners have encouraged developers to include high-density mixed land use, the use of infill developments and more development at bus, metro and rail stations. The 2009 Smart Growth Initiative focused on attracting economic sectors such as biosciences, green technologies and agriculture. The county is well positioned to do so, as it is home to the National Institute of Health, the FDA and a myriad of private firms whose research and work support these agencies.

Portland, Oregon, is a well-known model of smart growth and sprawl containment. The city established an "urban growth boundary" in 1980 that protects farmland surrounding the city and tightly limits development in outlying areas. Portland's approach has not been without controversy. For several years the urban growth boundary was accompanied by skyrocketing housing costs and discontent among those who resented restrictions on development. But the high costs of housing—which are in fact attributable to a host of factors, including a high rate of migration to Portland from other states, particularly California—have since declined to the point that they are roughly equivalent to those of other West Coast cities.

Because of the urban growth boundary, Portland has assimilated a sharply rising population without encroaching on its valuable land resources. Portland's urban design provides affordable and accessible public transit located close to schools, businesses and residential communities. In addition, walking and bike paths connect the entire community.

Smart Growth is a possible answer for municipalities facing heavy development pressure and looking for principles and policies to halt the abandonment of urban infrastructure and the building of greenfield sites. But it is still too early to say whether smart growth will become an effective policy to halt the seemingly relentless expansion of the suburban fringe into open spaces.

New Urbanism

The history of urban planning is full of attempts to reorganize the city along principles of rational efficiency, good design and encouragement of community. The latest in a long line of urban design movements is New Urbanism. It is a response to suburban sprawl that emphasizes revitalizing old urban centers; creating mixed-use centers where residences are located close to commercial and office sectors; planning for walkable, high-density, low-rise residential areas that are socially diverse communities; minimizing the speed of autos through urban areas and making cities more attractive to walking and casual social interaction.[11] There is not anything particularly "new" about these ideas; the century-old Garden City Movement promotes most of these ideals.

One example of the New Urbanism is Seaside in Florida, designed by Andres Duany and Elizabeth Plater-Zyberk, where neotraditional houses conform to a strict code, the houses are at a high density, car traffic is kept to the edges and the walkways and porches are constructed in order to foster community inter-action. Duany, in particular, is now a high-profile advocate of New Urbanism. Seaside was the community shown in the movie *The Truman Show*.

Perhaps the most cited example of New Urbanism is Celebration, a commu-nity initially planned and funded by the Disney Corporation. Walt Disney initially had a plan for an Experimental Prototype Community of Tomorrow (EPCOT). However, EPCOT was integrated instead into Florida's Walt Disney World theme park and was never realized as a community. Celebration opened in 1996 just outside Orlando, Florida. With its Charles Moore civic center and Cesar Pelli cinema, the place has attracted big-name architects and lots of attention.[12] Celebration has all the design elements of the New Urbanism: low-rise, high-density residential areas where garages are at the back of the residences, walkways and porches allow pedestrian movement and a mixed-use downtown. It is also emblematic for the exclusivity that has characterized New Urbanism. The lowest rents are $800 per month, and most of the population is upper-income. While the New Urbanism proclaims social heterogeneity, in practice it tends to be restricted to the middle- and upper-income groups.

In Europe, the equivalent of New Urbanism is the "urban village." In Copenhagen, Freiburg, Vienna, Zurich, Heidelberg and Barcelona urban redesign has focused around pedestrian streets, sidewalks, public squares and parks. The dense pedestrian center is then connected to the rest of the city with bicycle lanes and well-integrated public transport.[13] The result is more walkable cities that offer diverse attractions that aims to foster a sense of urban community.

Underlying New Urbanism and the urban village model is a nostalgic sense of community and neighborliness, a longing for a lost community. The older high-density cities of the half-remembered, half-created past are often portrayed as places of tight community, while more recent suburban growth of the present day is seen as a cause of the decline of community. New Urbanism is a catch-all phrase that in principle captures the discontent with contemporary developments, especially the nature of low-density sprawl and the alienation felt by many residents. In practice, it means developments are high-density, pedestrian-friendly and often socially exclusive.

New Urbanism is not all that new and is not all that urban. On closer inspection it looks like the latest version of up-market suburban communities. New Urbanism as it has been practiced so far does little to discourage suburban sprawl, since it still produces densities too low to support public transportation and truly mixed communities, creating instead homogenous enclaves. It is repackaged suburbanization, a useful marketing strategy, playing an important role in stimulating debate but with little practical effect on creating community.[14]

Few would argue the need for some kind of alternative to the standard forms of suburban sprawl. It is wasteful of resources, lacks aesthetic appeal and produces a series of edges rather than centers. New Urbanism is, at the very least, proposing alternatives to a city dominated by the auto, the highway and the parking lot. New Urbanism as a design guide is a step in the right direction. However, the claims made for New Urbanism as a source for "recovery of community" are based on a series of assumptions (community is declining, urban form can resuscitate this community) that are asserted rather than demonstrated. The New Urbanism is a source of interesting design ideas, but as a method of recreating community, the verdict is still out.

Slow growth

The Slow City Movement has member cities in Italy, Germany, Norway and England. It was founded in October 1999 in Italy by three mayors as a policy framework and network of small cities—they have to have populations of less than 50,000—seeking to connect the three E's of environment, economy and equity. There are now numerous cities with a designated list of slow-growth environmental policies and urban designs. A study of two German cities in the network points to the protection of city-owned pastures and apple trees and revitalization of houses as community public spaces.[15] Cities in the slow-growth network tend to be not only small but also homogenous, with shared political agendas. In larger, more heterogeneous cities with a less interventionist political culture, slow growth may be politically untenable.

However, it is suggestive of how groups of cities can network their ideas and share practices of urban sustainability.

Preservation

There has always been some appreciation of cultural and architectural heritage, but only since the mid-twentieth century have societies become increasingly aware of the significance of our urban historic structures and sites. In the 1950s and 1960s the prevailing attitude was that "old" was bad and "new" was superior. By 1970, that idea was changing.

In the US the preservation movement evolved from two distinct paths.[16] The *private sector* path focused on important historical figures and landmark structures. It has been called the "George-Washington-Slept-Here" approach. The *public sector* path was involved with establishing national parks and this also included historic buildings. In the 1930s and 1940s the public sector established some historic districts, such as Charleston, South Carolina and the Vieux Carré (French Quarter) section of New Orleans and Alexandria, Virginia. In 1949, several organizations evolved into the National Trust for Historic Preservation, inspired by the British version. The purpose was to link the preservation efforts of the private side with the federal government/ National Park Service activities. The most important piece of historic preservation legislation was the National Historic Preservation Act of 1966. It established new laws, authorized funds for preservation activities and encouraged locally regulated historic districts. It went beyond merely protecting landmarks to recognizing a variety of historically and architecturally significant buildings, sites, structures, districts and objects. Preservation no longer focused on saving single landmarks; instead entire areas were delineated as historic districts and became important tools of urban revitalization during the 1970s and 1980s. Today there are thousands of local preservation associations and thousands of designated historic sites, buildings and other structures in cities. Historic preservation is composed of a variety of strategies: preservation, restoration, reconstruction and rehabilitation.

Preservation refers to the maintenance of a property without significant alteration to its current condition. When preservation is the guiding strategy, the only intervention is normal maintenance or special work needed to protect the structure against further damage. An example of innovation in preservation is Pike Place Market in Seattle. The old city market was threatened with demolition to make way for an urban renewal project. In a city-wide vote, however, residents voted to save the market as an important part of their city's life and culture. In order to prevent the loss of its original character as a working everyday market run by local farmers, fishermen and small entrepreneurs, the city developed an ordinance that not only protected the structure, but also the activities within.[17]

Restoration refers to the process of returning a building to its condition at a specific time period. However, this often means changing the natural evolution of a building, and creating a more contrived picture of its "original condition." This aspect to preservation is more common in historic homes, farms or churches. This strategy, however, is not without criticism, as it casts doubt on the authenticity of restored structures.

The term reconstruction refers to the building of a historic structure using replicated design and/or materials. This approach is taken when a historic structure no longer exists. The earliest and best example is Williamsburg, Virginia. In 1926 John D. Rockefeller was persuaded to fund the restoration of the entire colonial town of Williamsburg. The primary problem was that much of the original town had been lost over the centuries, and while many of the historic buildings remained, a few central buildings from the town's original layout were missing. Planners decided to reconstruct the Governor's Palace (which had been destroyed by a fire in 1781). The efforts to reconstruct Williamsburg were not without controversy—as some buildings were removed to make way for reconstructed ones.[18] Yet it remains one of the most visited historic districts in the US. In addition to both restoration and reconstruction, Colonial Williamsburg presents live recreations of historic events by actors in period costumes (Figure 16.1). This way of presenting historical places and artifacts is often referred to as a "living history museum," and is increasingly popular.

Lastly, many buildings no longer perform their original function or use, but retain their architectural integrity. For these structures, a common strategy is rehabilitation, sometimes referred to as adaptive reuse. The purpose is to modify or update portions of the structure and adapt the building for a new purpose. Numerous examples abound: abandoned factories that are adapted into microbrewery pubs or museums or residential lofts. Increasingly rehabilitation is a strategy employed by cities seeking to revitalize old areas of the city. Figure 16.2 shows the adaptive reuse of an old power plant in Baltimore Harbor.

Historic preservation is increasingly at the center of urban redevelopment efforts as cities search for a way to celebrate their past while looking to the future. It can be used to preserve neighborhoods and ecologically important areas within the city.

Figure 16.1 Colonial Willamsburg is an example of historic reconstruction, complete with the recreation of historic events by actors in period costumes

Source: Photo by Lisa Benton-Short

Planning for sustainability

Recently cities have begun to develop sustainability plans. Comprehensive sustainability plans are documents that present municipalities' overarching sustainability visions and priorities. Such plans often explain and explore current problems and barriers to sustainability, identify priorities and present indicators that can measure progress over time.

Kent Portney notes that it is a logical step for cities to begin developing sustainability plans. He notes that this reflects a "new localism" in environmental policy, whereby municipal governments take primary responsibility in promoting sustainable development.[19] This makes sense as municipal governments have considerable control over land-use planning and zoning, transportation and other infrastructure investments, waste management, municipal operations and a variety of factors which impact social issues such as public education. It is also the local governments that are in many ways best equipped to educate the public, promote sustainable decision making to individuals and respond to concerns of their citizenry. Already cities undertake many kinds of planning exercises, all of which have the potential to include elements of sustainability: zoning and comprehensive plans, natural

Figure 16.2 A power plant gets a new life as mixed-use development that includes a Barnes & Noble, a Hard Rock Café and an ESPN Zone

Source: Photo by Lisa Benton-Short

hazards plans, stormwater management plans, ecosystem management plans, transportation and smart growth plans. Portney has shown that some city governments now see sustainability policies as giving them an edge for economic development, and others may even see sustainability as a way of saving money.[20]

But comprehensive planning can be a major challenge. Eric Zeemering suggests that sustainability poses significant challenges for local policymakers because of the need for coordinated action across governmental departments, as well as coordinated action between government and the private and non-profit sectors.[21] In this respect a city's sustainability plan should not be under the control of just the department of environment, but should also involve numerous city offices and departments.

European cities have been at the forefront of global efforts to develop comprehensive sustainability plans. Paris, Freiburg, Helsinki, Oslo and London are pioneers in the area of sustainability plans. Malmö and Copenhagen, for example, often rank among the top of many "green cities" lists. Many European cities have initiated innovations in bike sharing and car sharing that were adopted by cities around the world. Freiburg, Germany, for example, has created the Freiburg Charter for Sustainable Urbanism (Box 16.3). There are also a range of established networks that encourage European cities to move toward sustainability. For example, the Aalborg Charter connects cities and provides information about sustainability initiatives. The organization also gives out a European Sustainability City Award (the first was issued in 1996), something that is now highly coveted and valued by politicians and city officials.[22] Cities in Europe can also compete for the designation of Green Capital City.

BOX 16.3

The Freiburg Charter for Sustainable Urbanism

The 12 guiding principles:

1. Diversity, safety and tolerance;
2. city of neighborhoods;
3. city of short distances;
4. public transport and density;
5. education, science and culture;
6. industry and jobs;
7. nature and environment;
8. design quality;
9. long-term vision;
10. communication and participation;
11. reliability, obligation and fairness;
12. cooperation and partnership.

Source: Academy of Urbanism (2010) "The Freiburg Charter for Sustainable Urbanism: learning from place." http://www.scribd.com/doc/79197159/AoU-Freiburg-Charter-for-Sustainable-Urbanism (accessed July 2012).

Developing a comprehensive sustainability plan can be a daunting challenge. However, Timothy Beatley and colleagues note that European cities represent important sources of ideas and inspiration about green urban development and policies.

Sustainability planning in US cities has lagged behind European cities, in part because Agenda 21 efforts are not as uniformly adopted at the local level in the US as they are in other countries. A second reason for the delay in creating sustainability plans in US cities is financial: the development of a comprehensive sustainability plan demonstrates a significant investment of resources. Third, the literature suggests that sustainability policy and management is not a well-coordinated effort within the US. Instead, there is considerable evidence that it is left up to environmental organizations to initiate campaigns that draw cities into networks and alliances that promote sustainability planning. Even given these obstacles, more than 50 US cities have developed sustainability plans in the last ten years.

Yet because there is no single template from which cities create their sustainability plans, they vary in size and shape and quality. For example, PLANYC, New York City's sustainability plan, is more than 160 pages long, and covers numerous issues including housing, open space, brownfields, water, transportation, energy, air and climate change. Cincinnati, Chattanooga, Portland and Baltimore also have extensive sustainability plans. On the other end of the spectrum are those sustainability plans that are not particularly comprehensive. Kansas City's sustainability plan is 16 pages long and focuses mostly on water; Cleveland's plan is about 23 pages, while New Orleans is 55 pages, but has a strong focus on rebuilding more sustainably in the wake of Hurricane Katrina. Regardless of size, these plans serve as a blueprint for implementing sustainability, and they can be built on in the future.

An interesting case is Houston, the oil capital of America. In 1999 it overtook Los Angeles as the most polluted city. When Houston bid to compete to host the 2012 Olympic Games, not a single person on the US Olympics Committee voted for it. But there have been big changes in just a few years. The city now has a director of sustainability and a plan. Its transport authority oversees light rail and is adding more than 20 miles of track; most of the traffic lights are LED bulbs, and more than half of the cars in the city's fleet are hybrid or electric. It has implemented strong energy codes for buildings, and is one of the biggest municipal buyers of renewable energy; about one-third of its power comes from Texan wind farms. Houston's sustainability plan is credited with making the public take more interest in sustainability. In 2012, a survey found that 56 percent of Houstonians think a better public transport system is "very important" for the city's future; 51 percent liked the idea of a smaller house in a more interesting district.[23] These are big changes for a city whose leaders have been wary of environmental regulation, and big on cars and McMansions.

Implementing sustainability

In this section, we look at the many ways in which sustainability is being implemented in cities—from the small-scale efforts directed at single

---

BOX 16.4

Sustainability plans

Sustainability plans come in all shapes and sizes. We offer a few examples of recent sustainability plans and highlight the table of contents to provide a sense of some of the topics, themes and elements cities are addressing.

City	Year	Table of contents of sustainability plan
Baltimore, MD	2009	• Introduction: what is sustainability? • Cleanliness ◦ Litter ◦ Transform vacant lots • Pollution prevention • Resource conservation ◦ Energy use ◦ Reduce GHG ◦ Reduce water pollution • Greening the city ◦ Double tree canopy • Transportation ◦ Improve public transport ◦ Increase bicycles • Education and awareness • Green economy

New York City	2007	• Housing • Open space • Brownfields • Water quality • Water network • Transportation congestion • Energy • Air quality • Climate change
Charleston, NC	2007	• Measuring emissions • Better buildings • Cleaner energy • Sustainable communities • Improved transportation • Zero waste • Green education • Moving forward
New Orleans	2008	• Green buildings and energy efficiency • Alternative energy • Waste reduction and recycling • Transportation and clean fuels • Environmental outreach and justice • Flood risk reduction

Source: compiled by authors

buildings, residences or neighborhoods, to larger-scale efforts that are more city-wide.

Green houses

Every house is a vortex of environmental consumption—of materials, energy, and water. In the twentieth century, planners tended to view housing within a vast infrastructure grid of gas, electrical power, water supply and sewerage. It was easy to expand this basic pattern with more suburbs and new towns. Suburbs and sprawl ultimately encouraged forms of urban growth that were oblivious to their effects on the environment. In contrast, sustainability would localize the impacts of growth by insisting that as many resources as possible (water, energy and food) are sourced, processed and disposed of locally.[24]

A "green house" is one in which the building or rebuilding attempts to make the house self-sustaining. Table 16.4 lists the objectives of green housing. There are three main systems that contribute to sustainability: (1) the waste system; (2) the drinking water system; and (3) the renewable energy system. For example, one type of waste treatment system relies on a "biolytic filter" that takes sewage, good waste and other organic material. The filter consists of a concrete tank with several layers of filter beds that contain microorganism and worms that sift, sort, digest and treat the solid waste and wastewater. The toilet, shower, bath, dishwater and sinks drain into a single sewer pipe that empties into the top filter bed of the tank. Once the microorganisms and worms have done their job, any remaining water is pumped up to an ultraviolet lamp to kill any remaining bacteria. The water can be used in the garden.

It is also possible to use rainwater to satisfy water needs. Rainwater falling on the roof can be collected and filtered for drinking water, watering gardens and other fresh water needs. Solar systems can consist of solar panels of photovoltaic cells that produce electricity from sunlight. In some instances, there is enough energy power to sell back to the utility. There are also small-scale designs for generating wind power for the home.

Green houses are also made using building materials from re-growth timber or materials produced by pollution-free manufacturing processes and use no materials that discharge toxic chemicals. For example, thick straw bale can be used for insulation for walls. The use of concrete floors can also help keep

Table 16.4 Objectives of green housing and green communities

Minimize the use of resources (water, land, energy)

Minimize production of waste

Minimize use of toxic materials

Integrate open space and green space into plans

Minimize need for travel and maximize low-energy modes of transport (pedestrian, bike and public)

Avoid privatized space, no gated communities (wasteful of land)

Design public space for personal safety

Insist on affordability and inclusiveness

Produce some of the food consumed

the house warm in the winter and cooler in the summer. It is possible to replace natural gas furnaces with stoves that burn corn kernels or wood pellets. Because corn kernels consume carbon dioxide as the plant grows, burning doesn't release new GHGs.

One of the most important elements of a green house is the emphasis on design for the local climate and the orienting of the house so that main windows face south (in the Northern Hemisphere). This helps to maximize the use of the sun during the winter to provide light and warmth.

The impact of a green house can be considerable. For example, by using rainwater, a house can save 26,500 gallons of water that would have been consumed from a river or reservoir. By treating waste on the property the house keeps 26,500 gallons of sewage from flowing into a treatment plant (or being discharged without treatment during rain). Composting food scraps and other organic material can cut the solid waste that might otherwise be taken to a landfill. A house using solar energy reduces carbon dioxide emissions from electricity generated at a coal-fired power station by eight tons.

Individuals and families that live in green houses are said to "live lightly on the grid" in that they have succeeded in creating homes that require only minimal energy from power plants and fossil fuels. The multiple benefits of a green house include avoiding high utility costs, electrical outages and combating climate change. The climate change connection is not inconsequential; the US EPA estimates that the average American house contributes more than twice as much GHG as the average automobile.

Green buildings

For many years architecture was rarely concerned with issues of sustainability. Sustainability often ranked below considerations of style and cost. More recently, however, green builders, architects and interior designers have developed designs for "green" buildings.

Buildings consume enormous quantities of the Earth's resources in both their construction and daily operation. The conceptual framework behind green buildings is the incorporation of features that support the conservation of the environment. Green buildings are often designed and oriented to minimize summer afternoon solar heat gain and optimize winter solar heat gain. Some may have solar energy as an alternative to fossil fuels. Many green builders select materials that do not have formaldehyde and have minimal or non-toxic properties to improve indoor air quality. Even interiors can incorporate materials and products that have high levels of renewability or reusability, such as bamboo flooring or cork tiles.

Another example is in the installation of landscaping on the tops of buildings. Green roofs reduce energy costs and soak up rainwater. Green roofs are partially or completely covered with vegetation and soil, or a growing medium, planted over a waterproof membrane. They were developed in Germany in the 1960s, and have since been adopted in many cities. Today, it's estimated that about 10 percent of all German roofs have been "greened."[25] Between 1989 and 1999, German roofing companies installed nearly 350 million square feet of green roofs, and the rate is increasing. Although green roofs have become increasing popular in Europe, the adoption of green roofs

---

BOX 16.5

LEED buildings

The US Green Building Council is a non-profit organization that promotes green buildings through its Leadership in Energy and Environmental Design (LEED) certification program, which is now a benchmark standard for measuring sustainability in buildings. The certification has five design categories: sustainable sites; water efficiency; energy and atmosphere; materials and resources; and indoor environmental quality. There are four categories of certification, in ascending order of sustainability: certified, silver, gold and platinum. One building that received the highest award of platinum was a 16-story building completed in 2006 in Portland, Oregon, to house a Center for Health and Healing. The building was constructed on an abandoned site close to the river. There is on-site sewage treatment, while water used for toilets and landscaping comes from rainwater and wastewater. Energy is provided by solar shades that are also power generators. While conventional buildings seal out air and light, the design of this building harnesses them.

LEED-certified buildings have wider benefits to the urban community, such as reducing waste sent to landfills, conserving energy and water, reducing harmful GHG emissions, creating compact and walkable communities with good

> access to neighborhood amenities and transit. They also have benefits to developers as they have lower operating costs and higher asset values than non-green buildings, and lead to healthier and safer buildings for occupants. LEED-certified buildings qualify for tax rebates, zoning allowances and other incentives in hundreds of cities. Building green is no longer just a laudable thing to do; it now makes good business sense.
>
> Source: Cidell, J. (2009) "A political ecology of the built environment: LEED certification for green buildings." *Local Environment* 14: 621–33.

in the US has been slower. Green roofs can impact the environment in the following ways: reduce carbon dioxide, reduce summer air-conditioning needs, reduce winter heat demands, reduce stormwater run-off, provide songbird habitat and remove nitrogen and other pollutants from rainfall. In addition, green roofs also reduce the urban heat island effect. Traditional building materials soak up the sun's radiation and reflect it back as heat, making cities at least 7°F hotter than surrounding areas. Figure 16.3 shows a green roof in Vienna, where by contrast temperatures on a hot day are typically many degrees cooler than they are in traditionally roofed buildings.

Seattle, Portland and Vancouver have been lauded as leaders in green buildings in the US; there are over 30 green buildings in Seattle alone. Chicago and Toronto have won numerous awards for advancing green roofs. Toronto initiated a "Green Roofs for Healthy Cities" program and has proposed a series of "green walls" where vegetation will grow on the sides of buildings.

In 2012 the city of Brooklyn announced it would soon be home to the world's largest rooftop farm. Bright Farms, a company that builds greenhouses, unveiled plans to build a multi-acre farm on 100,000 square feet of rooftop space atop the Liberty View Industrial Plaza in the Sunset Park area. Construction is set to begin soon and the farm is scheduled to open in 2013. Bright Farms predicts it will produce over one million pounds of vegetables each year, including tomatoes, lettuces and herbs. All of the produce will be grown hydroponically, meaning no soil will be used as mineral nutrients are absorbed directly from the water in which the plants are grown. The company plans to sell the food in the community.

Figure 16.3 Green roof in Vienna. This building, called the Global Yard, is public housing that comprises 50 percent Vienna-born residents and 50 percent immigrants

Source: Photo © Rob Crandall

Green infrastructure

Green infrastructure (GI) is a term given to cover a range of natural resources in cities—small parks, golf courses, preserves, riparian corridors and open spaces. In contemporary cities, one of the most prominent features in the marketing of residential areas is open space and natural features. Residential properties that line Central Park in New York, or even row houses that boast private gardens, are often the most desirable and sometimes the highest priced properties. Trees, plants and other vegetation are not merely cosmetic embellishment; they are basic infrastructure that makes important contributions to the city aesthetically and ecologically.[26] For example, trees and shrubs shade the walls of houses and buildings, thereby reducing indoor temperatures and hence the need for air-conditioning. One study suggests that merely viewing natural landscapes can have a positive effect on people's sense of physical and psychological well-being.[27] Another study, published in *Science*, showed that hospital patients whose rooms had window views of trees and gardens had significantly higher rates of post-surgical recovery and shorter hospital stays than those whose rooms did not have views of vegetation.[28] As we noted in Chapter 11, many cities now approach planning for stormwater management by expanding and developing their GI. Such approaches reduce stormwater while also improving the urban aesthetic.

Sustainable approaches to green spaces can include restoration of creeks and waterways, planting street and garden trees and the restoration of native plants and animals. Efforts to restore creeks not only help restore indigenous riparian plant species, but also help manage urban stormwater. Many creek and river restoration projects have the added element of providing residents and tourists with opportunities to experience these restored natural settings with the addition of new or improved walking trails. The greening of such small urban spaces is often initiated by urban community groups.

In New York City, the local grassroots organization Green Guerillas assists neighborhoods in creating community gardens and in developing garden coalitions to resist development of small urban open spaces. The organization dates back to 1973, when a Lower East Side artist, Liz Christy, organized her friends and neighbors to clean out a vacant lot on the corner of Bowery and Houston Streets. They created a vibrant community garden, thus establishing the modern community gardening movement in New York City. In other deindustrialized cities such as Pittsburg and Detroit, abandoned, derelict lands have been given over to food growing by unemployed workers. There is a revival of urban farming and agriculture.

Urban agriculture is not new to cities: it can be traced back to the earliest cities; even medieval cities in Europe grew crops within the walls. But it has been "rediscovered." Today, local supplies of fruit and vegetables are part of sustainability. Community gardens that grow food for local residents reduce the energy required to transport food from hundreds or thousands of miles away. In many developing cities, the growing of food in and around cities contributes to food security and poverty alleviation. For example, in Dar-es-Salaam, Tanzania, one of the fastest-growing cities, 67 percent of the families are engaged in urban farming, compared with 18 percent in 1970.[29] In Accra, Ghana, urban farmers supply 90 percent of the vegetables. In the slums of Lima, Peru, inhabitants produce a variety of crops from sweet potatoes and artichokes to chicken, fish and pork. Many of these urban farmers recently immigrated from the Andes mountains, where agriculture has been a way of life and their farming skills are put to good use in the slums.[30] Urban farming and community gardens are examples of locally organized and locally managed green projects. They also link to environment and equity issues; urban farms and community gardens can provide fresh food in low-income communities.

Increasingly, urban planners incorporate the principles of landscape and urban ecology into the siting and design of small patches of land, larger areas such as parks and even wider park systems. Many cities are currently planning to connect habitat patches through environmental corridors and networks of corridors. In Southern California, in 2009 the city of Irvine unveiled a new master plan for the site of the former El Toro Marine Corps Air Station, which closed in 1999. At 1,347 acres, the Great Park features a constructed two-and-a-half mile canyon, a "daylighted" (restored and removed from concrete ducts) stream, a large lake, a great lawn, an aviation museum, a conservatory/botanical garden, a promenade and a sports park

---

BOX 16.6

Urban agriculture in Havana, Cuba

"Cuba is the only country in the world that has developed an extensive state-supported infrastructure to support urban food production and urban growers. With the demise of the Soviet Bloc in 1989 all food imports were lost, resulting in the Cuban population experiencing immediate food shortages. Cuba also lost critical agricultural imports upon which its national food production system had become dependent—fertilizers, pesticides, tractors and spare parts and petroleum to provide fuel energy. Reductions in access to petroleum brought the food distribution system to a halt; severe fuel shortages meant that food could not be refrigerated or transported by trucks from the peri-urban and rural areas where food was produced to the urban areas where the majority of the population resided.

By the end of 1992, food shortages had reached crisis proportions throughout Cuba, including in the capital city of Havana, home to 2.2 million Cubans and the largest city in the Caribbean. Like many large cities, Havana was a food consumer city, completely dependent upon food imports brought in from the Cuban countryside and abroad. Havana had no food production sector or infrastructure, almost no land dedicated to the production of food. Worsening food shortages motivated Havaneros to spontaneously begin to plant food crops in the yards, patios, balconies, rooftops and vacant land sites near their homes. In some cases, neighbors got together to plant crops—beans, tomatoes, bananas, lettuce, okra, eggplant and taro. If they had the space, many began to raise small animals—chickens, rabbits, even pigs. Within two years there were gardens and farms in almost every Havana neighborhood. By 1994 hundreds of Havana residents were involved in food production. The majority of these urban growers had little or no access to much needed agricultural inputs—seeds, tools, pest controls, soil amendments. Nor did they have knowledge about the small-scale, agro-ecological techniques that urban gardening requires.

The Cuban Ministry of Agriculture responded to people's need for information and agricultural inputs by creating an Urban Agriculture Department in Havana. The Department's goal was to put all of the city's open land into cultivation and provide a wide range of extension services and resources such as agricultural specialists, short courses, seed banks, biological controls, compost and tools. The Department secured land use rights for all urban growers by adapting city laws to gain legal rights for food production on unused land. Hundreds of vacant lots, public and private, were officially sanctioned as gardens and farms.

This state-supported infrastructure for urban agriculture has allowed thousands of Cubans to become involved in food production in the nation's capital. Currently, about 30% of Havana's available land is under cultivation and there are more than 30,000 people growing food on more than 8,000 farms and gardens in Havana alone. The size and structure of these urban farms and gardens varies considerably. There are small backyard and individual plot gardens cultivated privately by urban residents (*huertos populares*). There are larger gardens based in raised container beds by individuals and state institutions (*organoponicos*). There are workplace gardens that supply the cafeterias of their own workplace or institution (*autoconsumos*). There are small family-run farms (*campesinos*) and there are farms owned and operated by the State with varying degrees of profit sharing with workers (*empresas estatales*).

Although urban agriculture in Cuba came about as a response to an acute food shortage, the benefits have been far reaching. These advances are directly due to the Cuban government's commitment to food security, which, in Cuba, has come to mean not only providing people with access to food but, providing them with healthy food, produced without chemical inputs harmful to human and environmental health. We hope this case study encourages other city governments to develop strategies and policies that contribute to an urban agriculture infrastructure that promotes small-scale sustainable farming methods and inputs, allows urban growers to thrive, increases local food security, and promotes ecological sustainability."

Pinderhughes, R., Murphy, C. and Gonzalez, M. (2000) "Urban agriculture in Havana, Cuba." http://online.sfsu.edu/raquelrp/pub/2000_aug_pub.html (accessed August 2012); see also this video: *Havana Homegrown: Inside Cuba's Urban Agriculture Revolution* (http://www.youtube.com/watch?v=iGuipXzxPFY)

(Figure 16.4). Ecologically, the park will be a vital link in the chain of land reserves stretching from the mountains to the sea. The park will also create social connections to the communities throughout the county by knitting together riding, hiking and multi-use trails from all parts of the region, linking all neighborhoods to the park and communities beyond. The plan centers on sustainability, stating:

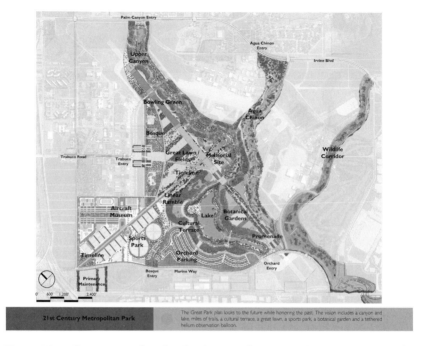

Figure 16.4 The master plan for the former El Toro Marine Base, soon to be the Great Park, in Irvine, California. This plan highlights a variety of sustainability features, from reusing materials from the old base, to restoring riparian water features

Source: http://www.ocgp.org

sustainability means utilizing the rich materials that already exist at the former El Toro Marine Corps Air Station and showcasing their use in the construction of the Great Park. The former base has existing resources that are far too beautiful to overlook. We plan to build bridges with the redwood now built into the military hangars. We will recycle the concrete runways to build the veterans memorial and roadways. We will clean existing storm water to be returned into the groundwater table. We will be energy-efficient.[31]

Green transportation

In many cities, transportation is by car. However, the costs are substantial. Building roads, bridges and highways takes enormous sums of money. The rise of the car has also come at the expense of public transportation and a diversity of transportation choices. It contributes to congestion, road deaths and injuries. Car dependency also has ecological costs—it impacts the atmosphere with GHGs and local pollution, such as photochemical smog. The car, however, is fundamentally not a city vehicle, but it has dominated many urban transportation priorities.

A sustainable city will involve a transportation system in which there is a network of paths connecting people safely on foot, bicycle or public transportation. In city centers, space will be allocated to pedestrian zones, street cafes and markets instead of parking garages and congested city streets. When people do use cars, they will travel in fuel-efficient vehicles, or low-emission vehicles that emit very few pollutants. A system of green transportation prioritizes foot travel, then cycling, then public transport and finally private motor vehicles, a reverse of the customary order.[32] As one architect says, "a sustainable city is compact, polycentric, ecologically aware and based on walking."[33]

The commercial success of pedestrian precincts created in the historic centers of many European cities is an example of a re-prioritization of transport. Figure 16.5 shows La Rambla in Barcelona, a tree-lined pedestrian mall that stretches for almost three-quarters of a mile (1.2 km). On La Rambla,

Figure 16.5 La Rambla is an example of a pedestrian precinct in the historic center of Barcelona. Such spaces are lively and important to counterbalance the automobile

Source: Photo by Lisa Benton-Short

visitors can find kiosks that sell newspapers and souvenirs, flowers and birds, street performers, cafes, restaurants and shops. In many cases, pedestrian precincts have become lively spaces. In the 1990s the city council in Birmingham, UK, decided that the historic nineteenth-century core of the city was not flourishing, in part because of a motorway that encircled the central business district. They demolished part of the motorway to allow full pedestrian access and today most of the city's central core is a walking precinct. In addition, the pedestrian areas are being linked to railway stations and major bus routes to allow foot travel to move easily from pedestrian areas to other forms of transportation.

In cities such as Hong Kong, Vienna, Zurich, Curitiba and Singapore, new tram and bus systems are convenient and efficient. In most Dutch cities such as Amsterdam there are expansive cycling networks that separate bicycles from automobiles, thus reducing the danger of accidents and providing fast and convenient pathways in and around the city. Bicycles are a dominant form in the visual landscape (Figure 16.6). Today many cities have established bike-share programs or dedicated bike lanes.

Investment in public transportation also addresses issue of social equity by providing transport for all those who do not have access to a car. Urban transportation that focuses on mass transit, walking and cycling means faster journeys, lower transportation costs and healthier people.

Sustainable cities are not necessarily ones where people own fewer cars, but they are ones where people use (or need) cars less.[34] The idea of moving away from cars to greener forms of transportation seems sensible in principle, but has proven complex in practice. In the interim, programs that encourage car sharing offer an alternative. In the US, many cities have FlexCar or ZipCar programs. Members pay a deposit and receive a key code that opens a variety of cars available all over the city (you can also locate them on the internet sites). Car-sharing plans start at around $8 an hour or $60 a day for the use of the car. Members share access to hundreds of vehicles, often within a five-minute walk of home or work. Members can reserve a car online or by phone and the hourly charge covers gas, insurance, monthly parking, and maintenance. Studies have shown that each car-share vehicle displaces up to 15 privately owned cars off the road.

For cities in the developing world, addressing transportation can raise a variety of other issues. Take the example of Bangkok, one of the world's most congested cities. In 2003, the Thai government announced a new measure for countering congestion in Bangkok: it was banning elephants from the

Figure 16.6 Bicycles in Amsterdam
Source: Photo by John Rennie Short

city's streets. This ban affected some 250 elephants that were wandering the city with their handlers, who were begging for food or selling trinkets. Often the elephants held up traffic, were hit by vehicles or fell into open drains.[35] Similarly, in Hyderabad, India, notorious for air pollution and congestion, sacred cows roam the city streets, an added obstacle for rickshaws, taxis and cars to negotiate. To compound issues of congestion, many cities in the developing world have invested little in mass public transportation. Megacities need a comprehensive regional transport plan to adequately deal with the congestion.

BOX 16.7

Participatory maps

Urban planners, designers and architects have their work cut out for them. The rate of urbanization around the world means that we have to reconfigure the way we think, design and plan cities. It is not like starting from scratch on a clean canvas. Reconstructing existing cities to fit the needs of a larger population will be especially difficult because it will require planners to accommodate the needs of current urbanites, while forecasting the needs of future urbanites. It will require a balancing act between the environment, public health and mobility, from the design stage all the way through implementation.

This is where participatory maps are an increasingly important aspect of studying and planning more inclusive cities. Inviting residents to participate in map-making gives them a voice in the spatial planning process. It also provides insights into how they use their cities—where they live, where they work, where they cycle. Take, for instance, the map made by Anton Polsky, a Russian artist who goes by the name "Make" and is the founder of a participatory urban re-planning website: www.partizaning.org.

In 2010, Make designed and shared a map, which he calls "USE/LESS" to bring to attention the dismal circumstances cyclists had to endure in his hometown of Moscow. Using the resources and opportunities available to him, Make created the map and designed a second online version to crowd-source the marking of cycling routes, dangerous roads and bicycle parking spaces. He made the first cycling map of the city and now uses his website to share these maps. Polsky encourages citizens to download, print, mark their favorite routes and drop off the maps at galleries across the city. What began as a personal art project has now expanded into a movement to create a participatory, informal bicycle map for Moscow. The maps include icons for bike parking spaces, cycle rental locations and shops. It is an example of citizen engagement and activism to make Moscow more bicycle-friendly. The importance of this project lies in that findings can be shared with local city authorities as they begin to create the city's bicycle infrastructure. This kind of activism and community involvement is a starting point for effective urban design. The map received a lot of support and attention from urban residents and the media, and it became part of an alternative vision for the city, called "Moscow 2020." This is just one example of participatory community mapping, which uses art as a method to engage citizens on important urban issues.

There are also efforts to map cycling routes in China, which is important given that many Asian cities—particularly in India and China—are seeing a reverse trajectory: a shift away from walkability, pedestrianism and cycling that instead promotes and facilitates car driving.

Participatory maps provide a unique way of engaging citizens, and they are a means to promote creative, relevant solutions to problems facing our urban future. Livable and humane cities require inclusive and participatory planning.

Source: Malhotra, S. (2012) "The city fix: participatory maps for inclusive cities." http://thecityfix.com/blog/participatory-maps-for-inclusive-cities (accessed July 2012). Malhotra, S. (2011) "DIY mapping: cycle routes in Moscow." http://patterncities.com (accessed July 2012).

Another challenge for developing cities is that as incomes rise many residents are determined to catch up with the rich world in terms of automobile ownership. In 2010 Chinese and Indian consumers bought almost 20 million new passenger cars. In China, from 1977 to 2008, vehicle ownership increased from one million to 51 million.[36] In cities such as Shanghai and Beijing, car sales are increasing dramatically; estimates are that 30 percent of residents in Beijing own a car, far higher than the national average of 3 percent. The forecast is that China will see a ten-fold increase by 2035 to nearly 300 million vehicles.

As urban planners deal with greening transportation, they are challenged to remodel and regenerate existing cities and to reintegrate the diverse forms of transportation in urban life. The campaign for walkable cities offers the opportunity to reconnect urban residents in new pedestrianized inner-city areas, reduce the need for the private car and allow public transport and urban village models to prevail.

Curitiba: a green city in the developing world

Much of our discussion of cities in the developing world has stressed "brown" issues such as poor sanitation, water quality, air pollution and housing problems. However, all is not doom and gloom. Curitiba, Brazil, is located near the coastal mountain range in the southern part of the country. Like many of Brazil's cities, it developed rapidly in the second half of the twentieth century, growing from 500,000 inhabitants in 1965 to 1.7 million in 2010. Such rapid growth brought the typical urban problems of unemployment, slums, automotive gridlock, pollution and environmental deterioration. Yet despite this, Curitiba is often cited as a model "green city." In the late 1960s and 1970s, Curitiba's political elites, led by its three-term mayor, Jaime Lerner, encouraged urban planners to think imaginatively, integrating social and ecological concerns. The planning process created innovative solutions in public transportation, recycling, garbage collection and green space expansion, many of them models of sustainable urban planning.

Curitiba was faced with an inefficient public transportation system and an increase in the number of private automobiles. Initially, planners leaned toward the development of a subway system, at the cost of some $60–70 million per kilometer. Instead, they turned to modifying the bus system, at $200,000 per kilometer, or 1 percent of the cost of the subway. To meet the growing needs for transportation, and to curtail the use of private automobiles, planners focused on encouraging Curitiba's physical expansion along linear axes, each with a central road that has a dedicated lane for express buses. Some call this a "surface subway." The aim is to reduce congestion while returning the central area to the pedestrian. The use of express buses was far cheaper than subways or light railways and highlights a more practical and affordable solution to public transportation in the developing world. The city was able to keep public transportation affordable; average low-income residents spend only 10 percent of their income on transport. In addition, modifications to the buses and boarding tubes have made the system very efficient. For example, along the bus routes are clear tubular structures that are level with buses, providing quick boarding and exiting and easier access for the handicapped. The buses themselves have five lateral doors along the side and can load three times as many passengers per hour as traditional buses. Although the city has more than 500,000 private cars (more cars per capita than most Brazilian cities), most residents use the bus system. In fact, the public transportation system is used by more than 1.3 million passengers each day, nearly two-thirds of the urban population.[37] One effect is improved air quality in the city.

An innovative solution to garbage is the city's recycling program. Introduced in 1989, the recycling program encourages city residents to separate organic and inorganic garbage. A city-wide environmental education program helps to reinforce the benefits of recycling. Over 70 percent of the community participates in the separation program, one of the highest rates for any city in the world. A second program is the Purchase Garbage program, aimed primarily at the slums in Curitiba. In the slums, known locally as *favelas*, there is no organized garbage collection. To help deal with potential health problems, the program encouraged favela residents to collect garbage in return for bus tickets and groceries. This program led to a decrease in city litter while helping to feed the urban poor. At one point there were 22,000 families involved in the Purchase Garbage program, feeding approximately 100,000 people while collecting some 400 tons of trash. In 1990 Curitiba received an award from the United Nations Environment Programme for these two successful waste management programs.

Small-scale, modest changes have also had positive impacts on the city. In the 1980s, Mayor Lerner distributed 1.5 million tree seedlings to neighborhoods to plant and care for, adding to green space improvement. Once planted, a tree cannot be cut without a permit, and for every tree that is felled, two must be planted. Green space has also expanded with construction of 17 new parks, 90 miles of bike paths and trees everywhere. Many of the parks were created when planners diverted water from the lowlands (which were prone to flooding) into lakes that are now the centerpieces of the new parks. Integrating flood control with park space development is an example of "design with nature." In addition, developers receive tax breaks if their projects include green areas. As a result, the city-wide ratio of open space to inhabitant has increased from 0.5 square meters to 52 square meters, which means that Curitiba has one of the highest averages of green space per urban inhabitant anywhere in the world.[38] Today, almost 18 percent of the city's land area is green space. This translates into 18 parks, 14 forests and more than 1,000 green public spaces.

Under Lerner's administration more than 30 percent of the city's budget went to providing healthcare (more than three times other Brazilian cities).[39] The infant mortality rate dropped more than 60 percent over a 20-year period. The city has also invested substantially in improving primary education, retraining teachers and renovating public school buildings. Mayor Lerner convinced private firms to build more than 30 neighborhood daycare centers, run by the city. This partnership works because the firms' employees can use the daycare centers, a service that would otherwise be costly for companies to provide on their own.

In 2007 Curitiba was third on the list of the 15 Green Cities in the World in the American magazine *Grist*. In 2010 the city was awarded the Globe Sustainability award for its sustainable urban development. The case of Curitiba provides valuable lessons not only to other developing cities, but to large cities everywhere. Curitiba rejected conventional wisdom that emphasized technologically sophisticated solutions to urban problems. Creative and labor-intensive ideas—especially where unemployment is a problem— can substitute for capital-intensive, high-technological solutions.[40] Planners have learned that the solutions to urban woes are not specific and isolated, but interconnected. Improving public transportation or refuse collection is inextricably bound up with issues such as poverty, crime, education and public health. The lesson is crucial: cities have the capacity to transform their environmental problems into creative and innovative solutions. Moreover, as the case of Curitiba shows, the integration of social justice and social services and environmental quality is central to urban sustainable development.

Conclusions

Strategies to achieve urban sustainability include planning for environmental issues related to air quality, energy efficiency, integrated water resources management, waste stream management and others. Through the efficient use of natural resources and switching from fossil to renewable energy in its various forms, more livable and stable environmental conditions can be provided to the community and its economy in the longer term. A sustainable city is better prepared for future global environmental resource conditions.

One challenge to achieving urban sustainability is the fragmentation of urban planning. Urban planners must abandon the current compartmentalization of planning specialties (e.g., housing, transportation, land use, etc.). Instead planning efforts need to be more holistically integrated. As we explored in earlier chapters, many nineteenth-century planners such as Ebenezer Howard, Frederick Law Olmsted, Patrick Geddes and Daniel Burnham created urban designs that factored in multiple elements such as natural resources, parks, transportation systems, regional planning and social equity.

Another challenge is political. The good news is that urban sustainability is increasingly tied into issues of governance and political control.[41] Aidan While and colleagues use the term "urban sustainability fix" to describe the incorporation of ecological perspectives into city government. With reference to English cities, they show how entrepreneurial urban regimes have incorporated the green agenda, as in the case of Leeds, while others have sought to insulate themselves from dissent, as in the case of Manchester. In effect, their work shows the greening of the urban growth machine.[42]

At the same time, moving toward sustainability is not simply "greening" the city. A genuinely sustainable city is a place where the environment is protected, the economy is sufficient to provide its residents with basic needs such as food and shelter, and where the residents who live there have opportunities to live good lives. A sustainable city is also a just city.

This means sustainability is without a doubt the most ambitious revisioning of the urban–nature dialectic to date. Can cities effectively deal with growth and urban sprawl? Can cities achieve true sustainability? Are current levels and projected levels of urbanization compatible with urban sustainability? These questions are among the most dramatic and, for the time being, unanswered questions facing us in the twenty-first century. Given that cities are the principal habitats of human life, the task at hand is a daunting one but unavoidable.

We end with this charge:

> The cities of the twenty-first century are where human destiny will be played out, and where the future of the biosphere will be determined. There will be no sustainable world without sustainable cities.
>
> Herbert Girardet[43]

Guide to further reading

Adams, W. M. (2008) *Green Development: Environment and Sustainability in a Developing World.* 3rd edition. London: Routledge.

Beatley, T. (ed.) (2012) *Green Cities of Europe.* Washington, DC: Island Press.

Birch, E. and Wachter, S. M. (eds.) (2010) *Growing Greener Cities: Urban Sustainability in the Twenty-first Century*. Philadelphia, PA: University of Pennsylvania Press.

Dresner, S. (2008) *The Principles of Sustainability*. 2nd edition. New York: Routledge.

Edwards, A. and Orr, D. (2005) *The Sustainability Revolution: Portrait of a Paradigm Shift*. British Columbia: New Society Publishers.

Fitzgerald, J. (2010) *Emerald Cities: Urban Sustainability and Economic Development*. Oxford: Oxford University Press.

Heynen, N., Kaika, M. and Swyngedouw, E. (eds.) (2006) *In the Nature of Cities: Urban Political Ecology and the Politics of Urban Metabolism*. New York: Routledge.

McManus, P. (2005) *Vortex Cities to Sustainable Cities: Australia's Urban Challenge*. Sydney and Canberra: UNSW Press.

Newman, P. and Jennings, I. (2008) *Cities as Sustainable Ecosystems*. Washington, DC: Island Press.

Portney, K. E. (2003) *Taking Sustainable Cities Seriously: Economic Development, the Environment and Quality of Life in American Cities*. Cambridge MA: MIT Press.

Risse, M. (2012) *On Global Justice*. Princeton, NJ: Princeton University Press.

Sarkar, S. (2012) *Environmental Philosophy: From Theory to Practice*. Oxford: Wiley.

Postscript
Ideas for a sustainable city

1 Combat climate change by adapting existing buildings.
2 Capture the energy of people going about their day.
3 Make a small difference in a community you know.
4 Map everything.
5 Use temporary structures to offer amenities to underserved communities.
6 Design for generational diversity.
7 Look for excess capacity everywhere and put it to good use.
8 Listen to your ecosystem.
9 Support a diverse economy.
10 Use public space to share ideas and start a conversation.
11 Understand food infrastructure so that we can create a healthy, resilient city.
12 Take care of nature wherever it is found.
13 Encourage the private sector to create public amenities.
14 Harness renewable energy from the tides, not just the sun and the wind.
15 Celebrate and activate the city's ecology.
16 Consider public transport options beyond road and rail.
17 Learn from the geologic material beneath your feet.
18 Design recreation space by watching how people play.
19 Stop sewer overflow with creative, preventative design.
20 Ask citizens how they want to improve their cities.
21 Turn toxic land into agriculturally productive lots.
22 Let citizens engage in the maintenance of infrastructure.
23 Think beyond shelter to help the homeless.
24 Appreciate what you have so you consume less.
25 Interrogate assumptions about what sustainability means.

Source: adapted from Urban Omnibus (2012) "50 ideas for the new city." http://urbanomnibus.net/ideas (accessed August 2012).

Notes

Chapter 1 The city and nature

1 Klineneberg, E. (2002) *Heat Wave: A Social Autopsy of Disaster in Chicago*. Chicago, IL: University of Chicago Press.

2 Booth, W. (2010, 18 January) "Haiti's elite spared much of the devastation." *The Washington Post*, A1, A6.

3 Solecki, W. and Rosenzweig, C. (2004) "Biodiversity, biosphere reserves and the Big Apple: a study of the New York metropolitan region." *Annals of New York Academy of Science* 1023: 105–28.

4 Robbins, P. (2007) *Lawn People: How Grasses, Weeds and Chemicals Make Us Who We Are*. Philadelphia, PA: Temple University Press.

5 Yates, J. S. and Gutberlet, J. (2011) "Reclaiming and recirculating urban natures: integrated organic waste management in Diadema, Brazil." *Environment and Planning A* 43: 2109–24.

6 Douglas, I. (1981) "The city as an ecosystem." *Progress in Physical Geography* 5: 315–67.

7 Wackernage, M., Kirtzes, J., Moran, D., Goldfinger, S. and Thomas, M. (2006) "The ecological footprint of cities and regions: comparing resource availability with resource demand." *Environment and Urbanization* 18: 103–12.

8 Eaton, R. L., Hammond, G. P. and Laurie, J. (2007) "Footprints on the landscape: an environmental appraisal of urban and rural living in the developed world." *Landscape and Urban Planning* 83: 13–28.

9 http://www.footprintnetwork.org/images/uploads/Calgary_Ecological_Footprint_Report.pdf (accessed May 12, 2012).

10 Cho, M. R. (2010) "The politics of urban nature restoration: the case of Cheonggyecheon restoration in Seoul, Korea." *International Development Planning Review* 32: 145–65.

11 Panayotou, T. (2001) "Environmental sustainability and services in developing global city-regions," in *Global City-Regions*, A. J. Scott (ed.). Oxford: Oxford University Press.

12 Evans, P. (ed.) (2002) *Livable Cities: Urban Struggles for Livelihood and Sustainability*. Berkeley and Los Angeles, CA: University of California Press.

13 Haughton, G. (1999) "Environmental justice and the sustainable city." *Journal of Planning Education and Research* 18: 233–43.

14 Hough, M. (2004) *Cities and Natural Process: A Basis for Sustainability*. 2nd edition. London: Routledge.

15 Boone, C. and Moddares, A. (2006) *City and Environment*. Philadelphia, PA: Temple University Press.

16 Cronon, W. (1991) *Nature's Metropolis: Chicago and the Great West*. New York: Norton.

17 Klingle, M. (2007) *Emerald City: An Environmental History of Seattle*. New Haven: Yale University Press.

18 Gandy, M. (2002) *Concrete and Clay: Reworking Nature in New York City*. Cambridge, MA: MIT Press.

19 Gandy, M. (2008) "Landscapes of disaster: water, modernity and urban fragmentation in Mumbai." *Environment and Planning A* 40: 108–30.

20 *International Journal of Urban and Sustainable Development* 2011 Special Issue, Volume 3, Urban Water Poverty.

21 Swyngedouw, E. (2004) *Social Power and the Urbanization of Water*. Oxford: Oxford University Press.

22 Cadenasso, M. L., Pickett, S. T. A. and Grove, M. J. (2006) "Integrative approaches to investigating human–natural systems: the Baltimore ecosystem study." *Natures Sciences Societes* 14: 4–14.

23 Keys, E., Wentz, E. A. and Redman, C. L. (2007) "The spatial structure of land use from 1970–2000 in the Phoenix, Arizona Metropolitan Area." *The Professional Geographer* 59: 131–47.

24 Paul, M. J. and Meyer, J. L. (2008) "Streams in the urban landscape." *Urban Ecology* III: 207–31.

25 The Baltimore Ecosystem Study at http://www.beslter.org and the Central Arizona-Phoenix Study at http://caplter.asu.edu.

Chapter 2 The pre-industrial city

1 Mumford, L. (1989) *The City in History*. San Diego, CA and New York: Harvest Books, p. 71.

2 Pfeiffer, J. (1980) "The mysterious rise and decline of Monte Albán," *Smithsonian*, February.

3 Hardoy, J. (1973) *Pre-Columbian Cities*. Judith Thorne (trans.). New York: Walker, p. 33.

4 Mumford, *The City in History*, p. 75.

5 Mumford, *The City in History*, p. 75.

6 Markham, A. (1994) *A Brief History of Pollution*. New York: St. Martin's Press.

7 It should be noted that not all cities built walls; some fortified cities only when political leaders felt they were in threatened frontiers.

8 Lazzaro, C. (1990) *Italian Renaissance Garden: From the Conventions of Planting, Design, and Ornament to the Grand Gardens of Sixteenth-Century Central Italy*. New Haven, CT: Yale University Press.

9 Merchant, C. (1980) *The Death of Nature: Women, Ecology and the Scientific Revolution*. London: Wildwood House.

10 Merchant, *The Death of Nature*, p. xviii.

11 Miller, I. (2002) *Washington in Maps 1606–2000*. New York: Rizzoli International Publications, p. 20.

12 Miller, *Washington in Maps*, p. 18.

13 Bowling, K. (1991) *The Creation of Washington, DC: The Idea and Location of the American Capital*. Fairfax, VA: George Mason University Press, p. 224.

14 Scott, P. (2002) "This vast empire: the iconography of the Mall, 1791–1848" in *The Mall in Washington, 1791–1991*, R. Longstreth (ed.). New Haven, CT: Yale University Press, p. 39.

15 Smith, H. (1967) *Washington, D.C: The Story of Our Nation's Capital*. New York: Random House, p. 13

16 Scott, P., "This vast empire," p. 43.

17 Karlen, A. (1995) *Man and Microbes: Disease and Plagues in History and Modern Times*. New York: G.P. Putnam's Sons, p. 52.

18 Karlen, *Man and Microbes*.

19 Quoted in Sjoberg, G. (1960) *The Preindustrial City*. New York: The Free Press, p. 93.

20 Littman, Robert J. (2009) "The plague of Athens: epidemiology and paleopathology." *Mount Sinai Journal of Medicine*, 76 (5): 456–67.

21 Boccaccio, G. (1972) *The Decameron*. G.H. McWilliam (trans.). New York: The Penguin Classics.

22 Bell, G. W. (1994) *The Great Plague of London*. London: Bracken Books.

23 Swanson, M. (1995) "The sanitation syndrome: bubonic plague and urban native policy in the Cape Colony, 1900–09," *Segregation and Apartheid in Twentieth-Century South Africa*, W. Beinart and Saul Dubow (eds.). New York: Routledge, pp. 25–42.

24 Swanson, "The sanitation syndrome," p. 26.

25 McFarlane, C. (2008) "Governing the contaminated city: infrastructure and sanitation in colonial and postcolonial Bombay," *International Journal of Urban and Regional Research* 32 (2): 415–35.

26 Powell, F. and Harding, A. (2011) "The renaissance of inner city living and its implications for infrastructure: a Wellington case study." Working paper. http://nzsses.auckland.ac.nz/conference/2010/papers/Powell-Harding.pdf (accessed February 2012).

27 Karlen, *Man and Microbes*, p. 62.

28 Borsos, E., Makra, L., Bécz, R., Vitányi, B. and Szentpéteri, M. (2003) "Anthropogenic air pollution in the ancient times." *Acta Climatologica et Chorologica* 36–37, 5–15.

29 Herlihy, D. (1970) *The History of Feudalism*, New York: Harper and Row, pp. 270–1.

30 Quoted in McKay, J., Hill, B. and Buckler, J. (1995) *A History of Western Society*. 5th edition. Boston, MA: Houghton Mifflin, p. 347.

Chapter 3 The industrial city

1 Engels, F. (1973) *The Condition of the Working Class in England*. Moscow: Progress Publishers, pp. 86–89, 92.

2 Mumford, M. (1989, first published 1961) *The City in History*. San Diego, CA: Harvest Books, p. 446.

3 Melosi, M. (1980) "Environmental crisis in the city: the relationship between industrialization and urban pollution," in *Pollution and Reform in American Cities, 1870–1930*, Martin Melosi (ed.). Austin, TX: University of Texas Press, p. 6.

4 Karlen, A. (1995) *Man and Microbes: Disease and Plagues in History and Modern Times*. New York: G.P. Putnam's Sons, p. 51.

5 Dennis, R. (1994) *English Industrial Cities of the 19th Century*. Cambridge: Cambridge University Press; Mumford, *The City in History*, pp. 467–8.

6 Melosi, "Environmental crisis in the city," p. 7.

7 Markham, A. (1994) *A Brief History of Pollution*. New York: St. Martin's Press, p. 20.

8 As quoted in Markham, *A Brief History of Pollution*, p. 16.

9 Markham, *A Brief History of Pollution*, p. 16.

10 Tarr, J. (1996) *The Search for the Ultimate Sink: Urban Pollution in Historical Perspective*. Akron, OH: University of Akron Press, p. 8–9.

11 Melosi, "Environmental crisis in the city," p 14.

12 Tarr, *The Search for the Ultimate Sink*, p. 324.

13 Tarr, *The Search for the Ultimate Sink*, p. 326.

14 Tarr, *The Search for the Ultimate Sink*, p. 14.

15 Tarr, *The Search for the Ultimate Sink*, pp. 14–15.

16 Markham, *A Brief History of Pollution*.

17 Robson, B. T. (1996) "The saviour city: beneficial effects of urbanization in England and Wales," in *Companion Encyclopedia of Geography*, I. Douglas, R. Hussert and M. Robinson (eds.). New York: Routledge, pp. 300–1.

18 Davison, G. (2008) "Down the gurgler: historic influences on Australian domestic water consumption," in *Troubled Waters: Confronting the Water Crisis in Australian Cities*, P. Troy (ed.). Canberra: Australian National University Press, pp. 37–65.

19 Galishoff, S. (1980) "Triumph and failure: the American Response to the urban water supply problem, 1860–1923," in *Pollution and Reform in American Cities, 1870–1930*, Martin V. Melosi (ed.). Austin, TX: University of Texas Press, pp. 35–58.

20 Boyer, M. C. (1986) *Dreaming the Rational City: The Myth of American City Planning*. Cambridge, MA: MIT Press, p. 17.

21 Galishoff "Triumph and failure," p. 36.

22 Galishoff "Triumph and failure," p. 48.

23 Kaika, M. (2006) "'Dams as symbols of modernization': the urbanization of nature between geographical imagination and materiality." *Annals of the Association of American Geographers* 96 (2): 276–301.

24 Davison "Down the gurgler," p. 40.

25 Tarr, *The Search for the Ultimate Sink*, pp. 344–5.

26 Tarr, *The Search for the Ultimate Sink*, pp. 344–5.

27 Tarr, *The Search for the Ultimate Sink*, pp. 12–13.

28 Melosi, "Environmental crisis in the city," p. 107.

29 Melosi, "Environmental crisis in the city," p. 113.

30 Grinder, R. D. (1980) "The battle for clean air: the smoke problem in post-Civil War America," in *Pollution and Reform in American Cities, 1870–1930*, Martin V. Melosi (ed.). Austin, TX: University of Texas Press.

31 Tarr, *The Search for the Ultimate Sink*, p. 16.

32 In Europe, royal parks and gardens provided only limited pubic access, usually on holidays, but there were exceptions. The Tiergarten in Berlin was opened to the public for "pleasure strolling" in 1649 and there were other public promenades.

33 Schulyer, D. (1986) *The New Urban Landscape: The Redefinition of City Form in Nineteenth Century America*. Baltimore, MD: Johns Hopkins University Press, p. 59.

34 This contrasts with existing urban parks in many European cities. There, existing gardens and parks were molded by earlier historical conditions and carried the imprint of different class–cultural relations. Hyde Park in London was open to the public (for a fee) in 1652 and in the1830s the palace and gardens of Luxembourg and Versailles became public.

35 Peterson, J. (1996) "Frederick Law Olmsted Sr. and Frederick Law Olmsted Jr.: the visionary and the professional," in *Planning the Twentieth Century American City*, Mary Corbin Sies and Christopher Silver (eds.). Baltimore, MD: John Hopkins University Press, pp. 37–54.

36 Gopnik, A. (1997, March 31) "Olmsted's trip," *New Yorker Magazine*, pp. 101.

37 Quoted in Cranz, G. (1982) *The Politics of Park Design: A History of Urban Parks in America*. Cambridge, MA: MIT Press, p. 29.

38 Cranz, *The Politics of Park Design*, p. 34.

39 Olmsted, F.L. (1996) "Public parks and the enlargement of towns," in *The City Reader*, Richard T. LeGates and Frederic Stout (eds.). London and New York: Routledge, p. 339.

40 Olmsted "Public parks and the enlargement of towns," p. 343.

41 Cranz, *The Politics of Park Design*, p. 41.

42 Schulyer, *The New Urban Landscape*, pp. 64–5.

43 Schenker, H. (2009) *Melodramatic Landscapes: Urban Parks in the Nineteenth Century*. Charlottesville, VA: University of Virginia Press.

44 Schulyer, *The New Urban Landscape*, p. 78.

45 Fishman, R. (2003) "Urban utopias: Ebenezer Howard, Frank Lloyd Wright and Le Corbusier," in *Readings in Planning Theory, Second Edition*, Scott Campbell and Susan Fainstein (eds.). Oxford: Blackwell, p. 22.

46 Howard, E. (1996) "Author's introduction from garden cities of to-morrow," in *The City Reader*, Richard T. LeGates and Frederic Stout (eds.). London and New York: Routledge, pp. 346–53.

47 Freestone, R. (1989) *Model Communities: The Garden City Movement in Australia*. Melbourne: Melbourne University Press.

48 Tuomi, T. (2003) *Tapiola: Life and Architecture*. Helsinki: Rakennustieto.

Chapter 4 Global urban trends

1 This paragraph draws heavily from the Introduction in Hall, T., Hubbard, P. and Short, J. R. (eds.) (2008) *The Urban Compendium*. London: Sage.

2 Kim, Y.-H. and Short, J. R. (2008) *Cities and Economies*. London: Routledge.

3 Au, C. and Henderson, J. (2006) "Are Chinese cities too small?" *Review of Economic Studies* 73: 549–70.

4 Yankson, P. W. K. and Gough, K. V. (1999) "The environmental impact of rapid urbanization in the peri-urban area of Accra, Ghana." *Geografisk Tidsskrift* 99: 89–100.

5 Deng, J. S., Wang, K., Hong, Y. and Qi, J. G. (2009) "Spatio-temporal dynamics and evolution of land use change and landscape pattern in response to rapid urbanization." *Landscape and Urban Planning* 92: 187–98; Su, S., Jiang, Z., Zhang, Q. and Zhang, Y. (2011) "Transformation of agricultural landscapes after rapid urbanization: a threat to sustainability in Hang-Jia-Hu region, China." *Applied Geography* 31: 439–49.

6 Li, Y., Zhu, X. Sun, X. and Feng Wang, F. (2010) "Landscape effects of environmental impact on bay-area wetlands under rapid urban expansion and development policy: a case study of Lianyungang, China." *Landscape and Urban Planning* 94: 218–27.

7 Shrestha, M., York, A. M., Boone, C. G. and Zhang, S. (2012) "Land fragmentation due to rapid urbanization in the Phoenix metropolitan area: analyzing the spatiotemporal patterns and drivers." *Applied Geography* 32: 522–31; Scolozzi, R. and Geneletti, D. (2012) "A multi-scale qualitative approach to assess the impact of urbanization on natural habitats and their connectivity." *Environmental Impact Assessment Review* 36: 9–22.

8 Huang, S.-L., Yeh, C.-T. and Chang, L.-F. (2010) "The transition to an urbanizing world and the demand for natural resources." *Current Opinion in Environmental Sustainability* 2: 136–43.

9 Paul, M. J. and Meyer, J. L. (2008) "Streams in the urban landscape." *Urban Ecology* III: 207–31.

10 Bulkeley, H. (2012) *Cities and Climate Change*. London: Routledge.

11 Martinez-Zarzoso, I. and Maruotti, A. (2011) "The impact of urbanization on CO_2 emissions: evidence from developing countries." *Ecological Economics* 70: 1344–54.

12 Katz, R., Mookherji, S., Kaminski, M., Haté, V. and Fischer J. E. (2102) "Urban governance of disease." *Administrative Sciences* 2:135–47.

13 Lui, L. (2008) "Sustainability efforts in China: reflections on the environmental Kuznets Curve through a locational evaluation of 'eco-communities,'" *Annals of Association of American Geographers* 98: 604–29.

14 Haiqing, L. (2003) "Management of coastal mega-cities: a new challenge in the 21st century." *Marine Policy* 27: 333–7.

15 Cave, D. (2012, 10 April) "Lush walls rise to fight a blanket of pollution." *New York Times*, A4.

16 Aguilar, A. G. (2008) "Peri-urbanization, illegal settlements and environmental impact in Mexico City." *Cities* 25:133–45.

17 The Blacksmith Institute (2011) *The World's Worst Polluted Places*. New York: The Blacksmith Institute.

18 Khanna, P., Jain, S., Sharma, P. and Mishra, S. (2010) "Impact of increasing mass transit share on energy use and emissions from transport sector for National Capital Territory of Delhi." *Transportation Research Part D: Transport and Environment* 16: 65–72.

19 Onodera, S. (2011) "Subsurface pollution in Asian Megacities," in *Groundwater and Subsurface Environments: Human Impacts in Asian Coastal Cities*, M. Tanigucho (ed.). Dordrecht: Springer, pp. 159–85.

20 Short, J. R. (2004) *Global Metropolitan*. London: Routledge.

21 Taylor, P. (2004) *World City Networks*. London: Routledge.

22 Nelson, A. C. and Lang, R. E. (2011) *Megapolitan America*. London: Routledge.

23 This section draws heavily on Short, J. R. (2007) *Liquid City: Megalopolis and the Contemporary Northeast*. Washington, DC: Resources for the Future.

24 Salvati, L., Munafoc, M., Morelli, V. G. and Sabbi, A. (2012) "Low-density settlements and land use changes in a Mediterranean urban region." *Landscape and Urban Planning* 105: 43–52.

25 Chorianopoulos, I., Pagonis, T., Koukoulas, S. and Drymoniti, S. (2010) "Planning, competitiveness and sprawl in the Mediterranean city: the case of Athens." *Cities* 27: 249–59.

26 Paul, V. and Tonts, M. (2005) "Containing urban sprawl: trends in land use and spatial planning in the metropolitan region of Barcelona." *Journal of Environmental Planning and Management* 48: 7–35.

27 Kucukmehmetoglu, M. and Geymen, A. (2009) "Urban sprawl factors in the surface water resource basins of Istanbul." *Land Use Policy* 26: 569–79.

28 European Environmental Agency (2006) *Urban Sprawl in Europe: The Ignored Challenge*. Copenhagen: European Environmental Agency.

29 Schneider, A. and Woodcock, C. E. (2008) "Compact, dispersed, fragmented, extensive? A comparison of urban growth in twenty five global cities using remotely sensed data, pattern metrics and census information." *Urban Studies* 45: 659–92.

30 Kunstler, H. J. (1993) *The Geography of Nowhere*. New York: Touchstone; Duany, A., Plater-Zyberk, E. and Speck, J. (2000) *Suburban Nation: The Rise of Sprawl and the Decline of the American Dream*. New York: North Point Press; Putnam, R. D. (2000) *Bowling Alone: The Collapse and Revival of American Community*. New York: Simon and Schuster.

31 Bruegmann, R. (2005) *Sprawl: A Compact History*. Chicago, IL: University of Chicago Press.

32 Galster, G., Hanson, R., Ratcliffe, M., Wolman, H., Xoleman, S. and Freihage, J. (2001) "Wrestling sprawl to the ground: defining and measuring an elusive concept." *Housing Policy Debate* 12: 681–718.

33 Soule, D. (ed.) (2006) *Urban Sprawl: A Comprehensive Reference Guide*. Westport, CT: Greenwood Press, pp. xv.

34 Wolman, H., Galster, G., Hanson, R., Ratcliffe, M., Furdell, K. and Sarazynski, A. (2005) "The fundamental challenge in measuring sprawl: which land should be considered." *Professional Geographer* 57: 94–105; Lang, R. E. (2003) "Open bounded places: does the American West's arid landscape yield dense metropolitan growth?" *Housing Policy Debate* 13: 755–78.

35 Frumkin, H., Frank, L. and Jackson, R. (2004) *Urban Sprawl and Public Health*. Washington, DC: Island Press.

36 National Research Council (2008) *Urban Storm Water Management in the United States*. Washington, DC: National Academies Press.

37 Kahn, M. E. (2000) "The environmental impact of suburbanization." *Journal of Policy Analysis and Management* 19: 569–86.

38 Volstad, J. H., Roth, N. E., Mercurio, G., Southerland, M. T. and Strebel, D. E. (2003) "Using environmental stressor information to predict the ecological status of Maryland non-tidal streams as measured by biological indicators." *Environmental Monitoring and Assessment* 84: 219–42.

39 Pearce, J. and Witten, K. (eds.) (2010) *Geographies of Obesity: Environmental Understandings of the Obesity Epidemic*. Burlington, VT: Ashgate.

40 Glaeser, E. L. (2011) *Triumph of the City*. New York: Penguin.

Chapter 5 The postindustrial city

1 Kerr, G., Noble, G. and Glynn, J. (2011) "The city branding of Wollongong," in *City Branding: Theory and Cases*, K. Dinnie (ed.). New York: Palgrave, pp. 213–20.

2 Gordon, D. (1997) "Managing the changing political environment in urban waterfront development." *Urban Studies* 34 (1): 61–83.

3 US EPA "Superfund." http://www.epa.gov/superfund/sites/npl (accessed August 7, 2012).

4 US EPA "Brownfields basic information." http://www.epa.gov/brownfields/basic_info.htm (accessed August 7, 2012).

5 Alberini, A., Heberle, L., Meyer, P. and Wernstedt, K. (2004) "The brownfields phenomenon: much ado about something or the timing of the shrewd?" Working Paper, Center for Environmental Policy and Management, University of Louisville.

6 Howland, M. (2003) "Private initiative and public responsibility for the redevelopment of industrial brownfields: three Baltimore case studies." *Economic Development Quarterly* 17: 367–81.

7 Oliver, L., Ferber, U., Grimski, D., Millar, K. and Nathanall, P. (2005) "The scale and nature of European brownfields." http://www.cabernet.org.uk/resourcefs/417.pdf (accessed August 7, 2012).

8 Garb, Y. and Jackson, J. (2010) "Brownfields in the Czech Republic 1989–2009: the long path to integrated land management." *Journal of Urban Regeneration and Renewal* 3: 263–76.

9 Deutz, P. (2004) "Eco-industrial development and economic development: industrial ecology or place promotion." *Business Strategy and the Environment* 13: 347–62.

10 De Sousa, C. (2003) "Turning brownfields into green space in the city of Toronto." *Landscape and Urban Planning* 62: 181–98.

11 De Sousa, "Turning brownfields into green space in the city of Toronto."

12 De Sousa, C., Wu, C. and Westphal, L. M. (2009) "Assessing the effect of publicly assisted brownfield development on surrounding property values." *Economic Development Quarterly* 23: 95–110.

13 Lee, S. and Mohal, P. (2012) "Environmental justice implications of brownfield redevelopment in the United States." *Society and Natural Resources* 25: 602–9.

14 Lee, A. G., Baldock, O. and Lamble, J. (2009) "Remediation or problem translocation: an ethical discussion as to the sustainability of the remediation market and carbon calculating." *Environmental Claims Journal* 21: 232–46.

15 Newton, P. W. (2010) "Beyond greenfield and brownfields: the challenge of regenerating Australia's greyfield suburbs." *Built Environment* 36: 81–104.

Chapter 6 The developing city

1 Aluko, O. (2012) "Environmental degradation and the lingering threat of refuse and pollution in Lagos state." *Journal of Management and Sustainability* 2: 217–26.

2 Central Pollution Control Board (2010) "Study of urban air quality in Kolkata for source identification and estimation of ozone, carbonyls, NOx and VOC emissions." http://cpcb.nic.in/upload/NewItems/NewItem_160_cups.pdf (accessed July 2012)

3 Gurjar, B., Jain, A., Sharma, A., Agarrwal, A., Gupta, P., Nagpure, A. and Lelieveld, J. (2010) "Human health risks in megacities due to air pollution." *Atmospheric Environment* 36: 4606–13.

4 Aluko, O. (2012) "Environmental degradation and the lingering threat of refuse and pollution in Lagos state."

5 Fashae, O. A. and Onafeso, O. A. (2011) "Impact of climate change on sea level rise in Lagos, Nigeria." *International Journal of Remote Sensing* 32: 9811–19.

6 Sherbinin, A. de, Schiller, A. and Pulsipher, A. (2007) "The vulnerability of global cities to climate hazards." *Environment and Urbanization* 19: 39–64.

7 Peterson, M. J. (2009) "Bhopal plant disaster: situation summary." International Dimensions of Ethics Education in Science and Engineering Case Study Series, University of Massachusetts, Amherst.

8 *Time* (2011) "The ten most polluted cities in the world." http://www.time.com/time/specials/packages/completelist/0,29569,1661031,00.html (accessed May 15, 2012).

9 Vennemo, H., Aunan, K., Lindhjem, H. and Seip, H. M. (2009). "Environmental pollution in China: status and trends." *Review of Environmental Economics and Policy* 3 (2): 209.

10 You can follow the progress of the new city at http://www.tianjinecocity.gov.sg (accessed May 15, 2012).

11 Castells, M. and Hall, P. G. (1994) *Technopoles of the World*. London: Routledge.

12 Daniels, P. W., Ho, K. C. and Hutton, T. A. (2012) *New Economic Spaces in Asian Cities*. New York: Routledge.

13 Nam, V. H., Sonobe, T. and Otsuka, K. (2009) "An inquiry into the transformational process of village-based industrial clusters: the case of an iron and steel cluster in northern Vietnam." *Journal of Comparative Economics* 37: 568–81.

14 http://www.aecom.com/deployedfiles/Internet/Careers/Student%20 Connections/Urban%20SOS/SOS_Urbanriver.pdf (accessed July 20, 2012).

15 Piercy, E., Granger, R. and Goodier, C. I. (2010) "Planning for peak oil: learning from Cuba's 'special period.'" *Proceedings of Institute of Civil and Building Engineering: Urban Design and Planning* 163: 169–76.

16 Klink, J. and Denaldi, R. (2012) "Metropolitan fragmentation and neo-localism in the periphery: revisiting the case of Curitiba." *Urban Studies* 49: 543–61.

17 There are numerous names for slum settlements or squatter settlements. They are called "ranchos" in Venezuela, "pueblo joven" in Peru, "favelas" in Brazil, "Barong-barongs" in the Philippines and "Kevettits" in Burma. The United Nations uses the term "slum," so have we.

18 UN Human Settlements Program (2003) *The Challenge of Slums*. London: Earthscan.

19 United Nations Centre for Human Settlements (2001) *Cities in a Globalizing World: Global Report on Human Settlements, 2001*. London and Sterling, VA: Earthscan, p. xxvi.

20 Hackenbroch, K. and Hossain, S. (2012) "The organized encroachment of the powerful: everyday practices of public space water supply in Dhaka, Bangladesh." *Planning Theory and Practice* 13 (3): DOI:10.1080/14649357. 2012.694265

21 Hardoy, J., Mitlin, D. and Satterthwaite, D. (2001) *Environmental Problems in Third World Cities*. London: Earthscan, p. 72.

22 Gruebner, O., Khan, A., Lautenbach, S., Muller, D., Kramer, A., Lakes, T. and Hostert, P. (2012) "Mental health in the slums of Dhaka: a geoepidemiological study." *BMC Public Health* 12:177.

Chapter 7 Urban sites

1 Solnit, R. (2010) *Infinite City: A San Francisco Atlas*. Berkeley and Los Angeles, CA: University of California Press.

2 Bettencourt, L. and West, G. (2010) "A unified theory of urban living." *Nature* 467: 912–13.

3 Colten, C. (2004) *An Unnatural Metropolis; Wresting New Orleans from Nature.* Baton Rouge, LA: LSU Press; Lewis, P. F. (2003) *New Orleans: The Making of an Urban Landscape.* 2nd edition. Sante Fe, NM: Center for American Places.

4 Banham, R. (1971) *Los Angeles.* Harmondsworth: Penguin.

5 Harvey, D. (2003) *Paris: Capital of Modernity.* London: Routledge.

6 Appadurai, A. (2000) "Spectral housing and urban cleansing: notes on millennial Mumbai." *Public Culture* 12: 627–51. See also Mehta, S. (2005) *Maximum City: Bombay Lost and Found.* New York: Vintage.

7 Graham, S. and Hewitt, L. (2012, April 25) "Getting off the ground: on the politics of urban verticality." *Progress in Human Geography.* doi: 10.1177/0309132512443147.

8 Meyer, W. B. (1994) "Bringing hypsography back in: altitude and residence in American cities." *Urban Geography* 15: 505–13.

9 Patz J. A, Martens, W. J, Focks, D. A. and Jetten, T. H. (1998) "Dengue fever epidemic potential as projected by general circulation models of global climate change." *Environmental Health Perspective* 106: 147–53.

10 Cusack, T. (2010) *Riverscapes and National Identities.* Syracuse, NY: Syracuse University Press.

11 Gumprecht, B. (1999) *The Los Angeles River: The Life, Death and Possible Rebirth.* Baltimore, MD and London: Johns Hopkins University Press.

Chapter 8 Hazards and disasters

1 Wisner, B., Blaikie, P., Cannon, T. and Davis, I. (2004) *At Risk: Natural Hazards, People's Vulnerability and Disasters.* London: Routledge.

2 Hartman, C. and Squires, G. (eds.) (2006) *There is No Such Thing as a Natural Disaster.* New York: Routledge.

3 Eakin, H. and Luers, A. L. (2006) "Assessing the vulnerability of social-environmental systems." *Annual Review of Environment and Resources* 31: 365–94; see also Cutter, S. L., Baruff, B. T. and Shirley, W. L. (2003) "Social vulnerability to environmental hazards." *Social Science Quarterly* 84: 242–61.

4 Chafe, Z. (2007) "Reducing natural disaster risk in cities," in *State of the World: Our Urban Future.* New York: Norton, pp. 112–29; see also, Bull-Kamanga, L., Diagne, K., Lavell, A., Leon, E., Lerise, F., MacGregor, H., Maskrey, A., Meshack, M., Pelling, M., Reid, H., Satterthwaite, D., Songsore, J., Westgate, K.

and Yitambe, A. (2003) "From everyday hazards to disasters: the accumulation of risk in urban areas." *Environment and Urbanization* 15: 193–203.

5 According to the Centre for Research on the Epidemiology of Disasters (CRED): http://www.cred.be.

6 Davis, M. (1998) *The Ecology of Fear*. New York: Holt.

7 Pelling, M. (2003) *The Vulnerability of Cities: Natural Disaster and Social Resilience*. New York and London: Routledge.

8 Tinniswood, A. (2003) *By Permission of Heaven: The True Story of the Great Fire of London*. London: Jonathan Cape.

9 Bell, G. W. (1994) *The Great Plague of London*. London: Bracken Books.

10 Sawislak, K. (1995) *Smoldering City: Chicagoans and the Great Fire, 1871–1874*. Chicago, IL: University of Chicago Press.

11 Maps of the fires can be found at http://map.sdsu.edu (accessed June 13, 2012).

12 Rintoul, S. (2010, February 15) "Ban development in fire-prone areas, experts tell royal commission." *The Australian*.

13 Otieno, Carren (2010) "Fire hazard in Kenyan slums." Slum stories, Amnesty International.   http://www.slumstories.org/episode/fire-hazard-kenyan-slums (accessed March 16, 2012).

14 Seng. L. K. (2008) "Fires and the social politics of nation-building in Singapore." Asia Research Center. Working Paper 149, Murdoch University, p. ii.

15 Barry, J. (1997) *Rising Tide: The Great Mississippi Flood of 1927 and How it Changed America*. New York: Simon and Schuster.

16 Kim, K.-G. (1999) "Flood hazard in Seoul: a preliminary assessment," *Crucibles of Hazards: Mega-cities and Disasters in Transition*, J. K. Mitchell (ed.). Shibuya-ku: United Nations University Press.

17 Collins, T. W. (2010) "Marginalization, facilitation, and the production of unequal risk: the 2006 Paso del Norte floods." *Antipode* 42: 258–88.

18 The World Bank (2010, November 11) "Project appraisal document on a proposed loan in the amount of US$250 million to the Corporación Autónoma Regional de Cundinamarca with a guarantee from the Republic of Colombia for a Río Bogotá Environmental Recuperation and Flood Control Project." http://www-wds.worldbank.org/external/default/WDSContentServer/WDSP/IB/2010/12/01/000333038_20101201222814/Rendered/PDF/543110PAD0REPL10BOX353792B01public1.pdf (accessed July 20, 2012); see also Corporación Autónoma Regional de Cundinamarca (2006) "Plan de Ordenación y Manejo de la Cuenca Hidrográfica del Río Bogotá: Resumen Ejecutivo." http://www.car.gov.co/?idcategoria=1375# (accessed June 13, 2012).

19 http://www.1906eqconf.org/mediadocs/BigonestrikesReport.pdf (accessed July 2006).

20 Davis, D. (2005) "Reverberations: Mexico City's 1965 earthquake and the transformation of the capital," in *The Resilient City: How Modern Cities Recover From Disasters*, L. J. Vale and T. J. Campanella (eds.). New York: Oxford University Press.

21 Davis, D. (2005) "Reverberations: Mexico City's 1965 earthquake and the transformation of the capital," in *The Resilient City: How Modern Cities Recover From Disasters*, L. J. Vale and T. J. Campanella (eds.). New York: Oxford University Press, p. 276.

22 Chen, B. (2005) "Resist the earthquake and rescue ourselves: the reconstruction of Tangshan after the 1976 earthquake," in *The Resilient City: How Modern Cities Recover From Disasters*, L. J. Vale and T. J. Campanella (eds.). New York: Oxford University Press, p. 251.

23 Oxfam (2012) *Haiti: The Slow Road to Reconstruction.* http://www.oxfamamerica.org/press/publications/haiti-the-slow-road-to-reconstruction (accessed April 15, 2012).

24 Select Bipartisan Committee to Investigate the Preparation for and Response to Hurricane Katrina (2006) *A Failure of Initiative: Final Repost of the Select Bipartisan Committee to Investigate the Preparation For and Response to Hurricane Katrina.* Washington, DC: US Government Printing Office. http://www.gpoacess.gov/congress/index/html (accessed April 15, 2012).

25 Dodson, J. and Sipe, N. (2008) *Shocking the Suburbs: Oil Vulnerability in the Australian City.* Sydney: UNSW Press.

26 Graham, S. (2011) *Cities Under Siege: The New Military Urbanism.* London: Verso.

27 Vale, L. J. and Campanella, T. J. (eds.) (2005) *The Resilient City: How Modern Cities Recover From Disasters.* New York: Oxford University Press.

28 Fallahi, A. (2006) "Mobilization of local and regional capabilities through community participation in the process of recovery after the Bam earthquake in Iran." Paper presented to the International Geographical Union Conference, Brisbane, Australia, July 5.

29 Horwich, G. (2000) "Economic lessons of the Kobe earthquake." *Economic Development and Cultural Change* 48: 521–42.

30 Skidmore, M. and Toya, H. (2002) "Do natural disasters promote long-run growth?" *Economic Inquiry* 40: 664–87.

31 United Nations Environment Program (2007) "Environment and reconstruction in Aceh: two years after the tsunami." http://postconflict.unep.ch/publications/dmb_aceh.pdf (accessed July 2012).

32 Asian Development Bank (2009) *Indonesia: Aceh-Nias Rehabilitation and Reconstruction*, pp. 21–5). http://www.adb.org/Documents/Produced-Under-TA/39127/39127-01-INO-DPTA.pdf (accessed July 2012).

33 McLennan, B. and Handmer, J. (2012) "Reframing responsibility-sharing for bushfire risk management in Australia after Black Saturday." *Environmental Hazards* 11: 1–15.

Chapter 9 Urban political ecology

1 Cadenasso, M., Pickett, S. T. A. and Grove, M. J. (2006) "Integrative approaches to investigating human–natural systems: the Baltimore ecosystem study." *Natures Sciences Sociétés* 14: 4–14.

2 Shrestha, M. K., York, A. M., Boone, C. G. and Zhang, S. (2012) "Land fragmentation due to rapid urbanization in the Phoenix Metropolitan Area: analyzing the spatiotemporal pattern and drivers." *Applied Geography* 32: 522–31.

3 Georgi, N. J. and Zafiriadis, K. (2006) "The impact of park trees on microclimates in urban areas." *Urban Ecosystems* 10: 195–209.

4 Blaikie, P. M. (1985) *The Political Economy of Soil Erosion in Developing Countries*. London: Pearson.

5 Elmhirst, R. (2011) "Introducing new feminist political ecologies." *Geoforum* 42: 129–32; Perkins, H. (2007) "Ecologies of actor-networks and (non) social labor within the urban political ecologies of nature." *Geoforum* 38: 1152–62.

6 Truelove, Y. (2011) "(Re-)conceptualizing water inequality in Delhi, India through a feminist political ecology framework." *Geoforum* 42: 143–52.

7 Wolman, A. (1965) "The metabolism of cities." *Scientific American* 213: 179–90.

8 Kennedy, C., Cuddihy, J. and Engel-Yan, J. (2007) "The changing metabolism of cities." *Journal of Industrial Ecology* 11: 43–59.

9 Gandy, M. (2008) "Landscapes of disaster: water, modernity and urban fragmentation in Mumbai." *Environment and Planning A* 40: 108–30.

10 Swyngedouw, E. (2004) *Social Power and the Urbanization of Water*. Oxford: Oxford University Press.

11 Ioris, A. A. R. (2011) "The geography of multiple scarcities: urban development and water problems in Lima, Peru." *Geoforum* 43: 612–22.

12 Ioris, A. A. R. (2012) "The neoliberalization of water in Lima, Peru." *Political Geography* 31: 266–78.

13 Cooke, J. and Lewis, R. (2010) "The nature of circulation: the urban political ecology of Chicago's Michigan Avenue Bridge, 1909–1930." *Urban Geography* 31: 348–68.

14 Graham, S. (ed.) (2010) *Disrupted Cities: When Infrastructure Fails*. New York and London: Routledge.

15 Marvin, S. and Medd, W. (2010) "Clogged cities: sclerotic infrastructure," in *Disrupted Cities: When Infrastructure Fails*, S. Graham (ed.) New York and London: Routledge, pp. 85–96.

16 Monstadt, J. (2009) "Conceptualizing the political ecology of urban infrastructures: insights from technology and urban studies," *Environment and Planning A* 41: 1924–42.

17 Wrigley, N. (2002) "Food deserts in British cities: policy context and research practice," *Urban Studies* 39: 2029–40.

18 Yates, J. S. and Gutberlet, J. (2011) "Reclaiming and recirculating urban natures: integrated organic waste management in Diadema, Brazil." *Environment and Planning A* 43: 2109–24; quote is p. 2120.

19 Gene Desfor, G. and Lucian Vesalon, L. (2008) "Urban expansion and industrial nature: a political ecology of Toronto's Port Industrial District." *International Journal of Urban and Regional Research* 32: 586–603.

20 Ley, D. (2012) "Waterfront development: global processes and local contingencies in Vancouver's False Creek," in *New Urbanism: Life, Work and Space in the New Downtown*, I. Helbrecht and P. Dirksmeir (eds.). Farnham and Burlington, VA: Ashgate, pp. 47–60.

21 The official website for the Highline is http://www.thehighline.org.

22 Forero, J. (2011, October 16) "Hit the highway, São Paulo is told." *The Washington Post*, A12.

23 Kitzes, J. and Wackernagel, M. (2009) "Answers to common questions in ecological footprint accounting." *Ecological Indicators* 9: 812–17; see the websites: http://www.footprintnetwork.org/en/index.php/GFN; http://www.footprintnetwork.org/en/index.php/GFN/page/footprint_for_cities.

24 Geis, S. (2006, August 15) "Northern Nevadans don't want to gamble with their water." *The Washington Post*, A3.

25 Sovocool, B. K. and Brown, M. A. (2010) "Twelve metropolitan carbon footprints: a preliminary comparative global assessment." *Energy Policy* 38: 4856–69.

26 Brown, M. A., Southworth, F. and Sarzynski, A. (2009) "The geography of metropolitan carbon footprints." *Policy and Society* 27: 285–304.

27 Novotny, V. (2010) "Urban water use and energy use from current use to cities of the future." *Proceedings of the Water Environment Federation* 23: 118–40.

28 Zhao, S., Lin, J. and Cui, S. (2011) "Water resource assessment based on the water footprint for Lijian City." *International Journal for Sustainable Development and World Ecology* 18: 492–7.

29 Galli, A., Wiedmann. T., Ercin, E., Knoblauch, D., Ewing, B. and Giljum, S. (2012) "Integrating ecological, carbon, and water footprints into a footprint family of indicators." *Ecological Indicators* 16: 100–12.

30 Kareiva, P. M., Tallis, H., Ricketts, T. H., Daily, G. C. and Polasky, S. (2011) *Natural Capital*. New York: Oxford University Press.

31 The study's website is http://www.teebweb.org.

32 Eigenbrod, F., Bell, V. A., Davies, H., Heinemeyer, A., Armsworth, P. R. and Gaston, K. J. (2011) "The impact of projected increases in urbanization on ecosystem services." *Proceedings of the Royal Society B: Biological Sciences* 278: 3201–8.

33 Robbins, P. and Sharp, J. (2003) "The lawn-chemical economy and its discontents." *Antipode* 35: 955–79.

34 Groffman, P. M., Law, N. L., Belt, K. T., Band, L. E. and Fisher, G. T. (2004) "Nitrogen fluxes and retention in urban watershed ecosystems." *Ecosystems* 7: 393–403.

35 Kaye, J. P., Groffman, P. M., Grimm, N. B., Baker, L. A. and Pouyat, R. V. (2006) "A distinct urban biogeochemistry?" *Trends in Ecology and Evolution* 21: 192–9.

36 Grimm, N. B., Faeth, S. H., Golubiewski, N. E., Redman, C. L., Wu, J., Bai, X. and Briggs, J. M. (2008) "Global change and the ecology of cities." *Science* 319: 756–60.

37 McKinney, M. (2008) "Effects of urbanization on species richness: a review of plants and animals." *Urban Ecosystems* 11: 161–76.

38 Blewett, C. M. and Marzluff, J. M. (2005) "Effects of urban sprawl on snags and the abundance and productivity of cavity-nesting birds." *The Condor* 107: 678–93.

39 Blair, R. (2004) "The effects of urban sprawl on birds at multiple levels of biological organization." *Ecology and Society* 9: 1–21.

40 Gregg, J. W., Jones, C. G. and Dawson, T. E. (2003) "Urbanization effects on tree growth in the vicinity of New York City." *Nature* 424: 183–7.

41 Hope, D., Gries, C., Zhu, W., Fagan, W., Redman, C., Grimm, N. B., Nelson, A. L., Martin, C. and Kinzig, A. (2003) "Socioeconomics drive urban plant diversity." *Proceedings of the National Academy of Sciences of the USA* 100: 8788–92.

42 Chai, Y., Zhu, W. and Han, H. (2002) "Dust removal effect of urban tree species in Harbin." *Ying Yong Sheng Tai Xue Bao* 13: 1121–6.

43 Domen, E. and Sauri, D. (2007) "Urbanization and class-produced natures: vegetable gardens in the Barcelona metropolitan region." *Geoforum* 38: 287–98.

44 Evans, J. P. (2007) "Wildlife corridors: an urban political ecology." *Local Environment* 12: 129–52.

45 For a fuller discussion of the context of the Chicago School, see Short, J. R. (2006) *Urban Theory*. Houndmills and New York: Palgrave.

46 http://www.umbc.edu/ges/research/sohn%27s_res_info/sohn_figure3.htm (accessed July 2006).

47 A recent study of gardens, in Hobart, Tasmania, is available in Kirkpatrick, J. B. (2006) *The Ecologies of Paradise: Explaining the Garden Next Door*. Hobart: Pandani Press.

48 Cielsewicz, D. (2002) "The environmental impacts of sprawl," in *Urban Sprawl: Causes, Consequences and Policy Responses*, G. Squires (ed.). Washington, DC: Urban Institute Press.

49 Volstad, J. H., Roth, N. E., Mercurio, G., Southerland, M. T. and Strebel, D. E. (2003) "Using environmental stressor information to predict the ecological status of Maryland non-tidal streams as measured by biological indicators." *Environmental Monitoring and Assessment* 84: 219–42.

Chapter 10 The environmental revolution

1 Adler, J. H. (2004, June 22) "Smoking out the Cuyahoga fire fable: smoke and mirrors surrounding Cleveland." *National Review*, n.p.

2 Mazmanian, D. (2006) "Achieving air quality: the Los Angeles experience,. USC Bedrosian Center, Working Paper. http://www.usc.edu/schools/price/bedrosian/private/docs/mazmanianairquality.pdf (accessed July 2012).

Chapter 11 Water

1 US EPA Office of Water (2009) "Water on tap: what you need to know." http://water.epa.gov/drink/guide/upload/book_waterontap_full.pdf (accessed June 10, 2012).

2 US EPA Office of Water (2004) "Drinking water costs and federal funding." www.epa.gov/safewater (accessed June 11, 2012).

3 Jarvie, J. (2007, November 4) "Atlanta water use called shortsighted." *Los Angeles Times*. http://articles.latimes.com/2007/nov/04/nation/na-drought4 (accessed Feburary 15, 2012).

4 Fox, P. (2012, February 13) "Alabama, florida appeal water ruling to High Court." *Atlanta Journal Constitution*. http://www.ajc.com/news/georgia-government/alabama-florida-appeal-water-1348441.html (accessed February 14, 2012).

5 Riverkeepers (n.d.) "Combined Sewage Overflows (CSOs)." http://www.river-keeper.org/campaigns/stop-polluters/sewage-contamination/cso (accessed July 2012).

6 Natural Resources Defense Council (n.d.) "Testing the waters: a guide to water quality at vacation beaches." http://www.nrdc.org/water/oceans/ttw (accessed July 2012).

7 Surfers against sewage campaign: http://www.sas.org.uk/campaigns/sewage-and-sickness (accessed June 5, 2012).

8 Lee, J. H. and Bang, K. W. (2000) "Characterization of urban stormwater runoff." *Water Research* 34 (6): 1773–80; see also Gromarie-Mertz, M. C., Garnaud, S., Gonazalez, A. and Chebbo, G. (1999) "Characteristics of urban runoff pollution in Paris." *Water Science Technology* 39 (2): 1–8.

9 US EPA (2008) "Clean water needs survey 2008, Report to Congress," EPA-832-R-10-002, p. ix. http://water.epa.gov/scitech/datait/databases/cwns/upload/cwns2008rtc.pdf (accessed February 21, 2012).

10 Karvonen, A. (2011) *Politics of Urban Runoff: Nature, Technology, and the Sustainable City*. Cambridge, MA: MIT Press.

11 American Society of Civil Engineers (2011) "Failure to act: the economic impact of current investment trends in water and wastewater treatment infrastructure." http://www.asce.org/Infrastructure/Failure-to-Act/Water-and-Wastewater (accessed July 2012).

12 Philadelphia Water Department (2012) "The CSO long term Control Plan Update." www.Phillyriverinfo.org (accessed June 5, 2012).

13 Natural Resources Defense Council (2011) "Rooftops to rivers II: Philadelphia, Pennsylvania—a case study of how green infrastructure is helping manage urban stormwater challenges," p. 4. http://www.nrdc.org/water/pollution/rooftopsii/files/RooftopstoRivers_Philadelphia.pdf (accessed March 2012).

14 Philadelphia Water Department (2011) "Green City, Clean Waters (GCCW): program summary," p. 12. http://www.phillywatersheds.org/what_were_doing/documents_and_data/cso_long_term_control_plan (accessed June 2012).

15 Philadelphia Water Department (2012) "Green Stormwater Infrastructure Programs." http://www.phillywatersheds.org/what_were_doing/green_infra-structure (accessed February 2012).

16 Biswas, A. (2010) "Water for a thirsty urban world." *The Brown Journal of World Affairs*. Volume 17 (1): 146–62.

17 World Health Organization (2010) "Progress on sanitation and drinking water update," pp. 16–18. http://www.who.int/water_sanitation_health/publications/9789241563956/en/index.html (accessed June 5, 2012).

18 Uitto, J. and Biswas, A. (eds.) (2000) *Water for Urban Areas: Challenges and Perspectives*. Tokyo and New York: United Nations University Press, p. xiii.

19 Uitto and Biswas, *Water for Urban Areas*, p. xiii

20 Gandy, M. (2008) "Landscapes of disaster: water, modernity, and urban fragmentation in Mumbai." *Environment and Planning A* 40: 108–30.

21 Ioris, A. (2012) "The geography of multiple scarcities: urban development and water problems in Lima, Peru." *Geoforum* 43 (3): 612–23.

22 Pezzoli, K. (2001) *Human Settlements and Planning for Ecological Sustainability*. Boston, MA: MIT Press, p 59.

23 Engel, K., Jokiel, D., Kraljevic, A., Geiger, M. and Smith, K. (2011) *Big Cities, Big Water, Big Challenges: Water in an Urbanizing World*. Berlin: WWF, pp. 18–24. http://www.wwf.se/source.php/1390895/Big%20Cities_Big%20Water_Big%20Challenges_2011.pdf (accessed February 2012).

24 Sletto, B. (1995) "That sinking feeling." *Geographical* 67 (7): 24.

25 Ezcurra, E., Mazari-Hiriart, M., Pisanty, I. and Aguilar, A. G. (1999) *The Basin of Mexico: Critical Environmental Issues and Sustainability*. New York: United Nations University Press.

26 Tortajada-Quiroz, C. (2000) "Water supply and distribution in the metropolitan area of Mexico City," in *Water for Urban Areas: Challenges and Perspectives*, J. Uitto and A. Biswas (eds.). Tokyo and New York: United Nations University Press, p. 120.

27 National Research Council (1995) *Mexico City's Water Supply*. Washington, DC: National Academy Press, p. 14.

28 Ezcurra *et al.*, *The Basin of Mexico*.

29 Tortajada-Quiroz, "Water supply and distribution in the metropolitan area of Mexico City," p. 113.

30 Ezcurra *et al.*, *The Basin of Mexico*.

31 Barkin, D. (2004) "Mexico City's water crisis," *NACLA Report on the Americas*. July/August, pp. 27–8.

32 Barkin "Mexico City's water crisis."

33 Biswas "Water for a thirsty urban world," pp. 150–3.

34 Wolf-Rainer, A. "Megacities as sources for pathogenic bacteria in rivers and their fate downstream." *International Journal of Microbiology* 2011. doi:10.1155/2011/798292.

35 Stille, A. (1998, January 19) "The Ganges' next life." *The New Yorker*, pp. 58–67. B.O.D. stands for biochemical oxygen demand and is a measure

of how much oxygen has been used by microorganisms in the biodegrada-
tion process. A river or body of water with high levels of BOD may be unhealthy
for aquatic life and if very high, may asphyxiate fish and other marine organisms.

36 World Bank (2012) "India National Ganga River Basin Project." http://www.
worldbank.org/en/news/2012/05/31/india-the-national-ganga-river-basin-project
(accessed July 2012).

Chapter 12 Air

1 American Lung Association (2012) "Key findings, State of the Air 2012." http://
www.stateoftheair.org/2012/key-findings (accessed July 2012).

2 Ozone in the stratosphere is naturally occurring and beneficial in that it forms
a protective layer that filters out the sun's harmful ultraviolet (UV) radiation.
In contrast, ground-level ozone is not beneficial and negatively impacts human
health and the environment.

3 American Lung Association, "Key findings, State of the Air 2012."

4 Gauderman, W. J., Avol, E., Gilliland, F., Vora, H., Thomas, D., Berhane, K.,
McConnell, R., Kuenzli, N., Lurmann, F., Rappaport, E., Margolis, H., Bates, D.
and Peters, J. (2004) "The effect of air pollution on lung development from 10 to
18 years of age." *New England Journal of Medicine* 351: 1057–67.

5 Bayer-Oglesby, L., Grize, L., Gassner, M., Takken-Sahli, K., Sennhauser, F. H.,
Neu, U., Schindler, C. and Braun-Fahrländer, C. (2005) "Decline of Ambient
air pollution levels and improved respiratory health in Swiss children."
Environmental Health Perspective 113: 1632–7.

6 US EPA (2011) "Air pollution trends: toxic air pollutants." http://www.epa.gov/
oaqps001/airtrends/2011/report/toxicair.pdf (accessed July 2012).

7 US Environmental Protection Agency (2010) "A citizen's guide to radon." http://
www.epa.gov/radon/pubs/citguide.html (accessed March 2012).

8 Brain, M. and C. Freudenrich (n.d.) "How stuff works: 'how radon works.'"
How Stuff Works: "Learn how Everything Works." http://home.howstuffworks.
com/home-improvement/household-safety/tips/radon.htm/printable (accessed
March 2012).

9 US Environmental Protection Agency (2012) "Why is radon the public health
risk that it is?" http://www.epa.gov/radon/aboutus.html (accessed March 2012).

10 US Environmental Protection Agency, "Why is radon the public health risk that
it is?"

11 US Environmental Protection Agency (2012) "State radon contact information."
http://www.epa.gov/radon/whereyoulive.html (accessed March 2012).

12 Chen, J. and Moir, D. (2010, February 19) "An updated assessment of radon exposure in Canada." *Radiation Protection Dosimetry*, pp. 1–5. http://www. snolab.ca/public/JournalClub/rpd.ncq046.full.pdf (accessed July 2012).

13 US Environmental Protection Agency. (n.d.) "EPA map of radon zones." http:// www.epa.gov/radon/zonemap.html (accessed March 2012).

14 US Environmental Protection Agency, "State radon contact information."

15 US Environmental Protection Agency, "A Citizen's Guide to Radon."

16 Brain and Freudenrich, "How stuff works: 'how radon works.'"

17 US Environmental Protection Agency (n.d.) "Homebuyers: basic techniques : radon-resistant new construction." http://www.epa.gov/radon/rrnc/basic_tech-niques_homebuyer.html (accessed March 2012).

18 Center for Disease Control (2012) "Frequently asked questions about CO." http://www.cdc.gov/co/faqs.htm (accessed July 2012).

19 US Environmental Protection Agency (2011) "Air pollution trends." http://www. epa.gov/oaqps001/airtrends/2011/report/highlights.pdf (accessed July 2012).

20 US Environmental Protection Agency, "Air pollution trends."

21 US Environmental Protection Agency, "Air pollution trends."

22 United Nations Environment Programme (2012) "Urban air pollution." http:// www.unep.org/urban_environment/issues/urban_air.asp (accessed July 2012).

23 Ostro, B. (2004) *Outdoor Air Pollution: Assessing the Environmental Burden of Diseases at the National and Local Level.* Geneva: World Health Organization.

24 OECD (2012) "Environmental outlook to 2050: the consequences of inaction." http://www.oecd.org/document/34/0,3746,en_21571361_44315115_ 49897570_1_1_1_1,00.html (accessed July 2012).

25 World Bank (2002) *Cities on the Move: A World Bank Urban Transport Strategy Review.* Washington, DC: World Bank.

26 Sayed, A. (2010) "Ulaanbaatar's air pollution crisis: summertime complacency won't solve the wintertime problem." http://blogs.worldbank.org/eastasiapacific/ ulaanbaatar-s-air-pollution-crisis-summertime-complacency-won-t-solve-the-wintertime-problem (accessed July 2012).

27 The Asian Clean Fuels Association (2011) "Clearing Indian skies." http://www. acfa.org.sg/newsletterinfocus02_01.php (accessed July 2012).

28 Elsom, D. (1996) *Smog Alert: Managing Urban Air Quality.* London and Sterling, VA: Earthscan, p 5.

29 Elsom, *Smog Alert*, p. 27.

30 World Resources Institute (1998) *World Resources 1998–99*. Washington, DC: World Resources Institute.

31 Fewtrell, L., Kaufmann, R. and Pruss-Ustun, A. (2003) *Lead: Assessing the Environmental Burden of Disease at the National and Local Level*. Geneva: WHO.

32 World Bank (1998) "Clean Air Initiative in sub-Saharan Africa cities." http://www.worldbank.org/wbi/cleanair/caiafrica/index.htm (accessed February 2003).

33 Harvard School of Public Health (2012) "Superfund research program: Mexico City." http://www.srphsph.harvard.edu/pages/projectsites_mex.html (accessed July 2012).

34 The industrialized north faces indoor air pollution as well, primarily from chemicals and compounds such as radon or carbon monoxide and tobaccos smoke.

35 World Health Organization (2012) "Indoor air pollution." http://www.who.int/indoorair/health_impacts/exposure/en/index.html (accessed July 2012).

36 World Health Organization (2006) "Indoor air pollution fact sheet." http://www.who.org.int/mediacenter/factsheets/en (accessed July 2012).

37 World Resources Institute (1999) "Rising energy use: health effects of air pollution." www.wri.org (accessed February 2003).

38 World Resources Institute (1998) *World Resources 1998–99*. Washington, DC: World Resources Institute.

39 World Health Organization (2012) "Indoor air pollution: intervention to reduce exposure." http://www.who.int/indoorair/interventions/en (accessed July 2012).

40 Clean Air Initiative for Asian Cities (CAI-Asia) Center (2010) "Air quality standards and trends." http://cleanairinitiative.org/portal/sites/default/files/documents/AQ_in_Asia.pdf (accessed July 2012).

For up-to-date information on ozone/smog and interactive maps, go to: www.epa.gov/air/ozonepollution. Also, www.scorecard.org allows you to profile the air, land and water pollution in your state, city or community in the US.

Chapter 13 Climate change

1 US EPA (n.d.) "Carbon dioxide." http://epa.gov/climatechange/ghgemissions/gases/co2.html (accessed August 2012).

2 US EPA (2012) "Methane emissions." http://epa.gov/climatechange/ghgemissions/gases/ch4.html (accessed August 2012).

3 US EPA (2012) "Inventory of U.S. greenhouse gas emissions and sinks: 1990–2010." http://epa.gov/climatechange/ghgemissions/usinventoryreport.html, pp. 1–2 (accessed August 2012).

4 US EPA, "Inventory of U.S. greenhouse gas emissions and sinks: 1990–2010."

5 US EPA (2012) "Future climate change: climate models." http://epa.gov/climatechange/science/future.html (accessed August 2012).

6 US EPA (2012) "Climate change impacts and adapting to change." http://epa.gov/climatechange/impacts-adaptation (accessed August 2012).

7 De Sherbinin, A., Schiller, A. and A. Pulsipher (2007) "The vulnerability of global cities to climate hazards." *Environment and Urbanization* 19 (1): 39–64.

8 America's Climate Choices: Panel on Advancing the Science of Climate Change, Board on Atmospheric Sciences and Climate, Division on Earth and Life Studies, National Research Council of the National Academies (2010) "Sea level rise and the coastal environment," in *Advancing the Science of Climate Change*. Washington, DC: The National Academies Press, p. 245. http://books.nap.edu/openbook.php?record_id=12782&page=245 (accessed December 2011).

9 Anderson, W. and Jones, S. (2008) "The Clean Water Act: a blueprint for reform." White Paper #802Es, Center for Progressive Reform.

10 *The Economist* (2012, March 17) "Asia and its floods: save our cities," pp. 48–50.

11 International Panel on Climate Change (2012) "Glossary of terms," in *Managing the Risks of Extreme Events and Disasters to Advance Climate Change Adaptation*, C. B. Field, V. Barros, T. F. Stocker, D. Qin, D. J. Dokken, K. L. Ebi, M. D. Mastrandrea, K. J. Mach, G.-K. Plattner, S. K. Allen, M. Tignor and P. M. Midgley (eds.). Cambridge and New York: Cambridge University Press, pp. 555–64. http://www.ipcc.ch/pdf/special-reports/srex/SREX-Annex_Glossary.pdf (accessed September 2012).

12 Keep in mind, however, that cap-and-trade programs are conducted at the federal level; cities do not have control over these types of responses.

13 Nordhaus, T. and Schellenberger, M. (2007) *Break Through: From the Death of Environmentalism to the Politics of Possibility*. Boston, MA: Houghton Mifflin Company.

14 Pacala, S. and Socolow, R. (2004) "Stabilization wedges: solving the climate problem for the next 50 years with current technologies." *Science* 13 (5686): 968–72.

15 Cities for Climate Protection Australia Adaptation Initiative (2008) "Local Government Climate Change Adaptation Toolkit." http://www.iclei.org/fileadmin/user_upload/documents/Global/Progams/CCP/Adaptation/Toolkit_CCPAdaptation_Final.pdf (accessed September 2012).

16 Cities for Climate Protection Australia Adaptation Initiative, "Local Government Climate Change Adaptation Toolkit."

17 Tanner, T., Mitchell, T., Polack, E. and Buenther, B. (2009) "Urban governance for adaptation: assessing climate change resilience in ten Asian cities." Institute of Development Studies at the University of Sussex, Brighton, Working Paper #315.

18 US Conference of Mayors (2011) "Clean energy solutions for America's cities: A summary of survey results prepared by GlobeScan Incorporated and sponsored by Siemens." www.usmayors.org/cleanenergy/report.pdf (accessed August 2012).

19 Interagency Climate Change Adaptation Task Force (2011) "Federal actions for a climate resilient nation." http://www.whitehouse.gov/sites/default/files/microsites/ceq/2011_adaptation_progress_report.pdf (accessed August 2012).

20 Interagency Climate Change Adaptation Task Force, "Federal actions for a climate resilient nation."

21 Interagency Climate Change Adaptation Task Force, "Federal actions for a climate resilient nation."

22 Center for Science in the Earth System (The Climate Impacts Group) Joint Institute for the Study of the Atmosphere and Ocean University of Washington and King County, Washington (2007) *Preparing for Climate Change: A Guidebook for Local, Regional and State Governments.* Seattle, WA: Center for Science in the Earth System, p. 29. http://www.iclei.org/fileadmin/user_upload/documents/Global/Progams/CCP/Adaptation/ICLEI-Guidebook-Adaptation.pdf (accessed April 23, 2012).

23 Bulkeley, H. and Betsill, M. (2005) *Cities and Climate Change: Urban Sustainability and Global Environmental Governance.* New York: Routledge.

24 Greater London Authority (2007) "Action today to protect tomorrow: the Mayor's Climate Change Action Plan." http://legacy.london.gov.uk/mayor/environment/climate-change/docs/ccap_summaryreport.pdf (accessed April 23, 2012).

25 London.gov.uk (2007) "Mayor unveils London Climate Change Action Plan." http://www.london.gov.uk/media/press_releases_mayoral/mayor-unveils-london-climate-change-action-plan (accessed August 2012).

26 City of Chicago (2008) "Chicago Climate Action Plan." http://www. chicagoclimateaction.org/filebin/pdf/finalreport/CCAPREPORTFINALv2.pdf (accessed March 15, 2012); and also City of Chicago (2010) "Chicago Climate Action Plan: Progress Report 2008–2009." http://www.chicagoclimateaction.org/filebin/pdf/CCAPProgressReport.pdf (accessed March 15, 2012).

Chapter 14 Garbage

1 Melosi, M. (2000) *The Sanitary City: Urban Infrastructure in America from Colonial Times to the Present.* Baltimore, MD: Johns Hopkins University Press, p. 339.

2 Melosi, *The Sanitary City*, p. 338.

3 Melosi, *The Sanitary City*, p 340.

4 US Environmental Protection Agency (2010) "Municipal solid waste generation, recycling, and disposal in the United States: facts and figures for 2010." http://www.epa.gov/osw/nonhaz/municipal/pubs/msw_2010_rev_factsheet.pdf (accessed July 2012).

5 Department of Sanitation New York (2011) *Annual Report, 2011.* http://www.nyc.gov/html/dsny/downloads/pdf/pubinfo/annual/ar2011.pdf (accessed July 2012).

6 US Environmental Protection Agency, "Municipal solid waste generation, recycling, and disposal in the United States."

7 Urban Development Unit, World Bank (1999) *What a Waste: Solid Waste Management in Asia.* Washington, DC: World Bank, p. 18.

8 Melosi, *The Sanitary City*, p. 343.

9 US Environmental Protection Agency (1999) "National source reduction characterization report for municipal solid waste in the United States." www.epa.gov/osw, p 15. (accessed February 2000).

10 National Solid Wastes Management Association (NSWMA) and the Waste Equipment Technology Association (WASTEC) (2012) "MSW (Subtitle D) Landfills" http://www.environmentalistseveryday.org/publications-solid-waste-industry-research/information/faq/municipal-solid-waste-landfill.php (accessed July 2012).

11 Lipton, E. (2000, February 21) "Efforts to close Fresh Kills are taking unforeseen tolls." *New York Times*, Section A, p. 1.

12 Melosi, *The Sanitary City*, p. 261.

13 Royte, E. (2005) *Garbage Land: On the Secret Trail of Trash.* New York: Little, Brown.

14 Fishbein, B. and Azimi, S. (1994) *Germany, Garbage and the Green Dot: Challenging the Throwaway Society.* New York: Inform, pp. 18–21.

15 Skidmore, S. (2011) "SunChips biodegradable bag made quieter for critics." *Huffington Post.* http://www.huffingtonpost.com/2011/03/01/sunchips-biodegradable-bag_n_829165.html (accessed June 2012).

16 US EPA (2012) "Energy recovery from waste." http://www.epa.gov/osw/nonhaz/municipal/wte/index.htm (accessed June 2012).

17 Urban Development Unit, World Bank, *What a Waste*, pp. 7–8.

18 Ecenbarger, W. (2011) "It's time to clean up: an out-of-control garbage crisis threatens the physical and economic health of much of Asia." *Readers Digest.* http://beta.readersdigest.com.sg/article/2874 (accessed June 2012).

19 Urban Development Unit, World Bank, *What a Waste*, p 3.

20 Urban Development Unit, World Bank, *What a Waste*, pp. 10–11.

21 Rogerson, C. (2001) "The waste sector and informal entrepreneurship in developing world cities." *Urban Forum* 12 (2): 247–9.

22 Wilson, D, Araba, A., Chinwah, K. and Cheeseman, C. (2009) "Building recycling rates through the informal sector." *Waste Management* 29 (2): 632.

23 Seng, B (2001) "Municipal solid waste management in Phnom Penh, capital city of Cambodia." *Waste Management & Research: The Journal of the International Solid Wastes and Public Cleansing Association*, 29 (5): 497–8.

24 Rathana, K. (2009) "Solid waste management in Cambodia." Cambodian Institute for Peace and Cooperation, Working Paper no. 29.

25 US Department of Commerce (2001) "Brazil: solid waste statistical data." http://srategis.ic.gc.ca/SSG/dd72515e.html (accessed September 25 2012).

26 Moore, S. (2008) "The politics of garbage in Oaxaca, Mexico." *Society and Natural Resources* 21: 597–610.

27 Scheinberg, A. and Anschutz, J. (2006) "Slim Pickins: supporting waste pickers in the ecological modernization of urban waste management systems." *International Journal of Technology Management & Sustainable Development* 5 (3): 263–4.

28 Afon, A. O. (2007) "Informal sector initiative in the primary sub-system of urban solid waste management in Lagos, Nigeria." *Habitat International* 31 (2): 194.

29 Afon, "Informal sector initiative in the primary sub-system of urban solid waste management in Lagos, Nigeria."

Chaper 15 Race, class and environmental justice

1 United Church of Christ, Commission for Racial Justice (1987) *Toxic Wastes and Race in the United States: A National Report on the Racial and Socio-economic Characteristics of Communities with Hazardous Waste Sites*. New York: Public Data Access.

2 Mohai, P. and Saha, R. (2007) "Racial inequality in the distribution of hazardous waste: a national-level reassessment." *Social Problems* 54: 343–70.

3 Martuzzi, M., Mitis, F. and Forastiere, F. (2010) "Inequalities, inequities, environmental justice in waste management and health." *European Journal of Public Health* 20: 21–6.

4 Boone, C. G. and Modarres, A. (1999) "Creating a toxic neighborhood in Los Angeles County." *Urban Affairs Review* 35: 163–87. See also Pulido, L. (2000) "Rethinking environmental racism: white privilege and urban development in Southern California." *Annals of the Association of American Geographers* 90: 12–40.

5 Bullard, R. (1993) *Confronting Environmental Racism: Voices from the Grassroots.* Boston, MA: South End Press. See also Bullard, R. D. (ed.) (2000) *Dumping in Dixie: Race, Class and Environmental Quality.* Boulder, CO: Westview.

6 Brulle, R. J. and Pellow, D. W. (2006) "Environmental justice: human health and environmental inequalities." *Annual Review of Public Health* 27: 103–24.

7 Lara-Valencia, F., Harlow, S. D., Lemos, M. C. and Denman, C. A. (2009) "Equity dimensions of hazardous waste generation in rapidly industrializing cities along the United States–Mexican border." *Journal of Environmental Planning and Management* 52: 195–216, quote from p. 212.

8 Crenson, M. (1971) *The Un-politics of Air Pollution: A Study of Non-Decisionmaking in Cities.* Baltimore, MD: Johns Hopkins University Press; see also Benton, L. M. and Short, J. R. (1999) *Environmental Discourse and Practice.* Oxford: Blackwell, ch. 6.

9 Martinson, K., Stanczyk, A. and Eyster, L. (2010) *Low-skill Workers' Access to Quality Green Jobs.* Washington, DC: The Urban Institute.

10 Bartley, A. (2011) "Building a 'community growth machine': the green development zone as a model for a new neighborhood economy." *Social Policy* 41: 9–20, quote from p. 15.

11 Heynen, N., Perkins, H. A. and Roy, P. (2006) "The political ecology of uneven urban green space." *Urban Affairs Review* 42: 3–25.

12 Boone, C. G., Buckley, G. L., Grove, J. M. and Sister, C. (2009) "Parks and people: an environmental justice inquiry in Baltimore, Maryland." *Annals of the Association of American Geographers* 99: 767–87.

13 http://www.epa.gov/environmentaljustice/index.html (accessed August 16, 2012).

14 Walker, G. (2009) "Globalizing environmental justice." *Global Social Policy* 9: 355–82.

15 Carmin, J. and Agyeman, J. (eds.) (2011) *Environmental Inequalities Beyond Borders: Local Perspectives on Global Injustices*. Boston, MA: MIT Press.

16 Gleeson, B. (2006) *Australian Heartlands*. Crows Nest, NSW: Allen and Unwin.

17 Spirn, A. (1985) *The Granite Garden: Urban Nature and Human Design*. New York: Basic.

18 Biehler, D. (2009) "Permeable homes: a historical political ecology of insects and pesticides in US public housing." *Geoforum* 40: 1014–23.

19 Berg, L. (2011) "Banal naming, neoliberalism and landscape of dispossession." *ACME: An International E-Journal for Critical Geographies* 2011: 13–22.

20 Short, J. R. (2012) "Representing country in the creative postcolonial city." *Annals of the Association of American Geographers* 102: 129–50.

21 Gibbs, L. (1993) "Foreword," in *Toxic Struggles: The Theory and Practice of Environmental Justice*, R. Hofrichter (ed.). Philadelphia, PA: New Society Publishers.

22 See Short, J. R. (1989) *The Humane City*. Oxford: Blackwell, esp. ch. 4.

23 Kegler, M. C., Norton, B. L. and Aronson, R. (2008) "Achieving organizational change: findings from case studies of 20 California healthy cities and community coalitions." *Health Promotion International* 23: 109–18.

24 Dooris, M. and Heritage, Z. (2011) "Healthy cities: facilitating the active participation and empowerment of local people." *Journal of Urban Health*. doi: 10.1007/s11524-011-9623-0.

25 Hartley, N. and Wood, C. (2005) "Public participation in environmental impact assessment; implementing the Aarhus Convention." *Environmental Impact Assessment Review* 25: 319–40.

Chaper 16 Urban sustainability

1 United Nations, World Commission on Environment and Development (1987) *Our Common Future*. Oxford: Oxford University Press, p. 8.

2 Wheeler, S. (2004) *Planning for Sustainability: Creating Livable, Equitable and Ecological Communities*. London and New York: Routledge, p. 23.

3 Wheeler, *Planning for Sustainability*, pp. 23–5.

4 Girardet, H. (1999) "Sustainable cities: a contradiction in terms?" in *The Earthscan Reader in Sustainable Cities*, D. Sattherwaite (ed.). London: Earthscan, p. 419.

5 S. Dresner (2008) "What does sustainable development mean?" in *The Principles of Sustainability*. London: Earthscan, p. 69.

6 As quoted by Dresner, "What does sustainable development mean?" pp. 69–80.

7 Rees, W. E. (1999) "Achieving sustainability: reform or transformation?" in *The Earthscan Reader in Sustainable Cities*, D. Sattherwaite (ed.). London: Earthscan, pp. 22–54.

8 Keil, R. (2007) "Sustaining modernity, modernizing nature," in *The Sustainable Development Paradox*, R. Krueger and D. Gibbs (eds.). New York: Guilford Press, pp. 41–65.

9 Haughton, G. (1999) "Environmental justice and the sustainable city," in *The Earthscan Reader in Sustainable Cities*, D. Sattherwaite (ed.). London: Earthscan, pp. 62–79.

10 Singer, M. (2003, August 11) "The haves and the haves." *The New Yorker*, pp. 56–61.

11 From the charter of New Urbanism:

> Neighborhoods should be compact, pedestrian-friendly, and mixed use. Many activities of daily living should occur within walking distance, allowing independence to those who do not drive, especially the elderly and the young. Interconnected networks of streets should be designed to encourage walking, reduce the number of and length of automobile trips and conserve energy. Within neighborhoods a broad range of housing types and price levels can bring people of diverse races and incomes into daily interaction, strengthening the personal and civic bonds essential to an authentic community.
> (http://www.cnu.org/charter.html, accessed November 9, 1999)

12 Frantz, D. and Collins, C. (1999) *Celebration USA*. New York: Henry Holt. See also Ross, A. (1999) *The Celebration Chronicles*. New York: Ballantine.

13 Girardet, H. (2004) *Cities, People, Planet: Livable Cities for a Sustainable World*. Chichester: Wiley.

14 For a range of opinions, see Talen, E. (1999) "Sense of community and neighborhood form: an assessment of the social doctrine of New Urbanism." *Urban Studies* 36 (8): 1361–79; Krieger, A. (1998) "Whose Urbanism?" *Architecture*, November, 73–7; and Ford, L. (1999) "Lynch revisited: New Urbanism and theories of good city form." *Cities* 16: 277–57.

15 Mayer, H. and Knox, P. L. (2006) "Slow cities: sustainable places in a fast world." *Journal of Urban Affairs* 28: 321–34.

16 Tyler, N. (2000) *Historic Preservation: An Introduction to Its History, Principles and Practice*. New York: W.W. Norton.

17 Tyler, *Historic Preservation*, pp. 23–4.

18 Tyler, *Historic Preservation*, pp. 27–8.

19 See Portney, K. E.(2003) *Taking Sustainable Cities Seriously: Economic Development, the Environment and Quality of Life in American Cities*. Cambridge, MA: MIT Press.

20 Portney, K. E. (2009) "Sustainability in American cities: a comprehensive look at what American cities are doing and why," in *Towards Sustainable Communities: Transition and Transformation in Environmental Policy*, D. A. Mazmanian and M. E. Kraft (eds.). Cambridge, MA: MIT Press.

21 Zeemering, E. (2009) "What does sustainability mean to city officials." *Urban Affairs Review* 45: 247–73.

22 Beatley, T. (ed.) (2012) *Green Cities of Europe*. Washington, DC: Island Press.

23 *The Economist* (2012, July 14) "Changing the plans," p. 29.

24 Low, N., Gleeson, B., Green, R. and Radović, D. (2005) *The Green City: Sustainable Homes, Sustainable Suburbs*. Oxford: Routledge, p. 44.

25 http://hortweb.cas.psu.edu/research/greenroofcenter/history.html (accessed August 27, 2012).

26 Low *et al.*, *The Green City*, p. 78–9.

27 Low *et al.*, *The Green City*, p. 81.

28 Ulrich, R. S. (1984) "Views through a window may influence recovery from surgery." *Science* 224 (4647): 420-421.

29 Girardet, *Cities, People, Planet*, p. 239.

30 Girardet, *Cities, People, Planet*, p. 247.

31 Orange County Great Park (2009) "Sustainability goals." http://www.ocgp.org/learn/sustainability (accessed July 2012).

32 Low *et al.*, *The Green City*, p. 135.

33 As quoted in Low *et al.*, *The Green City*, p. 138.

34 Low *et al.*, *The Green City*, p. 149.

35 Girardet, *Cities, People,* Planet, p. 131.

36 China Mike (n.d.) "China's Car Culture." http://www.china-mike.com/facts-about-china/facts-transportation-autos-car-culture (accessed July 2012).

37 Rabinovitch, J. (1997) "A success story of urban planning: Curitiba," in *Cities Fit for People*, U. Kirdar (ed.). New York: United Nations Press, p. 425.

38 Rabinovitch, "A success story of urban planning," p. 424.

39 Rabinovitch, "A success story of urban planning," p. 425.

40 Rabinovitch, "A success story of urban planning," p. 429.

41 Irazábal, C. (2005) *City Making and Urban Governance in the Americas: Curitiba and Portland*. Burlington, VT: Ashgate. See also Gilbert, R. (1996) *Making Cities Work: The Role of Local Authorities in the Urban Environment*. London: Earthscan.

42 While, A., Jonas, A. E. G. and Gibbs, D. (2004) "The environment and the entrepreneurial city: searching for the urban 'sustainability fix' in Manchester and Leeds." *International Journal of Urban and Regional Research* 28 (3): 549–69.

43 As quoted by Beatley, T. (2007) "Sustaining the city: urban ecology," paper presented at the Symposium on Framing a Capital City, National Building Museum, Washington, DC, April 11, 2007.

Index